AF880483

Recent Advances in Separation Techniques—III

Norman N. Li, editor

T. Alan Hatton, Sun-Tak Hwang, Donald M. LaRue
Stephen A. Leeper, Daryl L. Roberts, co-editors

B.S. Ainsworth
Richard W. Baker
Ivars Bemberis
Lloyd Berg
Subray N. Bhat
Robert G. Blanck
G.R. Boucard
V.C. Boucard
W.R. Brown
Gregory R. Carmichael
A.A. Cassano
Sherril D. Christian
H.C. Chu
Ravindra Datta
R.D. Doctor
B.R. Dunbobbin
Robert O. Dunn
Gerald D. Dupre
James L. Dwyer
Rex T. Ellington
William Eykamp
Timothy P. Fitzgerald
Edward W. Funk
A. Gelb
Arye Gollan
Alan J. Grodzinsky
Jeffrey H. Harwell
Myles H. Kleper
Sudhir S. Kulkarni
Norman N. Li
W.G. Light
Ben J. McCoy
Judith M. Mohr
L.A. Mooney
Karen Neely
C.B. Panchal
B.S. Parekh
P.L. Parise
Klaus-Viktor Peinemann
Barry W. Penney
J. Carl Pirkle, Jr.
L. Roettger
John F. Scamehorn
R.W. Serth
Jeffrey H. Siegell
A.X. Swamikannu
C.E. Swietlik
E.J. Taylor
Aryeh M. Weiss
F.W. Wilkins
S.K. Wood
Martin L. Yarmush
An-I Yeh
Randall A. Yoshisato

AIChE Symposium Series

Number 250 | 1986 | Volume 82

Published by

American Institute of Chemical Engineers

345 East 47 Street | New York, New York 10017

American Institute of Chemical Engineers
345 East 47 Street, New York, N.Y. 10017

AIChE shall not be responsible for statements or opinions advanced in papers or printed in its publications.

Library of Congress Cataloging-in-Publication Data
Main entry under title:

Recent advances in separation techniques III.

(AIChE symposium series ; no. 250, v. 82)
Includes index.
1. Separation (Technology) I. Li, Norman N. II. American Institute of Chemical Engineers. III. Series: AIChE symposium series ; no. 250.
TP156.S45R432 1986 660.2'842 86-26513
ISBN 0-8169-0392-1

Printed in the United States of America by
Twin Production & Design

FOREWORD

Separation processes make purified products for mixtures and purify waste streams. They are involved in virtually all processing industries. Separation methods are also key parts of many analytical procedures. Because of the importance of separation science and technology in a broad spectrum of industries, the Program Committee "Novel Separation Processes" of the American Institute of Chemical Engineers scheduled six sessions on separations at the 1985 Summer National Meeting of American Institute of Chemical Engineers in Seattle. Most of the papers in the symposium volume were the papers presented at this meeting. A few papers in the volume were presented in the separations sessions of the AIChE's Summer National Meeting in Boston.

All together there are 21 papers in this volume. They discuss a wide range of topics on chromatography, distillation, extraction, electrophoresis, extractive distillation, processes using magnetically stabilized fluidized bed, and various membrane processes. Some of them focus on theories while others concentrate on recently developed applications. In combination, these papers represent some important aspects of recent advances of separation science and technology.

All of the co-editors of this volume are the session chairmen at the above-mentioned National Meetings, whose sessions have contributed papers to this volume. They are T. Alan Hatton of Massachusetts Institute of Technology, Sun Tak Hwang of University of Cincinnati, Donald M. LaRue of EG&G Idaho, Stephen A. Leeper of EG&G Idaho, Inc., and Darly L. Roberts of SRI International.

My co-editors and I would like to express our sincere appreciation to the authors and the reviewers for their effort in making this volume possible. We also wish to thank Ms. Maura Mullen and her staff at the American Institute of Chemical Engineers for their invaluable editorial assistance.

Norman N. Li
Allied-Signal, Inc.
Des Plaines, Illinois

CONTENTS

ALCOHOL REMOVAL FROM BEER BY REVERSE OSMOSIS

W.G. Light, L.A. Mooney, H.C. Chu and S.K. Wood ■ Fluid Systems, A Unit of Allied-Signal Inc.
10124 Old Grove Road, San Diego, California 92131

Full-bodied beer with reduced or "no" alcohol content can be produced from regular beer using sanitary design, spiral-wound reverse osmosis membrane elements. The elements contain a FDA approvable thin-film composite membrane and are operated in a semi-batch or continuous feed-and-bleed mode. Efficient production of low-alcohol beer is achieved by operating systems at low net pressure.

In 1980, food processing accounted for about 13% of all U.S. manufacturing gross national product with a market value of $256 billion (1). In the last twenty years, value added in food processing has continued to increase independently of raw food prices, indicating a large potential market for membrane products capable of meeting food processing requirements. To infiltrate food markets effectively, membrane processes will have to reduce costs or give a product of superior quality, and they will have to satisfy regulations applicable to food processing.

For a technically feasible membrane food process to be commercially acceptable in many countries, all components of the membrane device must be made with raw materials which have been approved or are approvable for food contact by an appropriate governmental agency, such as the Food and Drug Administration (FDA) in the U.S. The device must also be constructed in a sanitary design with no stagnant areas and be easy to clean.

For certain food applications, membrane processes have been proven capable of giving lower cost, higher quality products, as a result of reduced energy requirements, decreased need for chemical preservatives, and lower operating temperatures. Since food processing using membranes does not require a phase change, heat sensitive fluids can be processed at lower temperatures, with reduced operating expenses and reduced heat-related damage. By exploiting these advantages, sanitary design reverse osmosis (RO) and ultrafiltration devices have been successfully used for solution sterilization, dewatering applications and component separations. Of these, a commercially important membrane application is the production of milk-based products (2), such as the recovery and concentration of proteins in cheese whey. Other processing applications include the isolation and recovery of proteins, the sterilization of beer and fruit juices, the dewatering of sugar solutions, and the clarification and removal of tartrate from wines.

Of the many possible food applications for membranes, the production of low-alcohol beer has been investigated because of recent commercial interest. In the brewing industry there are two types of products: low-alcohol beer containing 2 wt% ethanol and "non-alcoholic" beer containing less than 0.5 wt%. The goal in producing low-alcohol beer by RO is selectively to remove ethanol from beer without losing extract or flavor components. These essential components, including proteins, carbohydrates, acetates, and higher alcohols, give the beer its characteristic aroma, body, and taste.

Several methods have been tried unsuccessfully in an attempt to produce a full-bodied, low-alcohol beer with acceptable taste. Two major problems encountered in the production of low-alcohol beer have been the addition of unwanted flavors and loss of beer body. Low-alcohol beer by dialysis was not successful as an unacceptable percentage of the beer's extract passed through the membrane resulting in a very light-bodied beer (3). Distillation methods (4, 5) have had only moderate success, mainly because they are expensive and the flavor components are altered during the process. Heat sensitive proteins and carbohydrates in the beer are denatured at distillation temperatures giving the beer an unacceptable taste. Brewers have also tried to modify their brewing process to obtain a low-alcohol beer, but this is unsatisfactory because a separate, modified brewing formula is required, and the reduced wort concentration in the new formulation results in a light-bodied beer. In contrast, production of low-alcohol beer by RO has the advantage of being able to remove the ethanol from the beer after normal fermentation without losing extract or altering flavor (5, 6).

The objective of the work presented here is to demonstrate the use of commercially acceptable, sanitary design RO spiral-wound elements for the production of low-alcohol beer. Performance characteristics for elements tested in pilot-scale systems are presented and used as a basis for specifying the preferred process design and for elucidating a novel membrane separation mechanism.

MATERIALS AND METHODS

Membrane Elements

The thin-film composite (TFC) membrane contained in the RO elements is a membrane specially formulated for use in food processing applications. Its use is in compliance with applicable FDA regulations for food-contact surfaces. All non-membrane components of the element are made from FDA grade raw materials. A detailed description of element components and design is given elsewhere (7).

For the spiral-wound membrane element to have a sanitary design it was necessary to eliminate the stagnant areas between the element and the housing. This is accomplished by perforating the u-cup carrier to create a bypass around the u-cup seal which prevents feed from flowing between the element and housing in conventional elements. Small holes in the u-cup carrier allow 1 to 5% of the feed to leak past the element in the channel between the element and the housing. The controlled amount of feed bypassing the element can be modified by varying the number and size of the holes in the u-cup carrier. As can be seen in Figure 1, inlet cylindrical channels in the direction of flow have a diameter that is less than half that for the continuation channels through the u-cup carrier. This is done to prevent plugging of the channels downstream of the inlet. The inlet channels can also be designed to accommodate plugs containing holes of different sizes, thus allowing for adjusting the rate of bypass flow depending on the properties of the fluid being processed.

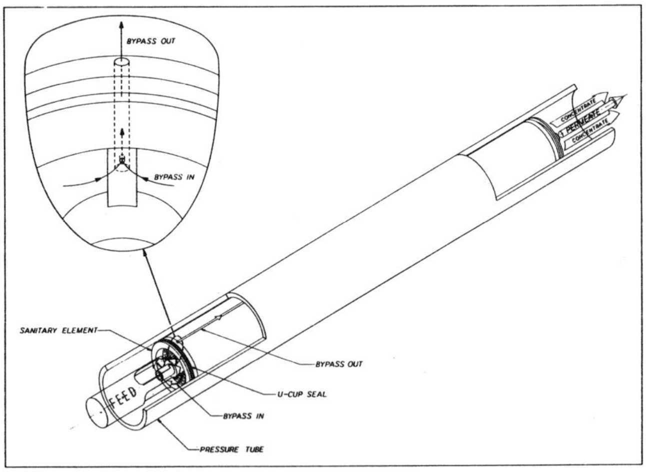

Figure 1. Sanitary design, spiral-wound element with perforated U-cup carrier for flow bypass.

Membrane System Process Design

Previous work has employed an approach for removing ethanol from beer using a membrane system operated in a batch-concentration mode whereby the beer being processed is either diluted with water before or after processing (5). Two disadvantages of this approach are that the concentrations of complex components in beer are altered and precipitates and other membrane foulants can form that affect the taste of the beer and reduce the membrane system productivity. Two approaches that were employed here that do not have these problems are the semi-batch and feed-and-bleed modes, which are shown schematically in Figures 2 and 3, respectively.

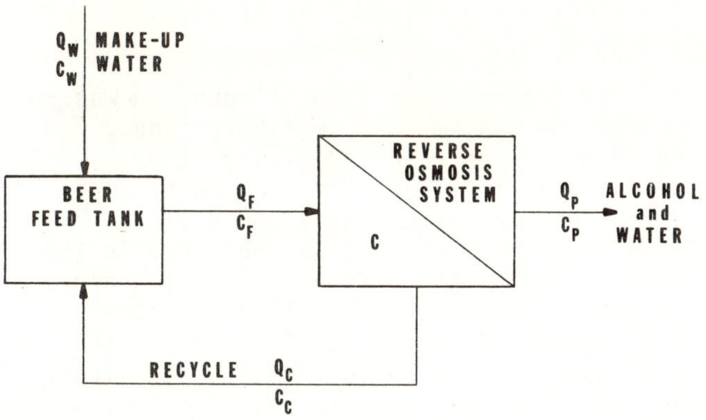

Figure 2. Flow schematic for semi-batch mode of operation for producing low-alcohol beer.

The fundamental process design equation for the semi-batch mode (8) for producing low-alcohol beer is the following

$$\frac{C_F}{C_B} = e^{\frac{1}{k}[(1-k)^{1-R}-1]S} \qquad (1)$$

where

C_F = Ethanol concentration in processed beer at time t

C_B = Ethanol concentration in unprocessed beer

k = Q_P/Q_F = Recovery per pass through the membrane system

R = $1-C_P/C$ = Membrane rejection for ethanol

S = Q_Pt/V = Volume of make-up water per volume of beer processed

Q_P = Permeate flow rate

Q_F = Membrane system feed flow rate

C_P = Concentration of ethanol in permeate

C = Concentration of ethanol in beer during processing

V = Volume of beer being processed

t = Time

Equation (1) can be used to calculate the fractional reduction in ethanol concentration (C_F/C_B) as a function of membrane rejection (R), recovery per pass (k), and the ratio of the volume of make-up water to the volume of beer processed (S). As S is determined by the flow rate of permeate (Q_P, equal to the flow rate of make-up water), the volume of beer being processed (V), and time (t), the fundamental equation can also be used to calculate the processing time required to give a specific reduction in ethanol concentration.

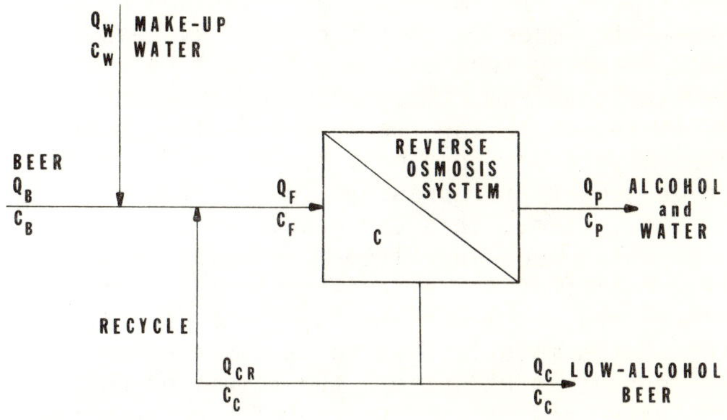

Figure 3. Flow schematic for feed-and-bleed mode of operation for producing low-alcohol beer.

In the continuous feed-and-bleed mode shown in Figure 3 for producing low-alcohol beer, the processed stream containing flavor components rejected by the membrane and reduced alcohol content is partially recycled (Q_{CR}) and mixed with fresh beer (Q_B) and make-up water (Q_W). The remaining processed stream (Q_C) is "bled" from the system and collected as product. To maintain a constant concentration of flavor components, volumetric flow rates are adjusted so that Q_C and Q_B are equal. The fundamental process design equation for calculating the reduction in ethanol concentration in processed beer (C_C) as compared to unprocessed beer (C_B) is given by

$$\frac{C_C}{C_B} = \frac{(1-k)-X}{(1-k)^R-X} \qquad (2)$$

where

C_C = Ethanol concentration in processed beer

X = Q_{CR}/Q_F = Fraction of feed that is recycle

Q_{CR} = Recylce flow rate

It can be seen from Equation (2) that the ethanol concentration change is a function of the recovery per pass (k), the membrane rejection (R), and the fraction of system feed that is recycled, processed beer (X). The fraction of the processed beer that is recycled is given as X/(1-k).

On comparing the two process designs, the semi-batch mode is an attractive design approach for small or pilot-scale units, but can be inappropriate for production or full-scale operation where a continuous mode of operation is preferred, such as a feed-and-bleed mode. An important disadvantage of the semi-batch mode is that for large systems, both incoming feed and recycle must be pressurized from atmospheric to operating pressure, while for the feed-and-bleed mode, the recycle pressure only has to be boosted to compensate for pressure losses incurred in the system. This can result in the energy cost for a feed-and-bleed mode being less than half that for a semi-batch mode. On the other hand, an important advantage for the semi-batch mode is that for a given ethanol concentration reduction, the volume of make-up water required per volume of beer processed is less than that for the feed-and-bleed mode.

Pilot-Scale Tests

Experiments were performed using sanitary design, spiral-wound elements operated in semi-batch and feed-and-bleed modes for reducing the ethanol content of beer. The pilot-scale system consisted of a positive displacement pump, six stainless steel pressure vessels with endplugs, two stainless steel reservoirs, and a propylene glycol cooling system. The cold feed was pumped from the feed tank through a manifold to three parallel arrays each containing two 2.5-in stainless steel pressure tubes. The concentrate was recycled back to the feed tank and the permeate containing ethanol, water, and carbon dioxide was discarded. The feed flow rate (Q_F) was adjusted to give acceptable recovery.

Beer is processed at 5 to 12°C as this temperature range allows for the greatest permeate flux without denaturing or precipitating beer proteins and other components. The pilot-system was operated at applied pressures of 690 to 4100 kPa (100 to 600 psig) while recovery per pass was varied from 1 to 10%. Feed and permeate flow rates were measured and the composition of samples was determined by chemical analyses. Data were collected for ethanol flux, ethanol rejection, and permeate flux as functions of pressure.

The general experimental approach was to operate the system in a semi-batch mode, recycling concentrate and mixing it with water added at a flow rate (Q_W) equal to that of permeate (Q_P) until the desired concentration of ethanol was obtained in the feed. Steady state performance measurements could then be made by operating in either a once-through or total-recycle mode, or the water make-up rate could be adjusted, fresh beer added to mix with the recycled concentrate, and a portion of the concentrate collected as product in a feed-and-bleed mode.

To verify that the alcohol content of the beer was actually reduced during the processing, samples of the processed and unprocessed beer were analyzed according to the protocol of the American Society of Brewing Chemists (9). The american pilsner beer used for the reduced-alcohol beer experiments was analyzed for original, real, and apparent extract, specific gravity, and ethanol content. The beer was also assayed for total protein and iso-α-acids. Beer samples were analyzed by gas chromatography for important flavor and aroma components including n-propanol and i-amyl alcohol.

RESULTS

In processing beer, the primary objective is to reduce the ethanol concentration in beer to a specified level with no loss of desirable components. The results of analyses listed in Table 1 show that by membrane processing the ethanol content of an american pilsner beer was reduced by 62 to 74% without loss of real extract, protein, or flavor components. The initial ethanol content of the beer was 3.4 wt%. After processing, the ethanol content of the "no-alcohol" beer was 0.9 wt%. Virtually all real extract was retained during processing. The value of the real extract was 3.7% for the initial sample, 3.9% for the low-alcohol sample, and 4.1% for the "no-alcohol" sample. The real extract increased during processing due to a proportional slight increase in volume. No substantial difference in the concentrations of taste and flavor components was observed between the processed and unprocessed beer. It is interesting to note that while the ethanol concentration decreased markedly, the concentrations of the higher alcohols remained nearly constant.

Table 1. Component analyses for American Pilsner beer as compared to low-alcohol and "no-alcohol" beer produced by RO from American Pilsner beer.

ANALYSES	REGULAR BEER	REDUCED ALCOHOL BEER BY REVERSE OSMOSIS	
		LOW-ALCOHOL BEER	"NO-ALCOHOL" BEER
ALCOHOL CONTENT (Wt %)	3.38	1.29	.85
REAL EXTRACT (Wt %)	3.74	3.92	4.11
TOTAL PROTEIN (mg/L)	.103	.108	.110
ISO-α-ACIDS (mg/L)	19.0	19.0	19.0
POLYPHENOLS (mg/L)	98	107	107
n-PROPANOL (mg/L)	19.7	19.2	21.8
i-AMYLALCOHOL (mg/L)	18.4	17.8	24.6

Figure 4 is a plot of TFC membrane ethanol rejection as a function of applied pressure ($P_{applied}$) using a beer feed at 10 to 12°C, which shows that ethanol rejection is not constant but increases with increasing applied pressure. For example, at an applied pressure of 1380 kPa the ethanol rejection is 34%, while at 3450 kPa the rejection is 68%. This increase in rejection with increasing pressure has an important effect on the selection of the preferred process pressure since at higher operating pressure more water will pass through the membrane with ethanol, adding to the processing costs. In contrast to the low, variable rejection for ethanol, the TFC membrane has a high rejection for chloride ion (>97% at 2760 kPa net), which is required to ensure that beer flavor components are not lost during processing.

A fundamental process parameter influencing flux and rejection for reverse osmosis systems is net transmembrane pressure differential (P_{net}). By expressing performance as a function of P_{net}, it is possible to relate results for different beers having different compositions. The term p_{net} is given as

$$P_{net} = P_{applied} - \Delta\pi \qquad (3)$$

where $\Delta\pi$ is the transmembrane osmotic pressure differential. For beer, an expression for $\Delta\pi$ consists of two parts, one for highly rejected components such as

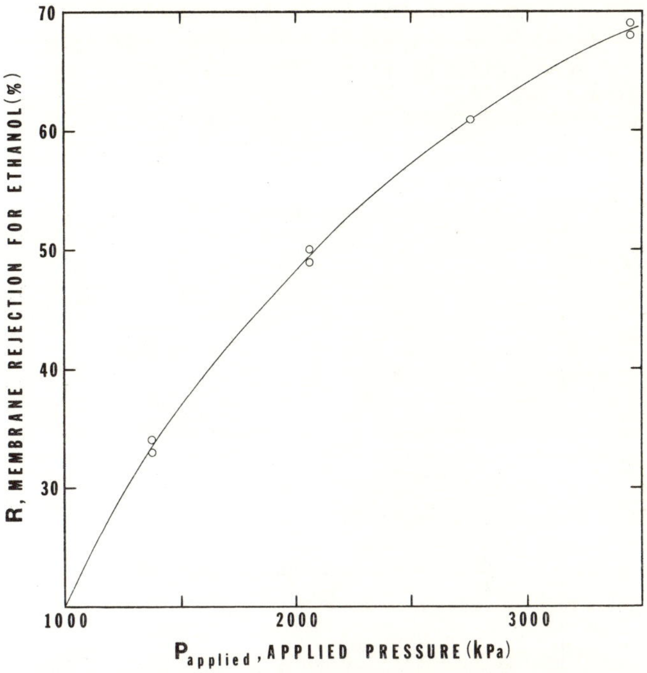

Figure 4. Membrane rejection for ethanol as a function of applied pressure for processing beer.

extract and one for modestly rejected species such as ethanol.

$$\Delta\pi = R\, C_{EtOH}\, k_{EtOH} + \pi_{EXT} \qquad (4)$$

where R = Membrane rejection at a given applied pressure

C_{EtOH} = Ethanol concentration in feed

k_{EtOH} = Osmotic pressure constant for ethanol

π_{EXT} = Osmotic pressure of extract

As data in Figure 4 show that R is a function of applied pressure, $\Delta\pi$ is also a function of applied pressure and must be calculated for each $P_{applied}$ used.

The osmotic pressure of the extract can be estimated using a graphical procedure, as shown in Figure 5, a plot of permeate flux as a function of applied pressure for a beer feed and an ethanol-water feed. For a totally rejected entity, extrapolation of the lines to zero permeate flux will give the osmotic pressure of feed solutions. For ethanol feed (3.5 wt%), the extrapolation to zero permeate flux gives an osmotic pressure value of zero for the feed solution because the rejection of ethanol approaches zero at zero pressure. Because the extract is highly rejected by the membrane, a value for the osmotic pressure of the extract can be deduced by extrapolating the beer feed line to zero permeate flux, shown on Figure 5 as about 500 kPa.

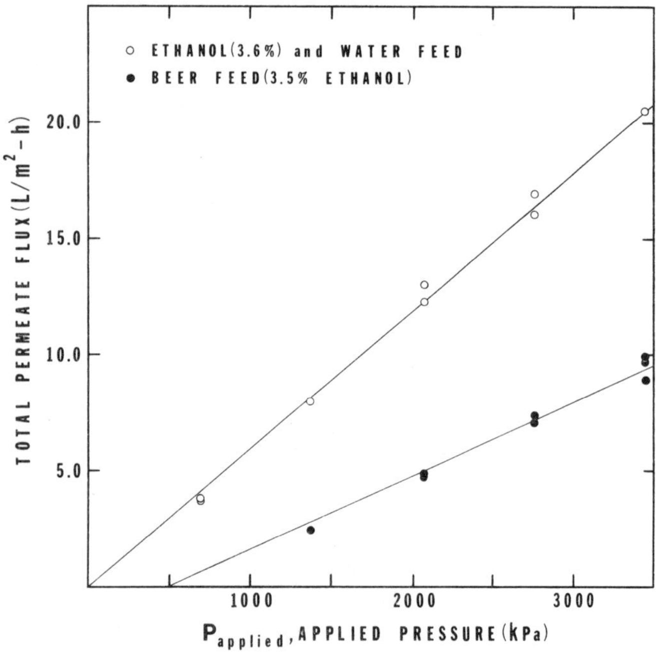

Figure 5. Permeate flux as a function of applied pressure for ethanol-water and beer feeds.

Figure 6 is a plot of ethanol flux and water flux as functions of net pressure, which was calculated using Equation (3). Water flux increases linearly with increasing net pressure. For example, at 250 kPa the water flux is 1.4 L/m^2-h while at five times that pressure (1250 kPa) the water flux has increased by five times to 7.0 L/m^2-h. On the other hand, it is interesting to note that while at low net pressure, ethanol flux also increases with increasing pressure, at high net pressures ethanol passage is nearly independent of net pressure. For example, at 250 kPa the ethanol flux is 0.05 L/m^2-h, but above 1500 kPa ethanol flux reaches a plateau and increases very slowly with further increases in net pressure. This indicates that there would be no benefit to operating a system at a net pressure greater than about 1750 kPa. It is also interesting to note that the shape of the ethanol curve is not due to concentration polarization since the water flux line shows no change in slope. The difference in slopes for the water and ethanol flux curves in Figure 6 accounts for the variable ethanol rejection as a function of pressure shown in Figure 4.

Figure 7 is a plot of the ratio of ethanol concentration in processed beer at time t (C_F) to ethanol concentration in unprocessed beer (C_B) as a function of time and S value for a semi-batch mode of operation at about 690 kPa net. Very good agreement is shown between actual empirical

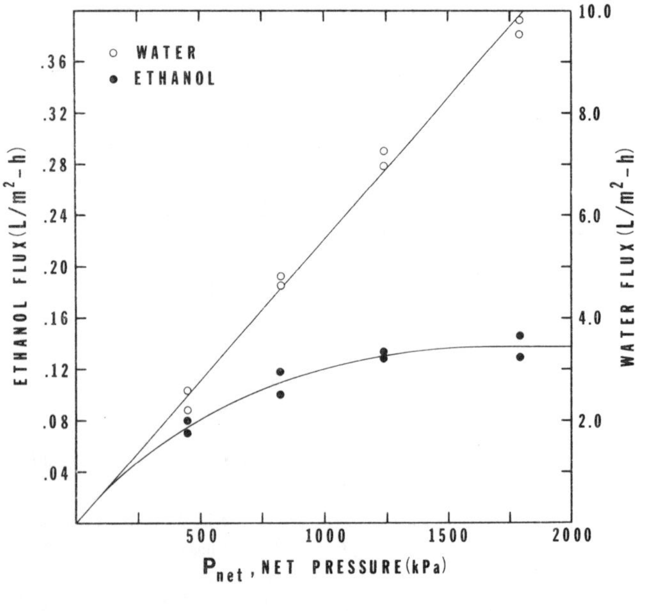

Figure 6. Flux for ethanol and for water as a function of net pressure.

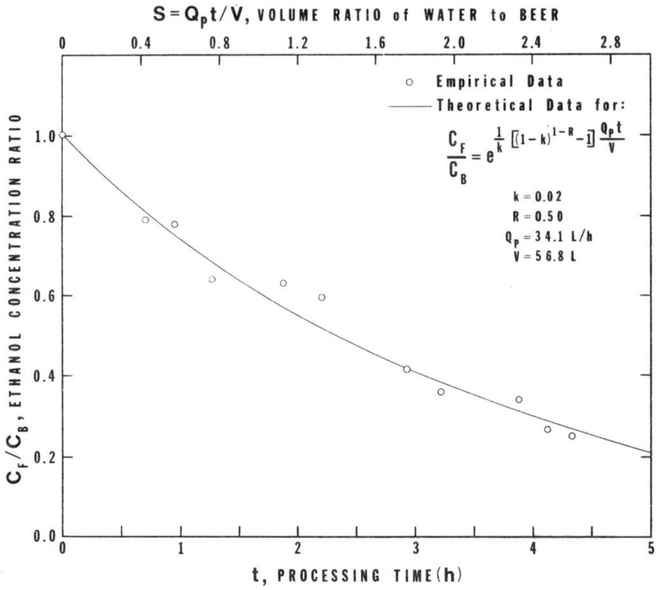

Figure 7. Ethanol concentration ratio for semi-batch operation as a function of time and S value.

data and theoretical predictions. For the conditions of the test, a reduction of 50% of the ethanol concentration required 2.3 h of processing. Had that been the goal of a commercial process either the semi-batch cycle would have been completed at this time or the process would have been converted to a feed-and-bleed mode for further processing in a continuous, steady state manner. The 50% ethanol reduction required 1.4 L water per L beer.

DISCUSSION

To design effective membrane systems, it is important to understand the factors that determine the flux and separating capability of the membrane. According to the widely held solution-diffusion mechanism, a distinction is made in aqueous membrane separations between the solution solvent (i.e., water) and the solution solute (e.g., salt). The driving force determining solvent flux is the net pressure differential across the membrane. For typical operating pressures, the solute flux is independent of net pressure differential and is determined by the transmembrane solute concentration gradient.

For the removal of ethanol from aqueous beverages, a TFC membrane has been developed that treats water as a conventional solvent, but treats ethanol as both a solvent (at low net pressure) and a solute (at high net pressure). This behavior is the basis for a novel approach for system design (6). A key additional feature of the TFC membrane is that it has an inherently high rejection for chloride ion (>97% at 2760 kPa net pressure). This is required to ensure that flavor components in beer are retained by the membrane.

The behavior displayed by the TFC membrane for ethanol separation could be due to several mechanisms. At low net pressure, ethanol flux is low and the transmembrane ethanol concentration gradient is relatively flat, with ethanol solubility being an important factor in determining the overall ethanol concentration in the membrane. For low net pressure, flux could increase with increasing pressure, like a solvent, due to pressure-effected displacement of ethanol in the polymer by ethanol in the feed. As pressure is further increased, water flux increases linearly, and the overall concentration of ethanol in the membrane could decrease due to the dilution affect of water. This dilution could result in establishing a steeper transmembrane concentration gradient for ethanol and causing the ethanol concentration in the membrane to no longer be solubility limited. With a steeper concentration gradient created, the ethanol flux would become less dependent on pressure and would become more dependent on transmembrane concentration gradient diffusion like a solute.

Selection of optimal process conditions is determined by cost factors, several of which depend on the net pressure and the influence it has on the relative passage of ethanol and water. The preferred pressure must be high enough that there is a substantial permeation rate for ethanol to give an acceptable membrane system size, but not so high that there is a relatively large flow rate for valuable water that permeates along with the ethanol. Also, the recovery per pass must be low enough so as not to cause fouling or precipitation of beer components, yet high enough to give an acceptable recirculation rate, pump size, and energy costs.

The sanitary design, spiral-wound elements described here have been shown to be capable of reducing the ethanol content of beer when operated in either a semi-batch or a feed-and-bleed mode. For the semi-batch mode, the reduction in ethanol concentration can be changed by varying the cycle time. For the feed-and-bleed mode, the ethanol concentration in processed beer can be varied by employing multiple stages in series. More than one stage is preferred when a reduction of more than about 70% of the ethanol in beer is desired. For a given application, selection of the preferred mode of operation will depend on cost factors and the preference for a continuous versus a semi-continuous process. In comparing modes, two process distinctions are worth noting. One is the higher costs for energy and pumps for repressurizing all of the recirculated flow for a semi-batch mode. The other is the higher cost for make-up water for a feed-and-bleed mode.

For a given net pressure, membrane rejection, and recovery per pass, more water is required in a feed-and-bleed mode than in a semi-batch mode because the ethanol feed concentration the membrane contacts is less in the feed-and-bleed mode than in a semi-batch mode. An equation for calculating the ratio of the volume of water required for a feed-and-bleed mode to that for a semi-batch mode to reduce the alcohol content by 50% (Z) is the following

$$Z = \frac{1.44[1-(1-k)^{1-R}]}{k-1+(1-k)^{R}} \qquad (5)$$

For example, for a recovery per pass of 4% and a rejection of 50%, Z is 1.47, meaning that 47% more water is required for a feed-and-bleed mode than for a semi-batch mode.

The cost of water has a major impact on the selection of the preferred process mode and the preferred applied and net pressures. If the cost of water is high, the preferred net pressure is low to reduce the amount of water expended during processing, the optimum net pressure being typically lower for the semi-batch than for the feed-and-bleed mode. For this case, overall annual costs can be lower for the semi-batch mode, primarily because the cost for expensive make-up water can be less than half that for the feed-and-bleed mode. This cost difference is due to the combined effect for the semi-batch mode of a lower rejection for ethanol at lower net pressure and a greater feed ethanol concentration driving force. Total costs for semi-batch can be lower than those for feed-and-bleed even though the electrical energy cost for semi-batch can be more than twice that for feed-and-bleed, and the pump capital cost is higher.

If the cost of water is low, the feed-and-bleed mode can be preferred because total costs are lower and because it is a continuous mode which can be more easily adapted to a commercial process. The optimum net pressure would be close to the plateau region on Figure 7 (e.g., 1380 kPa) to reduce the size of the membrane system, which would be the primary cost factor for the feed-and-bleed mode when the cost of water is low.

LITERATURE CITED

1. Greig, S.W., Economics and Management of Food Processing, AVI Publishing Inc., Connecticut (1984).

2. Delaney, R.A.M., and J.K. Donnelly, "Applications of Reverse Osmosis in the Dairy Industry," in Reverse Osmosis and Synthetic Membranes, Sourirajan, S. (Ed.), National Research Council of Canada, Ottawa (1977).

3. Moonen, H. and H.J. Neifind, Desal., 41, 327 (1982).

4. Raible, J.K., U.S. patent 3,291,613 to Aktiengesellschaft for Brauereiindustrie (Dec. 13, 1963).

5. British patent 1,447,505 to Lowenbrau (Sept. 4, 1972).

6. Light, W.G., U.S. patent application 736,047 to Allied-Signal, Inc. (May 8, 1985).

7. Mooney, L.A. and W.G. Light, "Food Applications for Sanitary Design Spiral-Wound Membrane Elements," International Membrane Technology Symposium, Tylosand, Sweden (May 1985).

8. Light, W.G., Nucl. Chem. Waste Manage., 1, 99 (1980).

9. Methods of Analysis of the American Society of Brewing Chemists, 7th ed., American Society of Brewing Chemists, Inc., Minneapolis (1976).

REVERSE OSMOSIS PROCESS FOR PRODUCING PHARMACEUTICAL-GRADE WATERS

B.S. Parekh, F.W. Wilkins, P.L. Parise, L. Roettger ■ Millipore Corporation
80 Ashby Road, Bedford, MA 01730

Pharmaceutical companies use a variety of high purity water such as USP Purified Water (PW) and Water for Injection (WFI). The processes utilizing the high purity water are drug manufacture, rinsing of autoclaves and process equipment, and formulations manufacture. Table 1 presents the minimum water quality requirements for the USP Purified Water and WFI as per criteria listed in the U.S. Pharmacopeia(1). The technologies employed to produce the Purified Water are ion exchange, reverse osmosis, and distillation; Water for Injection must be produced by distillation or reverse osmosis.

Currently, distillation is an overwhelming choice for producing the pharmaceutical grade waters. The reverse osmosis process is just beginning to make an impact on this industry, which demands sanitary design of system components, and a reliable, validatable process design. This paper will examine the use of reverse osmosis for the production of high purity water for the pharmaceutical industry. We will describe the development of sanitary design RO systems, present performance data from evaluation of a prototype unit and a field unit, discuss pretreatment options, and compare the operating and capital costs of comparable size distillation and reverse osmosis systems.

PHARMACEUTICAL RO SYSTEM DESIGN

An RO system designed to meet the stringent water quality requirements of the pharmaceutical industry must include features that minimize the growth of microorganisms (including pyrogenic by-products) and minimize the chemical contaminants that could leach into the water by system components. Also, "in-place" sanitization methods are essential to efficient and reliable operation.

Millipore has designed a series of single pass and two pass RO systems (Milli-RO S) specifically for the production of high purity waters(2). An example of the flow paths of two and single pass systems is given in Figures 1a and 1b. An illustration of a two pass system is shown in Figure 2.

A uniquely designed RO membrane module is a key component of the pharmaceutical RO systems. The RO module (spiral wound) has been designed to minimize stagnant areas within its structure and to be easily sanitized by chemical application.

The nominal module specifications are given in Table 2. The RO module contains materials selected to conform to appropriate FDA codes(3). A unique taper fit SS interconnector(4) is used to make all module connections (see Figure 3). The interconnector eliminates the use and potential for failure of O-ring seals. The module includes a controlled feed-to-reject bypass flow (see Figure 4) to maintain a water flow within the space behind the chevron seal. The outer wrap used is a heat shrinkable, fluorinated ethylene propylene which, unlike tape or wrapped fiberglass coverings, is smooth, crevice-free, and very low in extractables. Also, a flow channel has been incorporated between the outer wrap and cartridge surface.

System pressure vessels are 316L SS, hold three modules each, and are electropolished inside. All product piping is 316L SS, sanitary type (with no threaded or flanged fittings). The water flow velocity is maintained $\geq$ 1.5 m/s in the piping and tubing.

Multi-stage centrifugal submersible pumps are used in series on the two pass RO systems to provide up to 6200 kPa feed pressure to the first pass membranes. The pumps rely on the feed water for lubrication and thus there is no potential for oil leaks. All wetted metal parts are 316SS.

The RO unit is microprocessor controlled to aid in system operation and data storage. Validation of processes is a critical part of pharmaceutical operations. The microprocessor controller produces repeatable system operating conditions, monitors key parameters during operation, and provides a data base for validation procedures.

SYSTEMS PERFORMANCE

A variety of tests have been performed on a prototype two-stage RO system in our laboratory. These include an extractables test on the sanitary design RO module, a sanitization test on the system, and an operational test to produce WFI.

Testing was performed to demonstrate the efficacy of the module design (including the feed stream bypass) to facilitate flushing of residual sanitization agents (NaOCl, CH_3COOH, or HCHO). Flush water flow rates were from 0.28 to 0.41 l/s at 2756 kPa feed pressure. For modules with the feed flow bypass, all residual sanitizing agents were flushed to below detectable levels in the product water within thirty minutes. For modules without the sanitary design features, including no feed flow bypass, residual chemicals remained in the product stream after one hour of flushing.

Organic extractables tests were performed under normal service conditions and after repeated sanitization cycles (with either NaOCl or CH_3COOH). During all tests, the product water met the USP criteria for total oxidizables.

A sanitary design prototype RO system was operated in our laboratory for a period of nine months. During that time, biological and chemical data were collected in order to study the performance of the unit. Figure 5 shows the pretreatment used and the flow path of the entire system.

The operating conditions for the RO system were: Feed pressure 345 kPa, 1st stage pressure 4500 kPa, 2nd stage pressure 1900 kPa, feed flow 1.07 l/s, product flow 0.57 l/s, and reject flow 0.50 l/s. The water temperature was 20°C.

Samples for bacteria analysis were obtained three times a week. Samples for chemical and pyrogen analysis were obtained twice a week. Water samples for bacteria were taken after the sampling ports were flamed and water flushed through them for one minute. The MF technique using SPC agar was used during the test period for bacterial recovery.

Ionic contaminants were determined by either specific Ion electrodes or an inductive coupled plasma (ICP) apparatus. Total solids were obtained from conductivity readings in microsiemans.

Pyrogen levels were determined by the Limulus Amoebocyte Lysate assay.

To determine the salt rejection performance of the unit, a challenge test using 1500 mg/l NaCl was performed before and after the nine-month test under the same conditions. These results were 99.7% overall rejection before and 99.5% rejection after the test, indicating no loss of rejection during the test period.

Bacteria samples were taken at various times after chemical sanitization to determine if the levels increased as a function of time.

The chemical and microbial analyses of the feed and product streams (Table 3) indicate that over the nine months of testing, the prototype system consistently produced water quality that met the specifications outlined in Table 1.

During the test, two types of sanitizing agents were used, either 10 mg/l of NaOCl (pH = 5-6) or 100 mg/l of CH_3COOOH (pH = 4-5). The results indicated that right after sanitization, the bacteria levels were quite low, approximately 0-6 cfu/100 ml.

Bacteria and pyrogen levels in the product were compared to the feed in an attempt to correlate the results. A statistical analysis found the levels in the feed had no relation to the levels in the product. Such a correlation test may be useful in confirming module integrity. Since both sanitizing chemicals sanitize rather than sterilize, total bacteria kills will not be obtained at the stated concentrations.

Evaluation of a two pass unit that has operated at a customer location for five months indicates that USP chemical requirements for WFI have been easily met (data in Table 4). In addition, pyrogen concentrations have been below the limit of 0.25 EU/ml and bacteria counts in the 2nd stage product have been in the range of 1 to 2 cfu/ml (based on a weekly sanitization cycle).

PRETREATMENT CONSIDERATIONS

A proper pretreatment of feed waters is critical to maintaining an efficient and reliable operation of RO systems. A pretreatment process should be designed to accomplish the following: It must reduce the contaminants level to enable the RO systems to produce PW or WFI; it should protect RO system components (pump, membrane/module), and the treated water quality should improve with each treatment stage. A list of recommended treatment processes for the removal of commonly occurring RO feed water contaminants is provided in Table 5. A pretreatment capable of producing RO feed of enhanced quality should employ these processes in the right sequence.

COST OF PRODUCING PHARMACEUTICAL GRADE WATER

The reverse osmosis membrane-based process offers significant economic advantages over the distillation process for producing the pharmaceutical grade waters(2). The key advantages among these are energy costs and initial capital costs.

In distillation the greatest energy consumption takes place in vaporizing the water (latent heat of vaporization 540 cal/g @ 100°C.) The overall energy consumption of a still depends on its efficiency. The simplest (single effect) still designs are the least efficient and are used only on a laboratory scale. For large volumes, distillation columns are staged to provide increased flow rates at a higher level of efficiency. The staging of columns results in a multi-effect still.

The key factors which determine operating costs for distillation are: (1) Energy, (2) Feed water, (3) Cooling water, (4) Discharge water (sewer cost) and (5) Pretreatment.

Reverse osmosis utilizes a pump and a membrane to purify the feed water. The process occurs at ambient temperature. Unlike distillation, there is no phase change. The impurities are removed by passing water through a membrane, using pressure as the driving potential.

The key components of operating costs for a reverse osmosis system are:

1. Energy consumption by the high pressure pump
2. Annual membrane replacement
3. Tap water (product and reject)
4. Discharge water (sewer costs)
5. Pretreatment

Table 6 summarizes the capital and operating costs of a reverse osmosis system (Milli-RO S, 0.8 l/s) and a comparable size multi-effect distillation system. The capital costs include the purchase price of the systems and pretreatment equipment. The data show the capital cost of the distillation system is 52% higher than the Milli-RO S system ($5733/1000 liters/day for distillation vs. $3000/1000 liters/day for RO). The operating costs data indicate the Milli-RO S process costs $4.70/1000 liters while the distillation process costs $20.90/1000 liters. The total of the operating cost and capital cost suggests the Milli-RO S system costs approximately 28% of the distillation costs ($7.00/1000 liters vs. $25.00/1000 liters). The payback period for the Milli-RO S process is calculated to be 1.2 years.

DISCUSSION

The development of the pharmaceutical RO systems presented in this paper now facilitates the production of pharmaceutical grade water at a much lower cost than the traditional distillation process.

The lab evaluation of the prototype unit (data in Table 3) and testing of the field unit (data in Table 4) have demonstrated the ability of the pharmaceutical RO systems to produce the pharmaceutical grade waters (as per water quality standards in the U.S. Pharmacopeia).

In designing a process based on the pharmaceutical RO systems, it is very important to provide an adequate pretreatment for feed water, a proper sanitization protocol (chemical dosage, procedure) and a means for validating the system's performance. We have attempted to incorporate these features in the pharmaceutical systems. In order to establish the reverse osmosis process in the pharmaceutical industry, we also need to work with the users, develop a performance data base with long-term testing at customer sites and, based upon feedback from these tests, develop second generation RO systems.

ACKNOWLEDGMENTS

The authors gratefully acknowledge the contribution of Douglas Higgins, Fran Carlson, George Witham, Dave Marro and Barbara McLaughlin in the development work and manuscript preparation.

LITERATURE CITED

1. "The United States Pharmacopeia" 21st Rev. January 1, 1985, United States Pharmacopeia Convention, Inc., 12601 Twinbrook Parkway, Rockville, MD 20852

2. "The Development of Sanitary Design Reverse Systems for the Pharmaceutical Industry," P.L. Parise, B.S. Parekh, R. Smith, ACS-I&EC Symposium on Reverse Osmosis and Ultrafiltration, Philadelphia, August 26 - 31, 1984.

3. Code of Federal Regulations, Title 21-Food and Drugs, Parts 175.300, 177.2600, April 1, 1982.

4. Patent application filed.

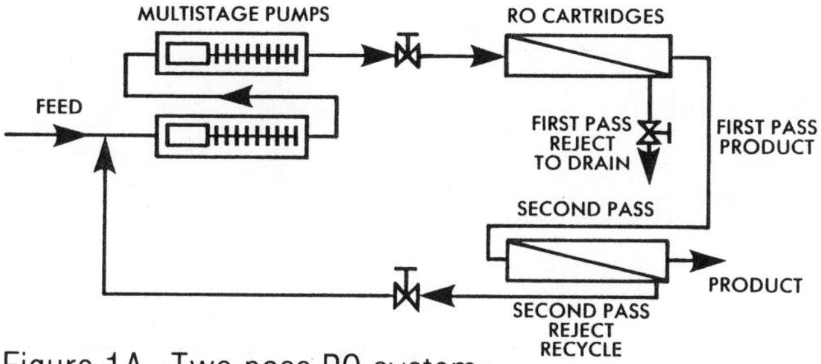

Figure 1A. Two pass RO system.

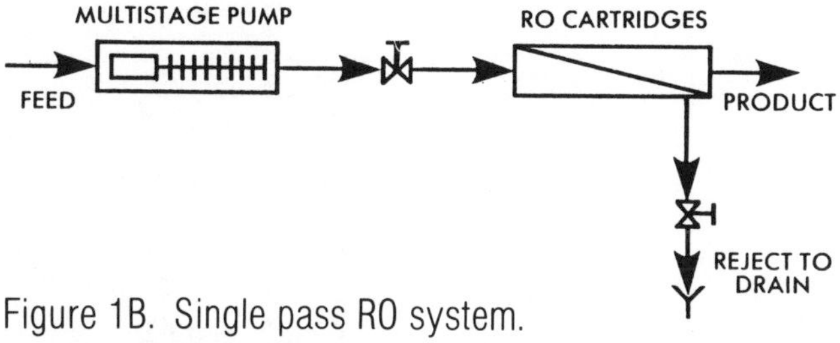

Figure 1B. Single pass RO system.

Figure 2. Milliro 2200 pharmaceutical RO system.

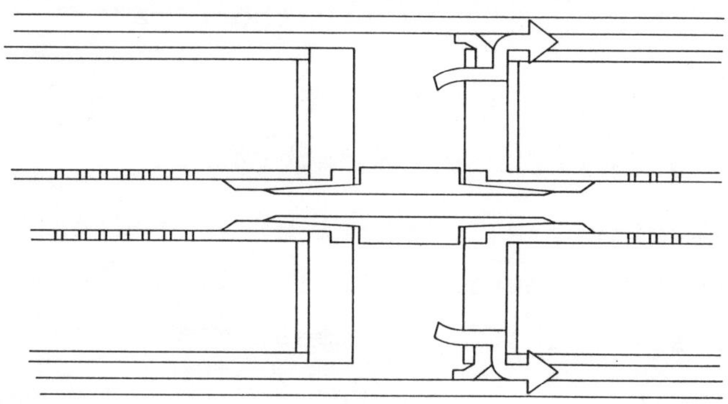

Figure 3. Taper-fit interconnector.

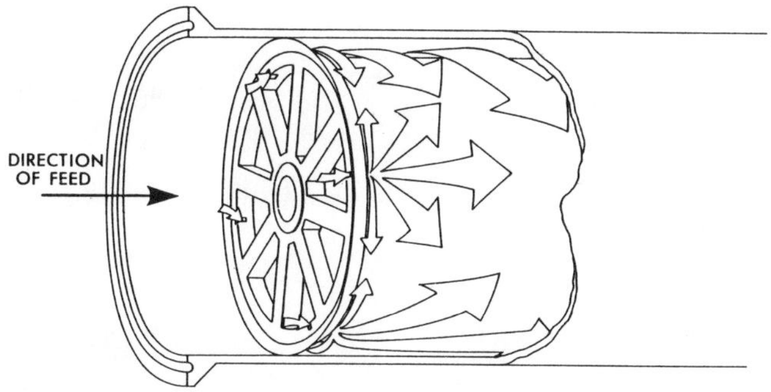

Figure 4. Feed stream bypass.

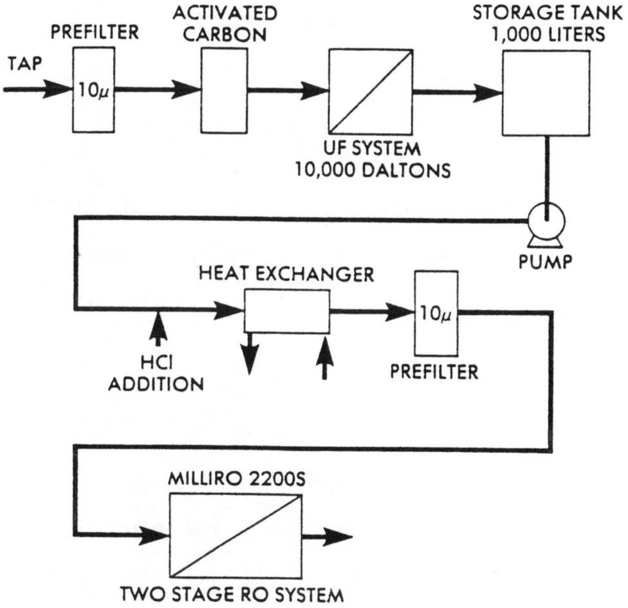

Figure 5. Pretreatment for lab system.

Table 1. Water Quality Standards

(Numerical Interpretation of USPC Standards)[1]

Constituent	Purified Water	Water for Injection
pH	5.0-7.0	5.0-7.0
Chloride	$\leq$0.5 mg/l	$\leq$0.5 mg/l
Sulfate	1.0 mg/l	1.0 mg/l
Ammonia	$\leq$0.1 mg/l	$\leq$0.1 mg/l
Calcium	$\leq$1.0 mg/l	$\leq$1.0 mg/l
Carbon Dioxide	$\leq$5.0 mg/l	$\leq$5.0 mg/l
Heavy Metals	$\leq$0.1 mg/l as Cu[2]	$\leq$0.1 mg/l as
Oxidizable Substances[3]	Passes USP Permanganate test	Passes USP Permanganate
Total Solids	$\leq$ 10 mg/l	$\leq$ 10 mg/l
Total Bacterial Count[4]	100 CFU/ml	50 CFU/ml
Pyrogen[5]	None specified	0.25 EU/ml (0.05

1 The USPC chemical test methods (except for pH and Total Solids) are quantitatively based on visual methods. The concentrations listed are the determined numerical equivalents for those tests.

2 Limits for other heavy metals may be determined.

3 Limits for specific oxidizable substances may be determined.

4 Proposed Action Guideline for the Microbial Control of Ingredien Water as issued by the USPC, effective November 1, 1983. It should be noted the manufacturers frequently impose more stringent internal guidelines.

5 As determined by LAL test.

Table 2. Cartridge Specification.

Configuration	Spiral Wound
Size	0.152 m diameter x 0.635 m length
Transmembrane Δ P	2756 kPa (normal)
	3100 kPa (max.)
Monovalent/Divalent Ion Rejection	93%/96% @ 2756 kPa, 25°C <20% recovery
Particulate, Bacterial and Pyrogen Rejection	>99%
Product Flow Rate	0.11 l/s minimum @ 2756 kPa, 25°C

Table 3. Analytical Results from Prototype Unit.

The following results are the average for the 9-month test:

	RO Feed	1st Stage	2nd Stage	Water for Injection
pH	5.7-6.4	6.4	6.3	5-7
Chloride	19.40 mg/l	367 μg/l	15.0 μg/l	≤0.5 mg/l
Sulfate	2.10 mg/l	19.0 μg/l	<10.0 μg/l	≤1.0 mg/l
Calcium	3.20 mg/l	9.0 μg/l	1.0 μg/l	≤1.0 mg/l
Carbon Dioxide	<2.0 mg/l	<2000 μg/l	<2000 μg/l	≤5.0 mg/l
Heavy Metals (as Cu)	.0025 mg/l	<1.0 μg/l	<1.0 μg/l	≤0.1 (as Cu)mg/l
Oxidizable Substances	fails USP test	passes USP test	passes USP test	passes USP Permanganate test
Conductivity	85μS	5.5μS	2.5μS	-
*Total Solids	40.4 mg/l	1430 μg/l	<150 μg/l	≤10.0 mg/l
Ammonia	0.05 mg/l	20.0 μg/l	19.0 μg/l	≤ 0.1 mg/l
Total Bacteria	283.3 cfu/ml	28cfu/100ml**	0-6cfu/100ml**	≤50 cfu/ml
Pyrogen	1.54 EU/ml	.126 EU/ml	0.067 EU/ml	≤ 0.25 EU/ml

Unit: cfu/ml = colony forming units/milliliter
EU/ml = Endotoxin units/milliliter

*Total solids values are calculated from conductivity after subtracting conductivity attributed to Carbon Dioxide

** Right after sanitization

Table 4. Typical Water Quality.
(Customer Location)

Constituent	City Water	Feed to RO*	RO Product Water
pH	7.7-8.6	5.6-6.0	5.7-6.0
Chloride	20-26 mg/ℓ	7-25 μg/ℓ	4-6 μg/ℓ
Sulfate	2-30 mg/ℓ	< 5 μg/ℓ	< 5 μg/ℓ
Ammonia	1.5 mg/ℓ	< 7 μg/ℓ	< 7 μg/ℓ
Calcium	21-31 mg/ℓ	<10-30 μg/ℓ	<10 μg/ℓ
Carbon Dioxide**	5-10 mg/ℓ	< 1 mg/ℓ	≤ 1 mg/ℓ
Heavy Metals	Pass USP Test	Pass USP Test	Pass USP Test
Total Solids	90-140 mg/ℓ	< 1 mg/ℓ	< 0.1 mg/ℓ
Total Bacterial Count	85-300 CFU/ml	2-3 CFU/ml	1-2 CFU/ml
Pyrogen	No data	No data	< 0.25 EU/ml
Oxidizable Substances	Pass USP Test	Pass USP Test	Pass USP Test

* Post Pretreatment
** Calculated

Table 5. Treatment Processes for RO Feed Water Contaminants.

Treatment Process	Contamination Removed
Depth Filtration	Particles
Ultrafiltration/synthetic adsorbent/electromedia filtration	Colloids
Softening	Scale forming salts of Ca^{++}, Mg^{++}, Sr^{++}, Ba^{++}
Green Sand	Fe^{++}. Fe^{+++}, Mn^{++}
H^{+} Addition	High pH
Ion Exchange	Dissolved ionic species
Carbon adsorption	Organic, Cl_2, chloramine
Degassification	Dissolved gases: CO_2, NH_3, low Mol.Wt. organics

Table 6. Cost Comparison: MilliRO System vs. Distillation

		MilliRO S	Distillation
I.	Capital Costs:		
	System	$115,500.00 [(a)]	$207,000.00 [(b)]
	Pretreatment [(c)]	$ 48,000.00	$ 43,000.00
	Total Purchase Price	$163,500.00 ($3000/1000 liters/day)	$250,000.00 ($5733/1000 liters/day)
II.	Operating Costs/Day		
	Energy Cost	$ 38.00	$ 862.00
	Water Cost	$ 80.00	$ 34.00
	Membrane Cost [(d)]	$ 123.00	
	Regeneration Chemicals	$ 14.00	$ 14.00
	Total Operating	$ 255.00 ($4.70/1000ℓ)	$ 910.00 ($20.9/1000ℓ)
	Capital Cost/Day [(e)]	$ 126.00	$ 192.00
III.	Total (Operating + Capital Cost) Per Day	$ 381.00 ($ 7.0/1000ℓ)	$1102.00 ($25.0/1000ℓ)
IV.	RO Savings Per Year	$137,300.	
V.	Payback Period	1.2 years	
VI.	5-Year Saving	$686,500	

(a) 0.94ℓ/s RO System

(b) 0.76ℓ/s Multi-Effect Still

(c) Automatic Separate Bed Deionizer

(d) Annual Membrane Replacement

(e) Amortized over 5 Years (260 days/Yr Operation)

THE SEPARATION OF ORGANIC VAPORS FROM AIR

Klaus-Viktor Peinemann, Judith M. Mohr,
and Richard W. Baker ■ Membrane Technology and Research, Inc., Menlo Park, California

Membranes used to recover low concentrations of organic solvents from air have extremely high membrane selectivities and are from 100 to 10,000 times more permeable to organic solvents than air. Because these selectivities are so high and because of the limitations of available vacuum pumps, it is not possible to operate these membranes at pressures that fully utilize their selectivities. As a result, membrane performance is determined by both selectivity and pressure. It can be shown that with membranes of equivalent solvent permeability there is an optimum membrane selectivity where the costs of recovering a fixed volume of solvent are at a minimum. Solvent recovery systems employing membranes with too low a selectivity consume too much energy, while systems employing membranes with too high a selectivity use too much membrane area. With membrane systems using liquid ring vacuum pumps able to economically draw a vacuum of 0.07 atm, the optimum membrane selectivity is in the range of 100 to 200.

INTRODUCTION

Thin film composite membranes have the potential of separating volatile organic solvents from contaminated air streams. (1,2) These streams are common by-products of industrial processes and represent a serious pollution problem. They also present a significant reuse and energy saving opportunity. It is estimated that total solvent emissions from industrial processes are on the order of 10 to 15 million tons per year(3). For their fuel value alone these solvents represent an annual loss of two billion dollars.

A membrane solvent recovery process is illustrated in Figure 1. The contaminated air stream from an industrial oven or other source flows past the surface of an organic-permeable, air- impermeable membrane. A vacuum pump is used to maintain a low partial pressure of solvent on the downstream side of the membrane, and solvent permeates the membrane preferentially. The solvent vapor on the permeate side of the membrane is then compressed, cooled, and condensed. The solvent depleted feed stream is recycled to the process oven. Because the solvent depleted stream is hot, recycling of the feed air to the process and allowing the heat content of the air to be recovered frequently lowers the total energy usage of the oven. For this process to be practical, membranes with high organic vapor permeabilities but low air permeabilities are required. This paper describes some theoretical aspects which are important for selection of these membranes.

THEORY

The solvent enrichment obtained with a membrane is determined by two factors. The first is membrane selectivity $\alpha_{2/1}$, defined as

$$\alpha_{2/1} = \frac{\text{Permeability of the membrane to organic solvent}}{\text{Permeability of the membrane to air } (P_1)} \quad (1)$$

The second is the pressure ratio, ϕ, across the membrane defined as

$$\phi = \frac{\text{total permeate pressure}(p'')}{\text{total feed pressure } (p')}. \quad (2)$$

The relationship between these parameters can be derived from the Fick's law expression for the membrane fluxes J_1 and J_2 given as

$$J_1 = \frac{P_1 (p_1' - p_1'')}{\ell} \quad (3)$$

and

$$J_1 = \frac{P_2\,(p_2' - p_2''}{\ell} \quad , \qquad (4)$$

where P_1 and P_2 are the permeabilities of components 1 and 2, is the membrane thickness, and p_1', p_2' and p_1'', p_2'' are the partial pressures of the two gases in the feed and permeate feed streams, respectively. The total gas pressure is equal to the sum of the partial pressures, i.e,

$$p' = p_1' + p_2' \qquad (a)$$

and (5)

$$p'' = p_1'' + p_2'' \qquad (b)$$

$$C_2' = \frac{p_2'}{p'} \ , \quad C_2'' = \frac{p_2''}{p''} \qquad (c)$$

$$\text{with } J_1/J_2 = \frac{C_1''}{1 - C_1''} = \frac{1 - C_2''}{C_2''} \qquad (6)$$

Combining Equations (1-6) then yields the expression

$$C_2'' = \frac{1}{2} \cdot \frac{1}{\phi} \left[C_2' + \phi + \frac{1}{\alpha - 1} \sqrt{\left(C_2' + \phi + \frac{1}{\alpha-1}\right)^2 - \frac{4\phi\, C_2'\, \alpha}{\alpha - 1}} \right] \qquad (7)$$

where C_2' and C_2'' are the feed and permeate volume fractions of the solvent gas.

Equation (7) is rather complex, but a plot of permeate solvent concentrations C_2'' versus ϕ can be shown to divide into three principal regions. At low pressure ratios, i.e. relatively modest permeate vacuums, when $\alpha_{2/1} >> 1/\phi$, it can be shown that the permeate concentration, C_2'', is proportional to the pressure ratio across the membrane and is independent of the membrane's selectivity, $\alpha_{2/1}$. We call this the pressure controlled region. In this region Equation (6) reduces to the expression

$$C_2'' = C_2'/\phi \ . \qquad (8)$$

At high pressure ratios, i.e. relatively low permeate vacuums, when $\alpha_{2/1} << 1/\phi$, the permeate concentration is proportional to the membrane selectivity and is independent of the pressure ratio across the membrane. We call this the membrane selectivity controlled region. In this region Equation (7) reduces to the expression

$$C_2'' = \frac{\alpha\, C_2'}{1 - C_2'\,(1 - \alpha)} \qquad (9)$$

There is, of course, an intermediate region between these two limiting cases when both the pressure ratio and the membrane selectivity affect the membrane system performance. These three regions are illustrated in Figure 2 which plots the calculated permeate organic vapor concentration, C_2'', against pressure ratio, ϕ, for a membrane with a selectivity $\alpha_{2/1}$ of 200.

RESULTS

We have measured the permeability properties of a number of rubbery polymers to oxygen, nitrogen and organic solvents. The details of these measurements are given elsewhere[1,2], but results for one polymer, neoprene rubber, are shown in Figure 3 by way of example. As this figure shows, high membrane selectivities are achieved for most organic solvents over air.

The selectivities of a number of membranes to one solvent, toluene, over air are plotted versus permeability in Figure 4. As shown for toluene, the most permeable materials are silicone polycarbonate, silicone rubber (poly(dimethyl siloxane)) and neoprene. Normally we might expect the most selective of these three materials, neoprene, to make the optimum membrane toluene recovery process. As shown below, this is not the case.

DISCUSSION

As shown in Figure 4, solvent selectivities from air of 100 to 10,000 are achievable with several membrane materials. If this selectivity is to be fully utilized, i.e. the membranes are to operate in the membrane selectivity controlled region, the pressure on the permeate side of the membrane must be maintained at a very low value, typically 1% of the feed pressure or less. The effect of pressure ratio across the membrane on the solvent concentration in the permeate for three typical values of $\alpha_{2/1}$ is shown in Figure 5.

The low permeate pressures required to operate selective membranes in the selectivity controlled region can be easily achieved in the laboratory. However, in an industrial process it is difficult and expensive to produce such low pressures, particularly with streams containing condensable vapors. These streams are best pumped by liquid ring vacuum pumps which do not operate efficiently at pressures below 0.07 atm. With a typical feed gas at a pressure of 1.2 atm this means the pressure ratio across the membrane is 0.06. Figure 5 shows that at this pressure ratio the permeate gas solvent concentration is the same for all membranes with selectivities greater than 100. With these membranes the permeate solvent concentration is only controlled by the pressure ratio. If we analyze the total membrane based solvent recovery process we can go one step further. Using a membrane with selectivities much greater than 100 at a moderate pressure ratio makes the process more expensive.

For these calculations the crossflow model (Weller and Steiner[4] and Pan and Habgood[5]) has been used. In the cross- flow model the gas on the high pressure side flows parallel to the membrane, while the permeate flows perpendicular to and away from the membrane. Spiral wound membrane modules operate in this crossflow pattern. The conditions used for our analysis are shown in Figure 6.

Using the fixed conditions shown in Figure 6, we can calculate the effect of changes in membrane selectivity of the solvent flux at a fixed pressure ratio across the membrane. Figure 7 shows the results of this calculation. The solvent flux is plotted versus membrane selectivity at pressure ratios of 0.1 and 0.01.

Even though the membrane solvent permeability is fixed, the solvent flux decreases drastically with increasing membrane selectivity. This is because at a constant solvent permeability a higher membrane selectivity means a lower air permeability. The permeate vapor pressure is therefore substantially higher, and this decreases the driving force for solvent permeation which is equal to the difference between the vapor pressure on the feed and the permeate side of the membrane.

As a result of the decrease in solvent flux with increasing membrane selectivity, the membrane area required to treat a fixed volume of solvent increases as the membrane becomes more selective, even though the membrane's solvent permeability is fixed. Figure 8 shows the membrane area necessary to recover 1000 liters of solvent per day versus membrane selectivity.

The results shown in Figure 7 and 8 would suggest that the optimum membrane selectivity is the lowest selectivity that would produce a permeate sufficiently enriched in solvent. However, this argument does not take the pumping energy used to produce the permeate vacuum into account. A low selectivity membrane requires a relatively small membrane area, but the energy requirement is larger. This is because the energy required at a fixed permeate pressure is proportional to the volume of permeate gas that has to be pumped. Since the concentration of solvent in the permeate is lower with less selective membranes, this volume is correspondingly higher. Figure 9 shows the volume of permeate in actual cfm versus membrane selectivity for a 1000 liters of solvent/day plant. The volume pumped and hence, the energy used increases drastically as the membrane selectivity falls.

ECONOMIC CALCULATIONS

Because of the opposite effects of membrane selectivity on the energy used and the membrane area required, there is a trade-off between these two factors. To calculate the optimum values of membrane selectivity which balance the capital costs of the membrane area against the cost of pumping energy, values must be assigned to these parameters. The assumptions we have used in our cost calculations are shown in Table I. The cost and operating efficiency of the vacuum pumps varies with the requirements and the pump used. Liquid ring pumps seem to be the only practical pumps for these condensable liquids. We have estimated the energy consumption of the pumps from manufacturers' literature. In their most efficient range at a vacuum of about 0.1 atm, a power of 1.1 brake hp is required to pump one standard cfm of gas. At lower pressures of 0.01 these pumps are less efficient and a power of 6 brake hp is required to pump one standard cfm of gas. The costs of the liquid ring pumps of varying capacities were obtained by graphical correlations given in references 6 to 8 These correlations are generally in good agreement with manufacturers' quoted prices.

Based on the assumptions given above, the cost of a 1000 liters/day solvent recovery system can now be calculated for membranes of varying selectivity. These costs are plotted in Figure 10. The total annual operating cost has a minimum at a membrane selectivity of between 100 to 1000 depending on the pressure ratio. The minimum cost decreases as the pressure ratio, ϕ decreases. However, liquid ring pumps able to generate pressure ratios less than 0.05 are unavailable. Therefore, the minimum practical ϕ is 0.05. The assumptions used to balance these energy and system considerations are illustrated in detail for this pressure ratio in Table II. The optimum membrane selectivity at this ratio is 150. As shown, the permeate gas is almost tenfold enriched in solvent. This solvent could be easily recovered by cooling the permeate stream. The total capital cost of the plant is $331,000 and the annual operating cost is $132,500. This operating cost could in part be offset by the value of the recovered solvent. If used for its heating value alone the solvent is not worth more than $.20/liter and its annual value is about $73,000/year. If the recovered solvent can be reused in the process, its recovered value would be higher and could easily offset the operating cost of the system completely. For example, the current value of perchloroethane (a common cleaning solvent) is $1.32 per liter. Reuse of this solvent would lead to savings of $481,000 per year in solvent costs and would quickly repay the cost of the system.

CONCLUSIONS

The costs of a membrane process to recover solvent vapors from air vary dramatically with the selectivity of the membrane even if the solvent permeability of the membrane remains unaltered. For liquid ring vacuum pumps operating in their most economical range of 0.1 to 0.05 atm, the membrane selectivity required to minimize total operating costs is in the region of 100 to 200. Less selective membranes are too energy intensive while more selective membranes require too much membrane area.

ACKNOWLEDGMENT

The authors wish to thank the U.S. Department of Energy, Office of Industrial Programs, for their financial support of this project.

REFERENCES

1. R.W. Baker, I. Blume, V. Helm, A. Khan, J. Maguire, and N. Yoshioka, "Membrane Research in Energy and Solvent Recovery from Industrial Effluent Streams," Department of Energy Report, DE84016819 (1984).

2. R.W. Baker, N. Yoshioka, J.M. Mohr, and A.J. Khan, "Separation of Organic Vapors from Air: Part I," paper submitted to J. AIChE, September (1985).

3. Control of Volatile Organic Emissions from Existing Stationary Sources, Vol. I, Control Methods for Surface Coating Operations, Environmental Protection Agency, Office of Air Quality Planning and Standards, EPA-450/2-76-028, November (1976).

4. S. Weller, W.A. Steiner, "Fractional Permeation Through Membranes," Chem. Eng. Prog. 46, 585-91 (1950).

5. C.Y. Pan, H.W. Habgood, "An Analysis of the Single Stage Gaseous Permeation Process," I&EC Fund. 13, 323-31 (1974).

6. J.L. Ryans and S. Croll, "Selecting Vacuum Systems," Chem. Eng. December 14, 1981, p. 72.

7. K.M. Guthrie, "Capital Cost Estimating," Chem. Eng. March 24, 1969, p. 114.

8. R.S. Hall, J. Matley, K.J. McNaughton, "Current Costs of Process Equipment," Chem. Eng. April 5, 1982, p. 80.

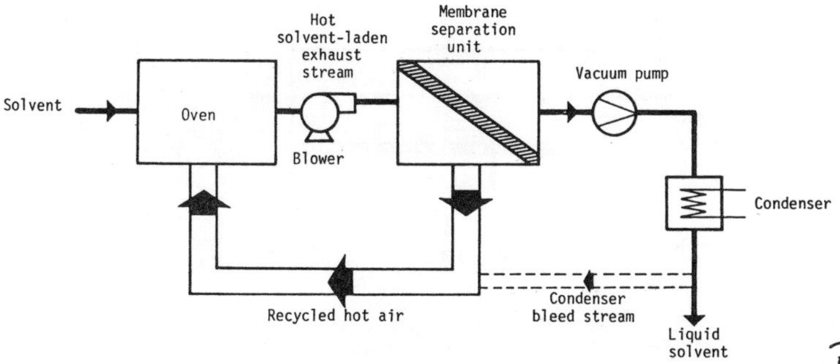

Figure 1. A solvent and waste heat recovery system.

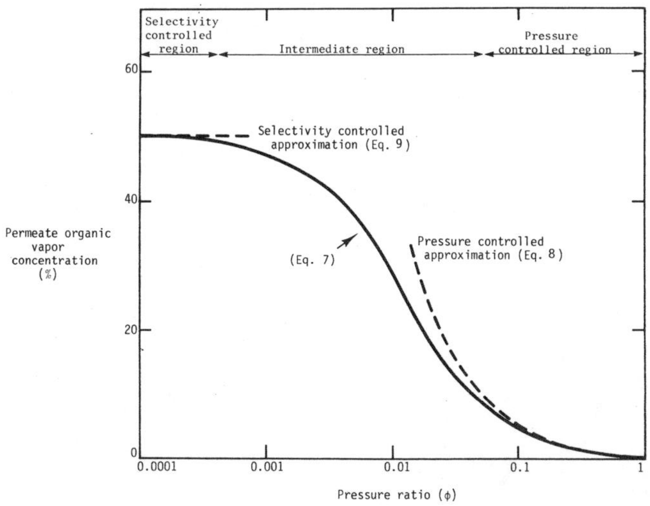

Figure 2. Calculated permeate solvent concentrations produced by a membrane with an α of 200 and a feed organic vapor concentration of 0.5%. The three different operating regions for this type of membrane are shown.

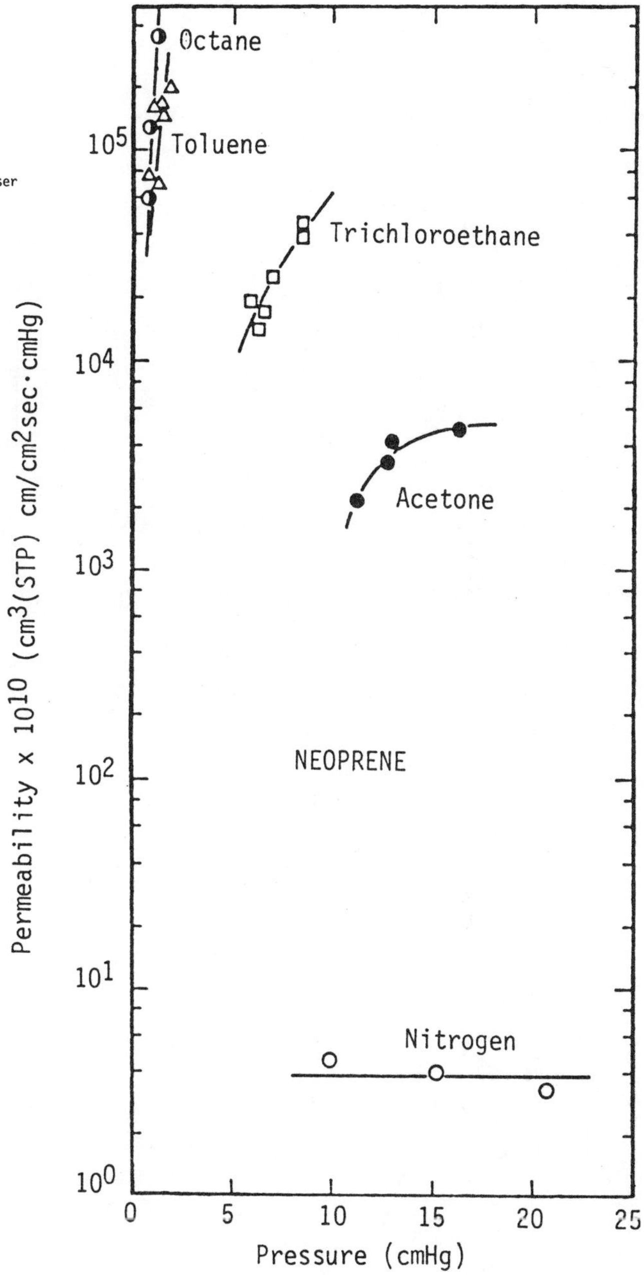

Figure 3. Permeability *vs.* pressure for nitrogen and various solvents through neoprene membranes (40°C).

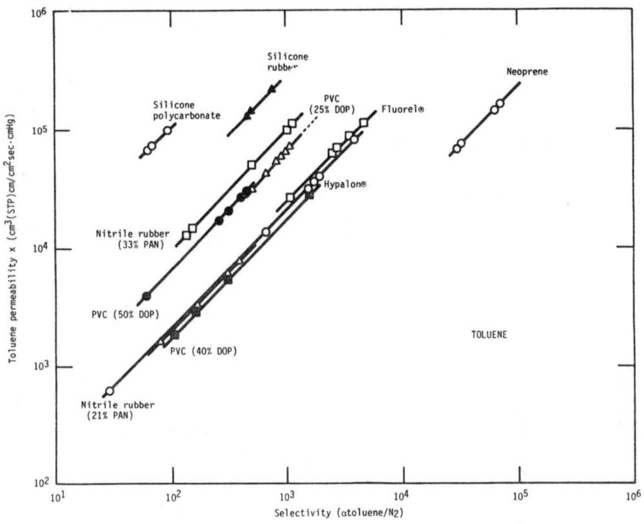

Figure 4. Permeability *vs.* selectivity for toluene in various rubber membranes (40°C).

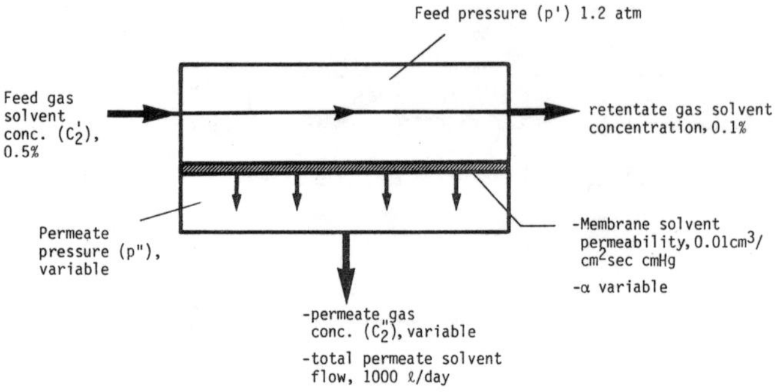

Figure 6. Schematic of a crossflow membrane module showing the fixed and variable parameters in our analysis.

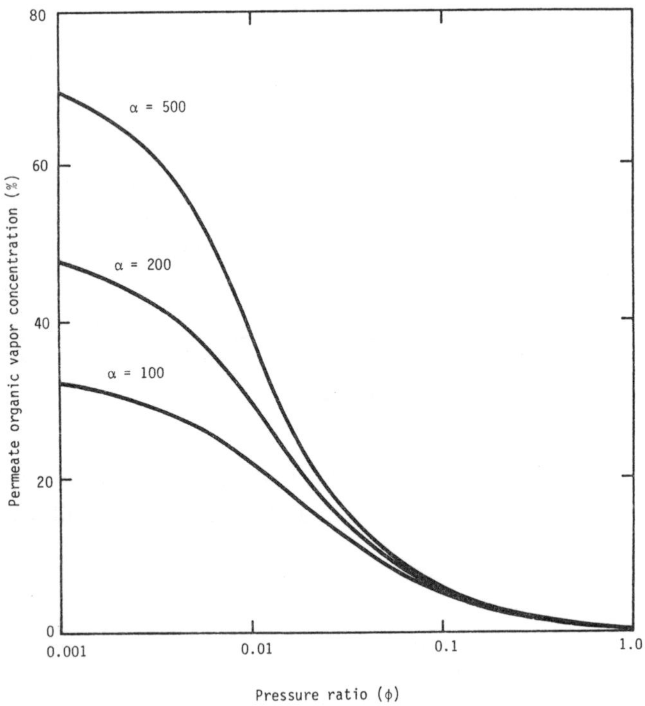

Figure 5. Permeate organic vapor concentration *vs.* pressure ratio across the membrane assuming a feed gas of 0.5% organic.

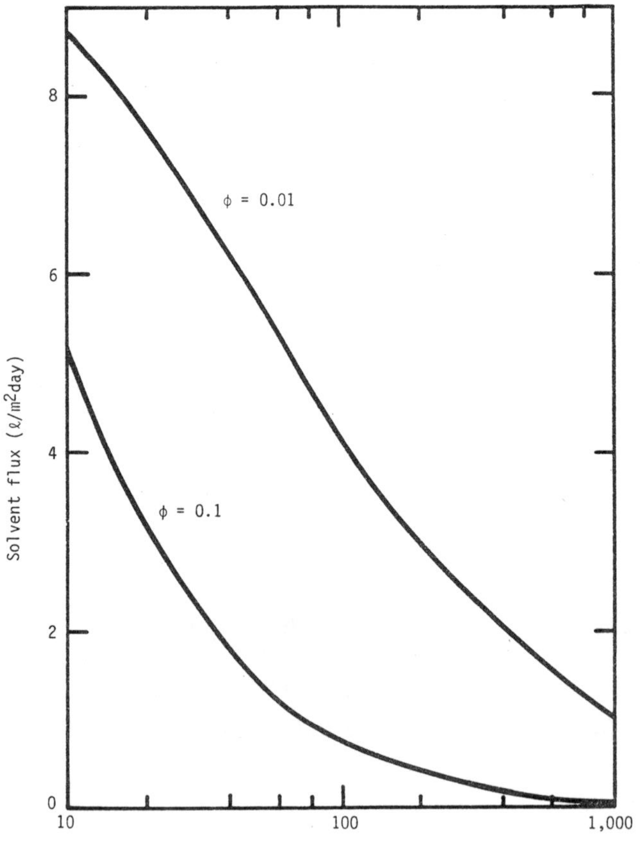

Figure 7. The effect of membrane selectivity on solvent flux for membranes of equal solvent permeability. As the membrane selectivity increases, the solvent flux through the membrane decreases.

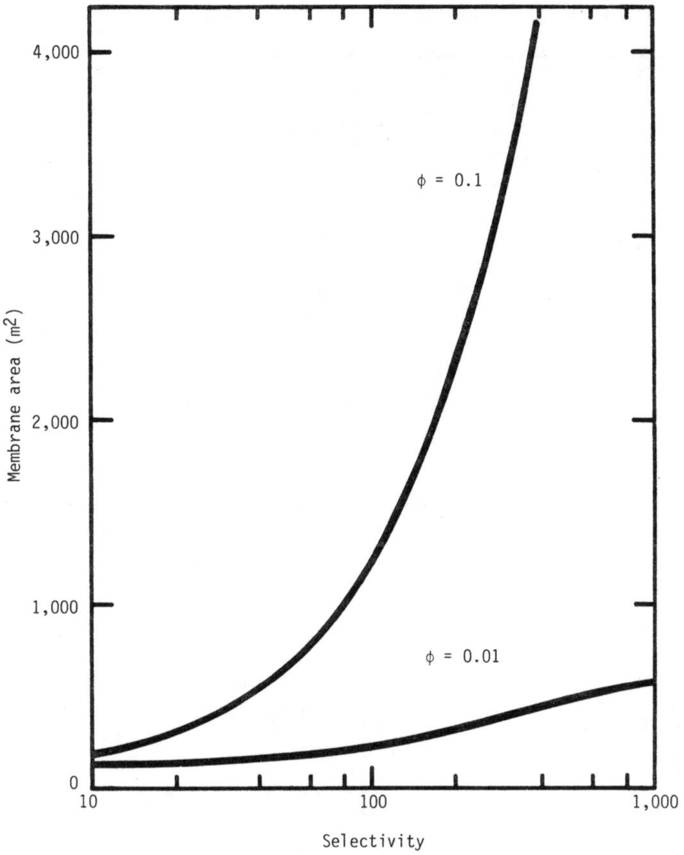

Figure 8. The membrane area of a plant designed to recover 1000 liters/day of solvent plotted against membrane selectivity. The value of the pressure ratio, **ϕ**, is indicated.

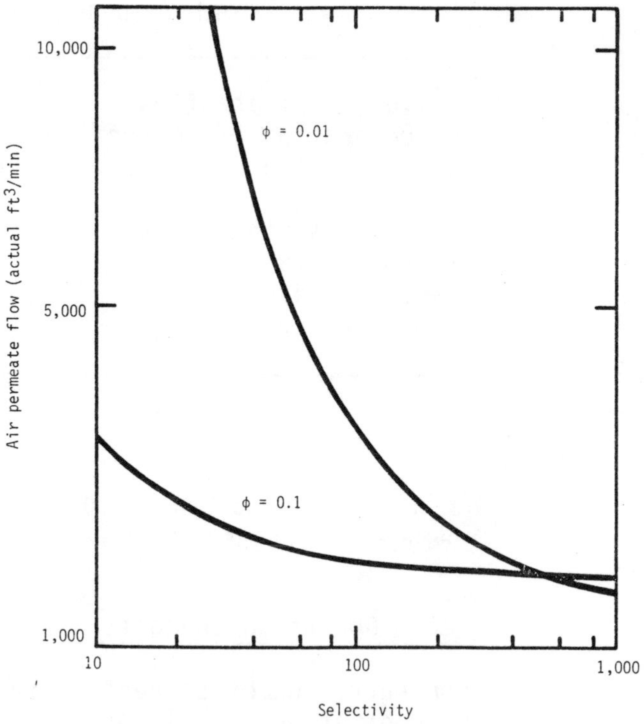

Figure 9. The effect of membrane selectivity on the total volume of permeate gas for membranes of the same solvent permeability. The total solvent flow through the membranes is fixed at 1000 liters/day. Other conditions used in these calculations are shown in Figure 6.

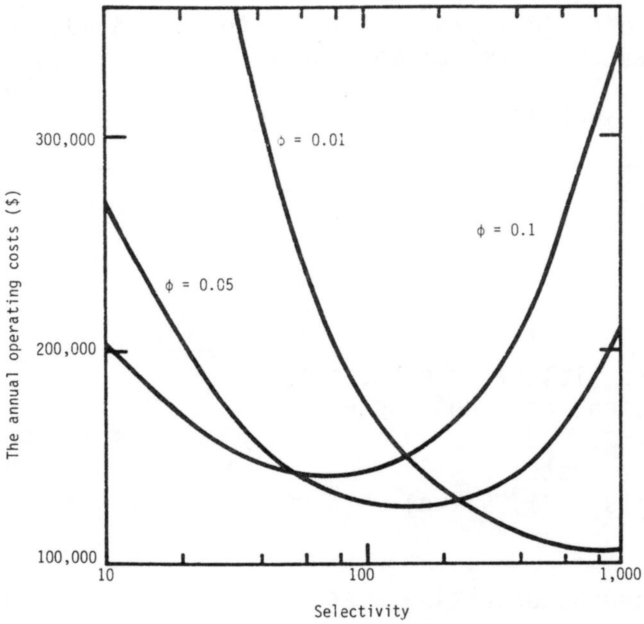

Figure 10. Plots of total annual operating costs calculated for a plant recovering 1000 liters/day of solvent *vs.* membrane selectivity for membranes of equal solvent permeability.

Table 1. Assumptions used to calculate system costs.

Solvent permeate flow	1000 liters/day
Concentration of solvent in feed air	0.5 vol%
Concentration of solvent in residue air	0.1 vol%
Membrane module costs	$40/m^2$
Membrane system costs	$40/m^2$
Energy costs	$.05/kwh
Membrane lifetime	3 years
Feed pressure	1.2 atm

Table 2. System and cost calculations for a solvent recovery plant operating at a feed pressure of 1.2 atm and a pressure ratio of 0.05.

Plant Operating Characteristics	
Concentration of solvent in feed air	0.5 vol %
Concentration of solvent in residue air	0.1 vol %
Concentration of solvent in permeate air	4.4 vol %
Volume flow of feed air	1400 scfm
Volume flow of product gas	130 scfm
Volume flow of recovered liquid solvent	1000 liters/day
Area of membrane required	1020 m^2
Pump required	200 brake hp
Blower required	17 brake hp
Membrane selectivity ( )	150
Capital Cost of Plant	
Membrane module costs (@ $40/m^2$) of membrane	$ 40,500
Other system costs (@ $40/m^2$ of membrane)	40,500
Blower costs	30,000
Pump costs (incl. installation)	220,000
Total Capital Cost	$331,000
Operating Costs	
Annual fixed costs (15% of investment for taxes, depreciation, etc. not including modules)	$ 48,000
Module replacement costs	13,500
Annual energy costs (@ $.05/kwh)	71,000
Total Annual Operating Costs	$132,500

EFFECT OF IMPURITIES ON CELLULOSE ACETATE MEMBRANE PERFORMANCE

E.W. Funk, S.S. Kulkarni and A.X. Swamikannu ■ Allied-Signal Engineered Materials Research Center
50 E. Algonquin Rd., Box 5016, Des Plaines, IL 60017-5016

This paper presents a systematic study of the effect of major acid-gas impurities on the performance of cellulose acetate membranes for the separation of CO_2 from CH_4. The focus is on H_2S, H_2O, hydrocarbons and phenol as major impurities in acid gas. The study combines membrane permeation studies of impurity-laden acid gas along with spectroscopic analysis of the used membranes. The results of this study can guide the rational design of pre-treatment processes for the control of H_2S and H_2O to the membrane element.

INTRODUCTION

The Allied-Signal Engineered Materials Research Center is making cellulose acetate (CA) membrane spiral modules for acid gas separations. Potential applications (1,2) include CO_2 recycle for enhanced oil recovery, natural gas sweetening, and upgrading of landfill gas. The CA membranes have attractive performance characteristics: CO_2 permeability of $\sim 7 \times 10^{-5}$(cm^3(STP)/cm^2.S.cmHg), [1 SCF/ft^2.D.psi(CO_2)] and CO_2/CH_4 selectivities of $\sim$25 at 24°C.

Impurities present in industrial gas streams could have significant effects on the physical and chemical characteristics of membranes. Harmful effects can be avoided by specifying pre-treatment or operating conditions.(3,4) An assessment of the effect of various impurities is important for designing a satisfactory, cost-effective pre-treatment or operating procedure. This paper discusses a laboratory study of the effect of impurities such as H_2S, H_2O, and polar as well as non-polar hydrocarbons on CA membranes. CA membranes as well as polymer resins were exposed to accelerated testing conditions. Membrane performance was examined and correlated with chemical, spectrophotometric, and microscopy studies.

It is well known in the literature (5,6) that penetrants that have strong affinity for the membrane are able to affect the permeation and sorption characteristics of other components in a gas mixture. As distinguished from this effect, the focus of the present paper is on whether there are any irreversible chemical or physical changes in the membrane structure as a result of exposure.

While the major focus of the paper is on the detailed chemical effect of gas impurities we also include some major process considerations and approaches for pretreatment of the feed gas.

EFFECT OF H_2S/H_2O

Experience with reverse osmosis has shown that acid hydrolysis can lead to decrease in salt rejection (7,8). Catastrophic failure can occur with cleaving of the ether link between the anhydroglucose units.

It is thus important to study the effect of H_2S both with and without the presence of moisture on the chemical nature of the membrane.

Membranes exposed to 10 psi H_2S for 14 days showed no evidence of deterioration (See Figure 2). Comparison of the CO_2/CH_4 selectivities before and after the run showed no significant difference, though the permeabilities for CO_2 and CH_4 were both 50% lower. Desorption of H_2S under vacuum for 10 days restored the original permeability to within 10%. With H_2S/H_2O the membranes showed an initial flux decrease before levelling off. As in the case with dry H_2S, no difference in CO_2/CH_4 selectivities could be detected. This membrane showed some surface discoloration; however, analysis by the infrared/attenuated total reflectance method was inconclusive as to whether any chemical change had occurred.

In order to sensitively detect chemical changes, finely divided (between 7.4×10^{-3} and 1.5×10^{-2}cm size) CA resin with 3 different degrees of acetyl substitution (2.4, 2.5 and 2.8 DS) was exposed to H_2S at 15-20 psi, saturated with moisture. Exposure was continued for 30 days with samples withdrawn at regular intervals for analysis.

Gel-permeation chromatography of these samples was performed using two different solvents (dioxane, tetrahydrofuran) and dextran as the molecular weight standard. The data are shown in Table 1 for the samples taken after 30 days exposure as compared to the unexposed resin. The molecular weights were relatively constant in THF and decreasec slightly in dioxane. This small decrease appears to be mainly due to loss of acetyl groups and not due to chain cleavage. IR analysis (KBr pellet) showed some evidence of deacetylation especially for the 2.8 DS case (Figure 3). The lower acetyl resins showed no experimentally detectable change in the acetyl content.

The deacetylation phenomena was further studied by the modified Eberstadt saponification method and also by C-13 NMR. The Eberstadt method gives a quantitative estimate of the remaining degree of acetylation, while C-13 NMR was used to study the substitution pattern of the primary (C_6) and secondary (C_2, C_3) acetyl groups (refer to Figure 1). The resonances of the acetyl carbonyl carbons between 169 and 171 ppm were measured and the assignment of the spectral peaks follows those made by Miyamoto et al.(10) No trend in the pattern of substitution was seen for the 2.4 and 2.5 DS CA. This indicates that the overall deacetylation is small which would be consistent with the IR results or that there is no detectable preference for hydrolysis of a particular acetyl group. It appears, however that in the case of the highly acetylated (2.8 DS) CA, primary acetyls are more susceptible to hydrolysis than the secondary groups. The primary/secondary ratio thus decreases with H_2S/H_2O exposure as shown in Figure 4.

Kinetic measurements indicate an initial rate of hydrolysis for the primary acetyl which is ~6 times faster than that for the secondary groups. Figure 5 shows the overall decrease in acetyl content over the 30 days.

Figure 6 shows two possible modes for the deacetylation of CA by H_2S/H_2O. Attack by H_3O^+ on the carbonyl carbon is probably the dominant mode, though it is possible that attack by H_2S/H^+ may also occur. X-ray photoelectron spectroscopy could not detect any carbon attached to S; however, the presence of a sulfide or a thiol species and oxygenated sulfur species were detected. Elemental analysis of 0.38% sulfur in the exposed CA also confirmed the presence of bound S.

The effect of moisture by itself on membrane morphology was also studied (see Figure 7). N_2 and CO_2 gases at 100 psi and at various humidity levels were continuously passed over small membrane samples. Moisture levels up to 20% RH appear tolerable but higher levels can cause irreversible membrane compaction. The probable cause for this is moisture condensation and re-evaporation in the porous substructure on the permeate side of the membrane. A transmission-electron microscopy (TEM) method involving a Co-doped embedding medium was used to reliably examine the membrane morphology. TEM photographs clearly show an increase in the thickness of the dense skin (Figure 8).

EFFECT OF HYDROCARBONS

Various polar impurities (phenols, mercaptans) as well as non-polar hydrocarbons present in the gas stream may affect membrane properties through plasticization and subsequent compaction.

A membrane performance test was run with the membrane exposed to the recycle stream from a hydrocracker pilot plant. The typical composition of this gas stream is shown in Table 2. The separation was good throughout the run, producing $\geq$ 99% H_2 as a permeate at a single-pass recovery of ∿50% and 200 psi. No compaction was observed under these conditions. The data are shown in Figures 9 (a-d).

Proton-NMR was used to quantify the H-bonding between polar impurities and CA. Figure 10 shows data for phenol; the shift towards a higher ppm value in the spectra corresponding to the hydroxyl proton of phenol indicates the strength of the H-bonding. This effect is much weaker in the case of ethyl or propyl mercaptan.

SUMMARY OF COMPONENT EFFECTS

The main objective in specifying pre-treatment conditions of the gas-stream before contact with a CA membrane is to prevent membrane compaction. Chemically, the membrane is fairly unaffected; no evidence of chain-cleavage due to H_2S/H_2O was found. A small degree of hydrolysis of acetyl groups can occur with primary groups being more susceptible than secondary acetyl. Dry H_2S did not cause any deacetylation.

The plasticizing effect of various penetrants on the physical structure of the membrane is summarized below in order of strongest to weakest plasticizer:

1. H_2O is a strong H-bond former and a powerful plasticizer. H_2O has been reported to reduce the tensile properties of CA.(11)

2. Phenols are also strong H-bond formers Decreases of 15% in the elastic and storage modulus of CA containing 10% phenol were measured.

3. H_2S is an intermediate H-bond former and a weak plasticizer. H_2S in the feed stream will decrease the permeability of other gases.

4. Mercaptans and nonpolar hydrocarbons have little effect on membrane strength.

In addition to the plasticizing effect, two impurities: H_2O and nonpolar hydrocarbons are candidates for condensing within the spiral module because of their typical feed concentrations. The difference between the two is that H_2O preferentially permeates through and condenses in the permeate side while hydrocarbons are concentrated in the reject stream.

PROCESS CONSIDERATIONS

The applicability of membrane technology is determined not only by the nature of the membrane but also by the operating conditions of the process. In the membrane design the extent of separation of a gas mixture depends upon the following factors:

a. Composition of the feed
b. Permeability of the membrane to the components in the feed
c. Temperature
d. Total pressure on both sides of the membrane (at the interface)
e. Durability (membrane degradation).

These factors determine the effectiveness of the separation process which will directly determine the capital and operating costs. This effectiveness is influenced by the membrane materials, the process design and the effect of impurities on the membranes. One important factor in the overall economic evaluation of the design is to determine the durability of the membrane. Frequent replacement of the membranes will decrease the attractiveness of the overall system. Thus the influence of impurities on the overall system is important in the economics of the design.

PRETREATMENT

One of the most limiting characteristics of membranes is the resistance of the membrane elements to temperature, mechanical stress or contaminants in the feed. These factors will effect the life of the membranes and are as important in the design as determining the separation ability of the membranes. Also, membranes are simple and easy to operate. However, upset conditions may cause harm to the membrane system. Thus, the design of the pretreatment system should take upsets into account. In effect, a reliable pretreatment step for the membranes is highly recommended to obtain optimum performance and service life. The feed should be evaluated to determine the presence of:

- particulates
- entrained liquids
- oil mist (after a compressor)
- condensable hydrocarbon
- all trace components
- water

The following pretreatment steps should be evaluated in order to determine if they should be included in the design to increase the overall reliability of the system:

A. A High Efficiency Solid-Liquid-Vapor Separator. This is necessary to remove any particulates or oil mist that may damage the membrane elements. In general, most vendors recommend that a filter coalescer be provided to remove particulates larger than 1 to 2 microns in size.

B. Knockout Drum. It is advisable to provide a knockout drum to remove liquid hydrocarbons or water before the feed gas flows to the membrane elements. This is also important in eliminating upset conditions before they reach the membrane elements.

C. A Reheat Step. It is prudent to heat the gas to about 20°F above its hydrocarbon and water dew point (at any point in the membrane element). The intent is to avoid the condensation of hydrocarbons or water and their possible adverse effect on the membrane. Insulation should be added on the feed line to insure that the temperature does not drop significantly before the gas enters the membrane. Also, it is important to design a system that insures that the feed temperature does not become too hot (from external electrical heaters, for example) because excessive temperatures will damage the membranes.

D. Component Removal Steps. It is necessary to remove certain components from the feed stream before the membrane elements. For example, in hydrogen recovery from ammonia and methanol purge streams, the ammonia and methanol contained in the purge stream is scrubbed out with water to avoid performance deterioration of the membrane units. Each component in the feed stream should be evaluated separately to determine if removal is required in the pretreatment step.

LITERATURE CITED

1. Kulkarni, S. S., Funk, E. W., Li, N. N., and Riley, R. L. AIChE Symp. Series, Electrochem Eng. and Tech.-II, p.172 (1983).

2. Goddin, C. S., "Hydrocarbon Processing," 61(5), 125 (1982).

3. Mazur, W. H. and Chan, M. C., Chem. Eng. Prog., 78(10), 38 (1982).

4. Knierem, H., Narayan, R. S., and Patton, C. J., Eur. DPat. Appl. EP 99868, A1, (1984).

5. Chern, R. T., Koros, W. J., Sanders, E. S. and Yui, R., J. Memb. Sci., 15, 157 (1983).

6. Stern, S. A. and Kulkarni, S. S., J. Memb. Sci., 10, 235 (1982).

7. Riley, R. L., Hightower, G. and Lyons,C.R., "Reverse Osmosis Membrane Research," H. K. Lonsdale and H. E. Podall eds., Plenum Publishing, (1971).

8. Vos, K. D., Burris, F. O. and Riley, R. L. J. Appl. Polym. Sci., 10, 825 (1966).

9. Genung L. B. and Mallatt, R. C., Ind. Eng. Chem., Analy. Ed., 13 (65), 369 (1941).

10. Miyamoto, Y., Sato, T., Shibata, T., Inagaki, H. and Tanahashi, M. J. Polym. Sci., Polym. Chem. Ed., 22 (1984) and ibid. 23, 1373 (1985).

11. Grant, J. N., "Cellulose and Cellulose Derivatives-Part IV," N. M. Bikales and L. Segal Eds., Wiley Interscience, p. 604, (1970).

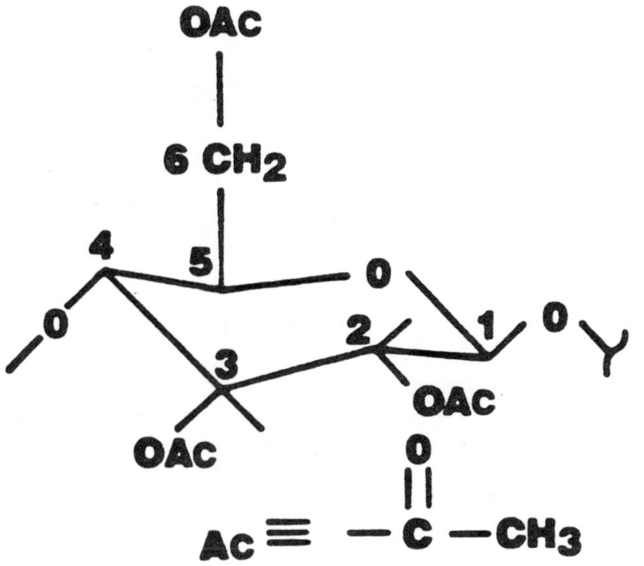

Figure 1. Chemical structure of cellulose acetate with primary (C_6) and secondary (C_2, C_3) acetyl groups.

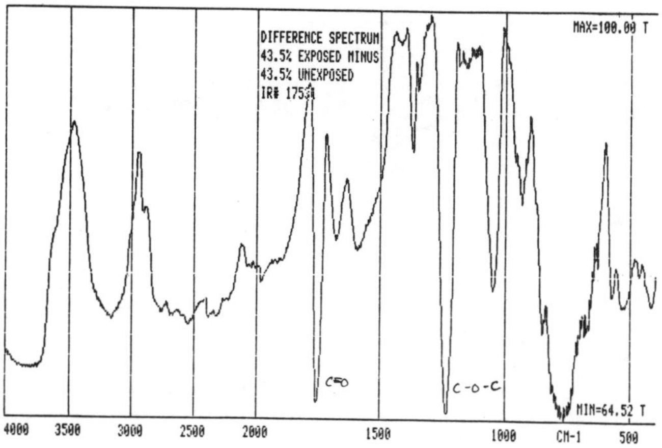

Figure 3. IR shows loss in acetyl character for 2.8 DS CA exposed to H_2S/H_2O.

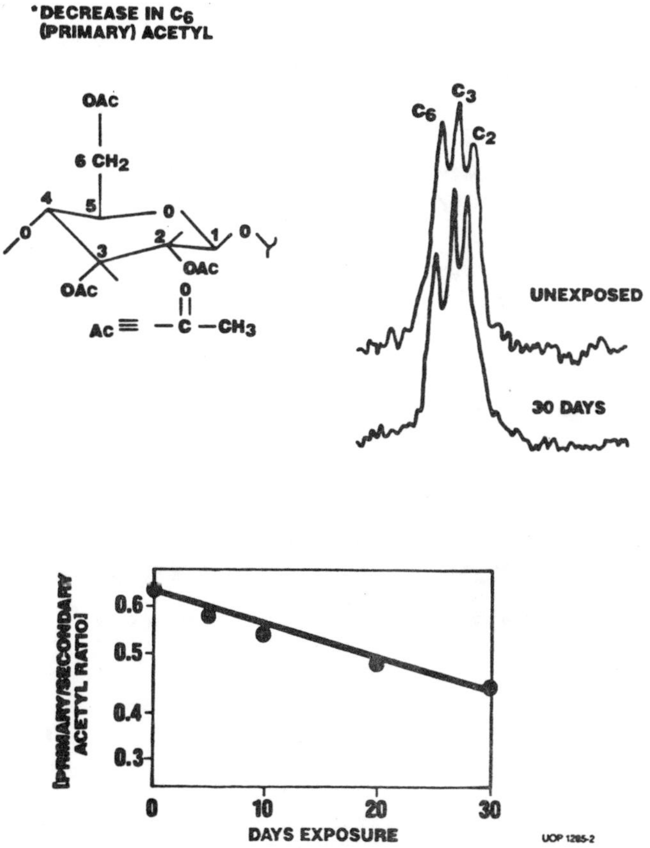

Figure 4. Primary/secondary ratio of 2.8 DS CA decreases with H_2S/H_2O exposure.

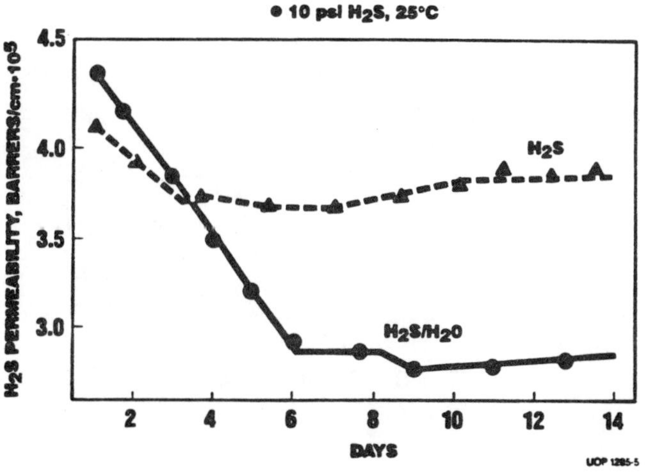

Figure 2. Permeation of H_2S and H_2S/H_2O through CA membranes at 10 psi H_2S and 26°C.

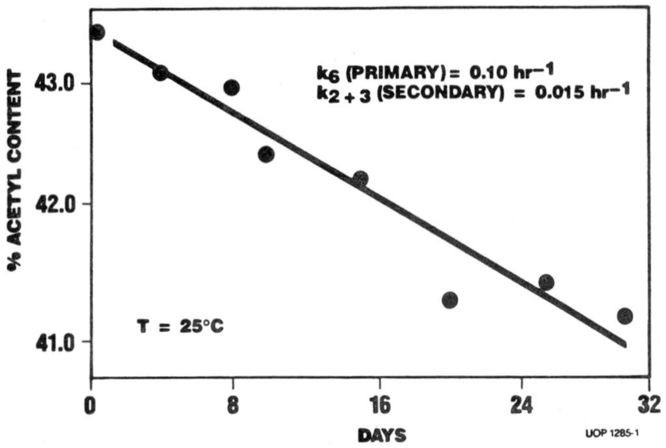

Figure 5. Decrease in overall acetyl content as a function of exposure time with 20 psi H_2S saturated with mositure.

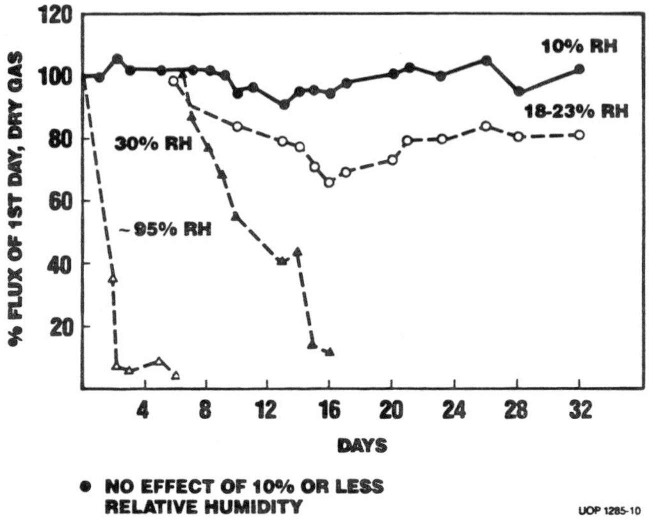

Figure 7. CA membrane permeability can be reduced by exposure to high humidity gas.

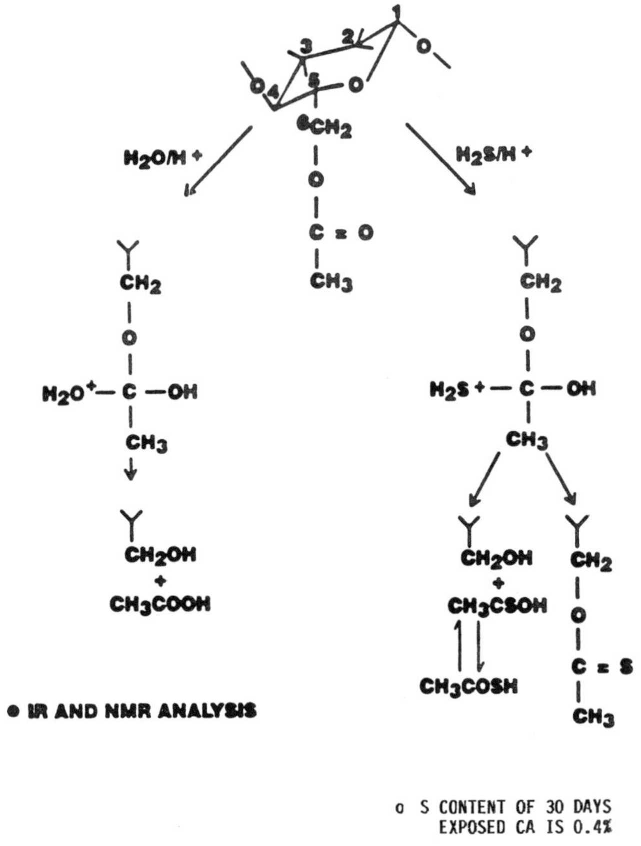

Figure 6. Two possible modes of CA deacetylation as a result of H_2S/H_2O exposure.

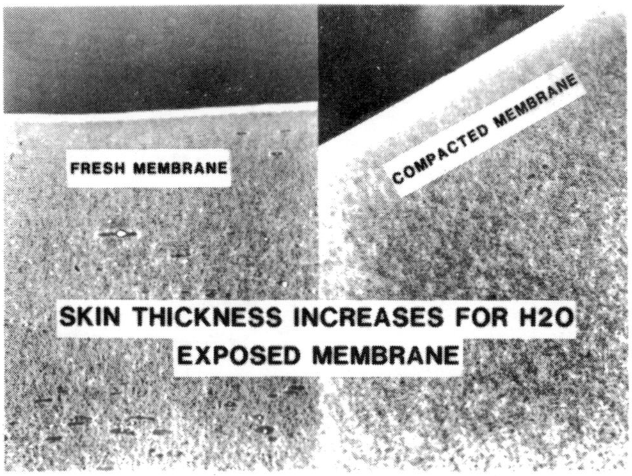

Figure 8. Thickness of dense skin increases for H_2O exposed membrane.

Figure 9A

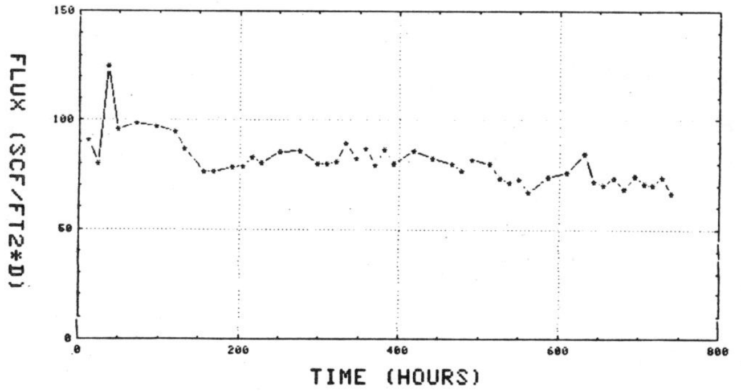

Figure 9B

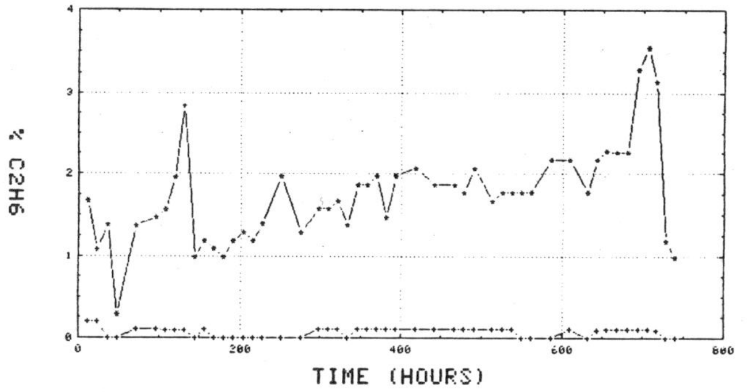

Figure 9C

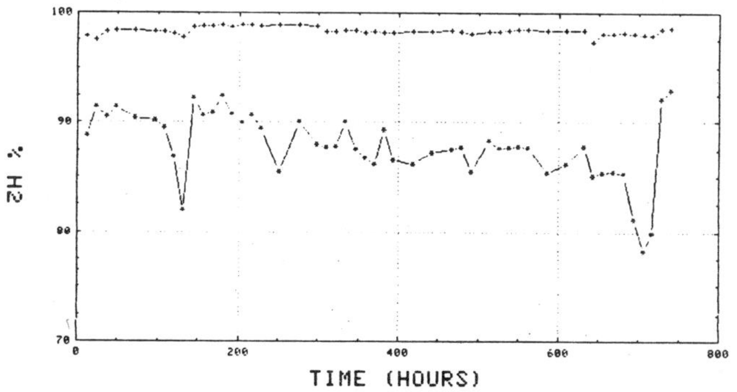

Figure 9D

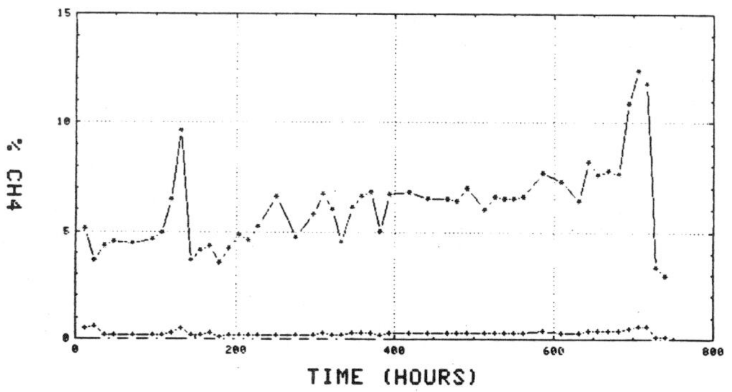

Figure 9A-D. CA membrane performance with hydrocracker recycle gas, 200 psi, 50% recovery.

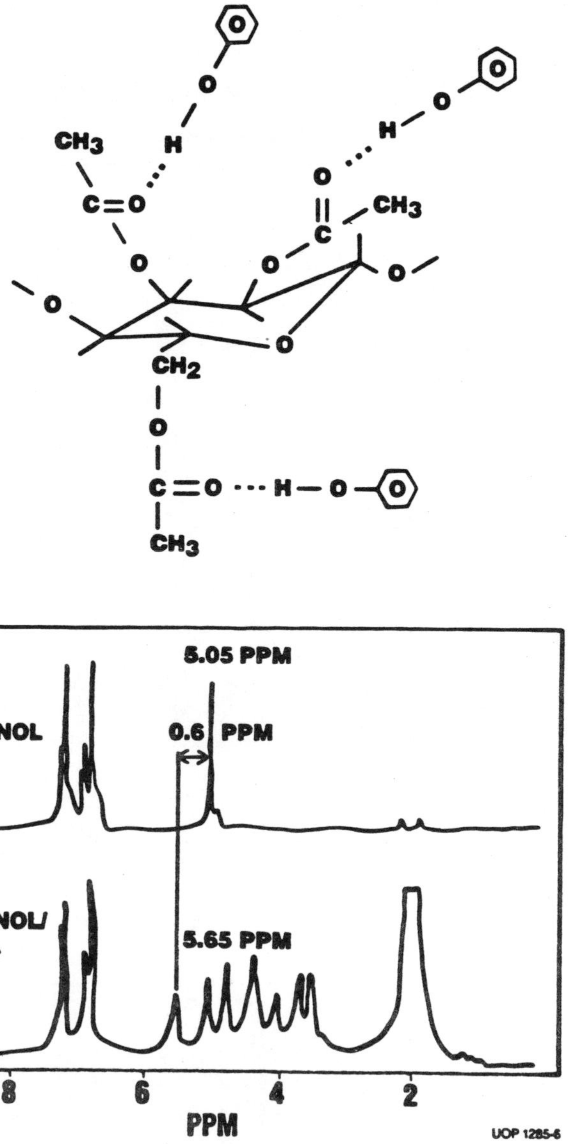

Figure 10. Proton shift in NMR spectra of phenol indicates strength of H-bonding with cellulose acetate.

TABLE 1

MOLECULAR WEIGHT Mw BY GPC ANALYSIS OF H_2S/H_2O-EXPOSED AND CONTROL CA, MEASURED IN TWO SOLVENTS

	$Mw \times 10^{-3}$			
	THF		Dioxane	
D.S. of CA	Control	Exposed	Control	Exposed
2.4	185	184	217	206
2.5	202	214	249	240
2.8	not soluble		411	359

TABLE 2

TYPICAL FEED COMPOSITION OF HYDROCRACKER RECYCLE STREAM

	Mole %
H_2	89.0
CH_4	5.6
C_2	1.9
C_3	1.2
i - C_4	0.4
n - C_4	0.4
i - C_5	0.1
n - C_5	0.1
C_6 +	0.1
H_2S	1.0
RSH	0.05 g/gmole

MEMBRANE-BASED AIR SEPARATION

Arye Gollan and Myles H. Kleper ■ A/G Technology Corporation, 34 Wexford Street, Needham, MA 02194

The separation of air into high purity oxygen and nitrogen streams has long been an important industrial process. Cryogenic fractionation of air is a well-established, widely practiced technology which is applied economically at unit capacities of 100 to several thousands of tons of gas per day. Pressure swing adsorption (PSA) for either oxygen or nitrogen generation has found a market niche in the medium capacity range of about 20 to 50 tons of gas per day and is broadening its market share to both larger and smaller systems.

For the very small volume user, requiring less than 5 to 10 tons per day of gas, on-site generation by cryogenic or adsorption technology has been, in general, economically prohibitive and liquid oxygen (LOX) or liquid nitrogen (LN) is purchased with delivery by truck or through pipelines.

On-site PSA and delivered LOX/LN each suffer drawbacks. On-site PSA is inherently a batchwise process requiring a complex valving and control system. PSA also requires heavy contact columns and is thus relatively bulky. In addition, PSA is energy-intensive for oxygen enriched air (OEA) generation. Delivered gas is subject to wide variations in price depending upon user location, is typically more expensive than on-site generation and is subject to interruptions in the delivery schedule. Delivered gas may appear to be more favorable economically if plant demand is non-continuous; however in this case, a small, cost-effective on-site system can be sized to fill storage tanks for delivery during subsequent peak demand periods. Cryogenic air separation is economically prohibitive for the small volume user and will not be discussed in this paper.

Membrane Separation Technology is an emerging air separation technique which is finding near-term niches in the small volume market (less than about 50 tons/day of gas). Membrane separation for production of either an OEA stream or a nitrogen enriched air (NEA) stream is particularly attractive when single-stage operation is employed yielding concentrations of up to 40% oxygen or 98% nitrogen. When higher purity gas is needed multi-stage membrane systems can deliver on-the-order-of 60% oxygen or over 99% nitrogen; however they are less attractive economically than single stage systems. Very high purity gas streams i.e., greater than 90% oxygen or greater than 99.9% nitrogen are impractical, and may be unachievable, with available membrane technology.

The simplicity of membrane systems is one of their overriding advantages versus competing technologies. The design, construction, operation and maintenance of membrane air separation systems are all straightforward, particularly in a single-stage unit. The modular nature of membrane systems makes them suitable for a wide range of capacities from several liters per minute to tonnage quantitites and allows system production capacity to be easily augmented by add-on modular assemblies. Furthermore, membrane systems are compact, particularly when the hollow fiber membrane geometry is used.

In this paper, the energy requirements, operating costs and capital costs for membrane systems are compared to corresponding costs for PSA sytems for both the generation of 35% OEA for typical combustion air applications and 95% nitrogen for typical blanketing applications (1). Where published data are available, the major commercial membrane systems for air separation are compared to the A/G Technology Corporation hollow fiber membrane system for both OEA and NEA generation. In addition, representative performance maps of A/G Technology membranes are presented, and the major factors affecting membrane performance are discussed. Further optimization of the A/G Technology membranes and the competitive systems are anticipated.

BACKGROUND

Intrinsic Membrane Properties

Gas separations in polymeric membranes, in the absence of any bulk reaction, occur due to different levels of solubility in, and diffusion through, the separating layer (2). Solubility is a thermodynamic property; diffusion is a kinetic quantity. The permeability, P_{ri}, of a given gas is an intrinsic property of the membrane. It is closely associated with Henry's Law for simple and non-interacting gases, and for low concentration levels is given by

$$P_{ri} = S_i \times D_i \qquad [1]$$

where,

P_{ri} = Intrinsic Permeability of gas i in the membrane, cm^3 (STP) cm/cm^2 sec cm-Hg.
S_i = Solubility of gas i in the membrane, cm^3(STP)/cm^3 cm-Hg.
D_i = Diffusivity of gas i in the membrane, cm^2/sec.

The mass transport of gas through a membrane is given by Fick's Law:

$$N_i = \frac{P_{ri}}{\delta} A \; \Delta P \qquad [2]$$

where,

N_i = Flowrate of gas i, cm^3(STP)/sec
A = Membrane area, cm^2
δ = Membrane separating barrier thickness, cm
ΔP = Transmembrane partial pressure difference of gas i, cm-Hg

For a given membrane system, the degree of separation between gases will depend upon the relative permeabilities of the gases to be separated. This ratio of gas permeabilites is termed the separation factor, or membrane selectivity, and the larger the value of this factor, the better the separation. For the separation to be economically feasible the flowrate, N, through the membrane must be sufficiently high.

By convention, the oxygen enriched air stream passing through the membrane is termed the **permeant**, while the oxygen depleted (nitrogen enriched) air stream remaining is termed the **retentate.**

It follows from Equation [2] that for a given membrane area the intrinsic permeability, P_{ri}, is not the only parameter contributing to the gas flowrate through the membrane. Equally important is the membrane thickness and the driving force across the membrane (Δ P). For economical and energy conservation reasons, the driving force must be kept as low as possible.

Thus, polymers which have high intrinsic permeability values and good separation factors are not necessarily suitable for commercial membrane systems if they cannot be rendered into sufficiently thin layers. What must be taken into account is the membrane effective permeability, Π_e , defined as

$$\Pi_e = \frac{P_{ri}}{\delta} \qquad \frac{cm^3\ (STP)}{cm^2\ sec\ cm\text{-}Hg} \qquad [3]$$

For air separation, an oxygen to nitrogen separation factor greater than 3 is necessary to achieve either 35% OEA or 95% NEA under realistic operating conditions. To perform these separations most economically, effective permeability values on the order of 10^{-4} cm^3(STP)/cm^2 sec cm-Hg are preferred to reduce the membrane area requirement which is directly related to the system capital cost.

Membrane System Operating Modes

Membrane air separation systems can be operated in either of two modes: **pressurized** (above atmospheric) or **vacuum** (subatmospheric), as depicted in the Figure 1 flow diagrams. In the pressurized mode, feed air is compressed, filtered, cooled and passed through a condensate trap before entering the membrane modules. The OEA permeant stream exits the modules at atmospheric pressure and the NEA retentate stream is available at essentially the same pressure as the compressed feed air.

In the vacuum mode, feed air is pressurized to slightly above atmospheric (about 1 psig). This prefiltered air is introduced into the membrane module and a vacuum pump on the permeant-side provides the driving force for gas permeation. The retentate is vented at atmospheric pressure.

The pressurized mode allows higher driving forces (differential pressure) than the vacuum mode resulting in reduced membrane area requirements for a given capacity system. This mode is, however, more energy intensive (especially for OEA generation) since larger volumes of air have to be compressed to higher pressures.

As a general rule, OEA systems are operated in the vacuum mode to take advantage of the lower energy requirements while NEA systems are operated in the pressurized mode to utilize the retentate at pressure.

Useful Definitions

The **recovery** of an air separation system is the ratio of the enriched gas flow to the feed air flow. For OEA generation the recovery is:

$$\text{OEA Recovery} = \frac{\text{Permeant Flow}}{\text{Feed Air Flow}} \qquad [4]$$

This definition is the same as that of the term "conversion" commonly used in liquid/solid separations.

For NEA generation,

$$\text{NEA Recovery} = \frac{\text{Retentate Flow}}{\text{Feed Air Flow}} \qquad [5]$$

The NEA recovery is therefore equal to (1 - OEA recovery).

In the following discussions, OEA and NEA membrane performance data are presented as plots of relative energy consumption factors versus system recovery over a range of compression ratios. For OEA generation the relative energy consumption factor takes into account the system compression ratio, the OEA recovery and the permeant oxygen concentration. For NEA generation this factor is a function of the system compression ratio and NEA recovery with no correction for gas concentrations less than pure (100%) nitrogen.

A further definition is required for accurate comparison of alternative OEA generation schemes since higher purity PSA or LOX must be admixed with air to achieve the desired OEA concentration for a given application. The entrained oxygen in air becomes available at essentially no cost in this blending step. The simple mass balance for calculating the contribution of the admixed air to the overall permeate oxygen content is shown in Figure 2. The actual quantity of oxygen available to the end user after blending may be termed: **Tons of Available Oxygen.**

Comparative Membrane Properties

Air separation systems based on membrane technology are offered commercially by Fluid Systems Division of UOP, Inc., Dow Chemical Company and Monsanto Company. Table 1 compares the best estimates for the intrinsic properties of the membranes offered by these companies with the air separation membranes developed by A/G Technology Corporation.

Low separation factor (<2) membranes are limited in their single-stage system's OEA concentration to about 30%, even at recoveries as low as 5% to 10% (3). Thus, these membranes are not suitable for OEA applications requiring over 30% oxygen. For 95% NEA generation, the product recovery for systems based on low selectivity membranes is about 10% (4), limiting their commercial usefulness versus alternative systems, primarily due to high energy consumption.

Membranes with low oxygen effective permeability values are not well-suited for OEA generation under the preferrable OEA vacuum operating conditions. The resultant low productivity for OEA adversely affects system economics by requiring excessive membrane areas for these membrane systems versus the UOP or A/G Technology membranes which have one to two orders-of-magnitude higher permeability. NEA generation by lower permeability membrane systems is more competitive with PSA since the NEA operating conditions, as will be discussed below, compensate for the low oxygen permeability.

The A/G Technology Corporation hollow fiber membranes offer a preferred combination of sufficiently high selectivity for oxygen to nitrogen and high oxygen effective permeability. Thus, oxygen concentrations of up to 40% or nitrogen concentrations of about 98% can be achieved in single-stage systems which operate at low energy requirements and have favorable capital and operating costs versus PSA or other competing membrane technology. The A/G Technology membranes require no break-in period for either OEA or NEA generation.

OXYGEN ENRICHED AIR GENERATION

General

Oxygen enriched air is potentially useful in a number of commercial and medical applications including:

* Industrial Burner Combustion Efficiency Improvement
* Fermentation Cell Growth Improvement
* Pulp Bleaching
* Inhalation Therapy

As an example of the benefits of using OEA, one can consider the energy conservation aspects of combustion with OEA versus air. The effect of oxygen concentration on the savings of natural gas for small torch applications is shown in Figure 3 (5). With 35% to 40% OEA combustion air, 40% to 50% natural gas savings can be achieved. It is noteworthy that above this OEA concentration range the increase in savings with increase in oxygen concentration diminishes and higher OEA concentrations are not warranted. In addition to these fuel savings, productivity improvements resulting from higher flame temperatures will further enhance the economics of OEA combustion.

In the production of OEA via membranes, both the separation factor and the effective permeability must be sufficiently high to achieve the desired oxygen concentration (e.g., 35%) and to permit the low energy vacuum mode of operation to be employed without excess membrane area requirements. Also as shown in Figure 4 for vacuum mode operation, membranes with high separation factors and low effective permeability values cannot compete on a productivity basis and will not be economical. Thus, for 35% OEA generation, systems based on membranes with an oxygen to nitrogen separation of <2 are unsuitable in terms of selectivity and systems based on oxygen effective permeability values $<1 \times 10^{-4}$ cm^3 (STP)/cm^2 sec cm-Hg are non-competitive in terms of productivity for vacuum mode operation. Membranes with high selectivity values may be more attractive in the pressurized mode for OEA generation; however this option has not been explored.

A/G Technology OEA Membrane Performance

Some representative performance data for A/G Technology Corporation OEA membranes are presented in Figures 5 and 6 for the vacuum and pressurized modes of operation, respectively. These data cover the realistic operating ranges for compression ratio and OEA recovery. It is important to note that for a given membrane (constant separation factor) the operating compression ratio and OEA recovery are the primary factors determining the oxygen concentration in the permeate.

The oxygen concentration data in Figures 5 and 6 are plots of specific energy factor versus OEA system recovery over a range of compression ratios. In the vacuum mode (Figure 5), 10% to 25% OEA recovery provides the minimum energy for a given oxygen concentration. In the pressurized mode (Figure 6), 40% to 50% OEA recovery results in minimum energy consumption. A comparison of the two figures shows, however, that for this particular membrane the vacuum mode is more energy efficient by a factor of about 3 over the pressurized mode. In applications such as OEA combustion, energy efficiency is critical and the vacuum mode of operation is preferred.

In designing a membrane system, one would elect operation on the right side of the energy minimum since this would lower the membrane area requirements (capital cost) with minimal penalty in energy consumption (operating costs). Changes in intrinsic membrane separation factor will not alter the relative relationships presented in these figures and subsequently in Figure 10 for NEA generation. Rather an increase in separation factor will shift the plots to the lower right (reduced energy), while a decrease in oxygen/nitrogen selectivity will shift the plots to the upper left (increased energy).

Comparative Analysis

For 35% OEA generation, the A/G Technology membrane system can be compared to PSA in 3 major categories: Power Requirement, Capital Cost, and Operating Cost. Delivered LOX can also be compared with regard to operating costs. Throughout this analysis, the alternative systems are compared on the basis of available oxygen at a 35% concentration as defined above and not on the basis of excess oxygen over air. An OEA plant capacity range of 1 to 20 metric tons/day available oxygen was selected to reflect the typical requirements for combustion applications. The OEA products from the membrane system and from LOX storage are typically delivered at atmospheric pressure while the PSA product can be available at low pressure.

The comparative power requirements for PSA and A/G Technology hollow fiber membrane systems are given in Figure 7. The membrane system power requirement is in the range of 180 to 200 Kwh/ton, while the PSA system power requirement ranges from about 250 to 400 Kwh/ton. The reduction in membrane system power as a function of system capacity is primarily due to improved compressor efficiencies at the larger scale. The range in power requirement for the membrane system at a given capacity indicates the trade-off available at currently achievable separation factors under realistic operating conditions. PSA power requirements were derived from vendor literature.

The capital costs for a membrane system are primarily a function of the membrane area requirement and the compressor (or blower/vacuum pump combination) cost. Both the membrane and PSA systems are envisioned to be skid-mounted and are costed as self standing units exclusive of interface with any other equipment (e.g., furnace interface and control). The capital equipment costs presented in Figure 8 are FOB, point of manufacture. The PSA system costs and the membrane system compressor costs are based on vendor contacts. Membrane module costs are estimated based on well-developed liquid separation membrane module pricing.

As can be observed from Figure 8, the A/G Technology hollow fiber membrane system has a capital cost advantage over current PSA systems for all capacities less than 20 ton/day available oxygen for production of 35% OEA. This capital cost advantage for membrane systems over PSA systems is about $15,000/ton per day throughout this capacity range.

The operating costs for a membrane air separation system are comprised of 6 major elements:

1. Power Requirement
2. Membrane Replacement Costs
3. Operator Labor
4. Maintenance and Taxes
5. Hardware Depreciation
6. Capital Charges

Assumptions used in developing the operating costs for both membrane and PSA systems are given in Table 2. Examples of the operating costs for 10 ton/day units are given in Table 3 and plots of operating costs over the capacity range of interest is shown in Figure 9 along with a comparative curve for delivered LOX.

In developing the operating costs, a 10 year life has been assigned to the membrane separation modules compared to a 15 year life for the PSA molecular sieve separation media. The molecular sieve has been assumed to be regenerable, with no negative effect on the operating cost, i.e., the molecular sieve cost is depreciated over the life of the equipment. The membrane life estimate is based on accelerated time/pressure tests. These ongoing tests have already simulated over 6 years OEA productivity with no performance deterioration at a differential pressure 4 times what the membranes would be exposed to under vacuum operation. Other factors, such as installation cost, operating costs and maintenance costs have been held constant for these two types of OEA systems.

The membrane system operating costs range from about $34/ton at the 1 ton/day capacity to about $24/ton at the 20 ton/day capacity. Throughout this capacity range the membrane system operating costs are lower than the costs for on-site PSA or delivered LOX. PSA system operating costs are estimated to be from $50/ton to $36/ton over this same capacity range, while delivered LOX costs are about $45 to $40/ton in the 1 to 5 ton/day range. Beyond a 5 ton/day capacity, on-site generation is typically preferred over delivered LOX.

NITROGEN ENRICHED AIR GENERATION

General

Blanketing operations in the storage and transportation of food and in the inerting of fuel tanks and stored chemicals are the major potential applications for nitrogen enriched air. Food storage typically requires between 88% and 95% NEA, while fuel tank inerting requirements are in the range of 91% to 96% NEA. For most blanketing applications the actual NEA requirements are higher than the theoretical requirements to account for potential air infiltration into the storage vessel.

Membrane systems for NEA production are typically operated in the pressurized mode to minimize membrane area requirements and to produce the NEA stream at pressure. In this mode, higher pressure operation (at a penalty of increased operating costs) can be used to compensate for membranes with low oxygen permeability values. The higher the membrane selectivity, the greater the NEA efficiency and the higher the NEA concentration which can be achieved with a given compression ratio.

For a typical inerting application, 95% NEA is required. Low selectivity membranes (separation factor <2) limit the NEA recovery to non-economical energy levels and hence, systems based on such membranes are not of interest for NEA generation. Membrane systems with accept-able separation factors but low permeability levels can provide NEA at acceptable recoveries. However, the low membrane effective permeability results in increased membrane area requirements and/or higher pressure operation.

A/G Technology NEA Membrane Performance

Figure 10 presents a representative performance map of A/G Technology Corporation membranes operated in the pressurized mode for 91% to 98% NEA generation (9% to 2% oxygen in the reten-tate stream). This plot is similar to the ones previously presented for OEA generation.

For 95% NEA, the optimum recovery (lowest energy case) is in the range of 20% to 30%. At a 30% recovery, a compression ratio of only 5 (60 psig inlet pressure) is required to achieve a 95% nitrogen concentration. The combination of reasonably high separation factor and high permeability allow the A/G Technology membrane system to provide 95% NEA under low pressure operating conditions without being penalized by excessive membrane area requirements.

Comparative Analysis

An economic projection of the A/G Technology membrane system for NEA generation is presented below. Also shown are best estimates for the Dow membrane system, the Monsanto membrane system and PSA for those conditions where sufficient information is available. The comparative power requirements for these four alternative systems are shown in Figure 11 for a nominal 3 ton/day capacity plant. The PSA power requirements assume high pressure (about 75 psig) operation, as most commonly used for low purity nitrogen production.

Using the PSA power requirement curve as the baseline, it is clear that the Monsanto system power exceeds PSA power at the 95% and 99% nitrogen levels (5% to 1% oxygen in the NEA) for which data are available (4). At the 95% level, the difference in power is about 50 Kwh/ton. In Figure 11 a straight line relationship is projected for intermediate NEA levels produced by this membrane system. It is of note that the economics of operating this membrane unit and other lower permeability membrane systems dictate operation at increased compression ratios, probably away from the optimum NEA recovery (see typical relationship, Figure 10). Lower energy could be achieved by operating at the optimum recovery. However this may cause an increase in membrane area which would adversely affect the system capital costs.

The energy requirements for the Dow system in the range of 91% to 97% NEA were calculated from published information (8). At 95% NEA, the Dow system has a similar power requirement to the Monsanto system.

The A/G Technology membrane power curve falls well below the alternative systems at NEA concentrations of 95% and lower. At 95% NEA, the comparative power requirements are:

System	95% NEA Power Requirement, Kwh/ton	
A/G Technology	240	(Projected, 70% compressor eff.)
PSA	280	(Vendor Information)
Dow	330	(Calculated from Reference (8), 70% comp. eff.)
Monsanto	330	Reference (4)

For the specific case presented, the A/G Technology power requirement is less than, or comparable with PSA at concentrations up to about 97% NEA. Above this concentration, the PSA system offers the lowest power alternative, although the total operating cost for a particular unit must be considered, as discussed below. Use of a higher selectivity membrane would lower the A/G Technology system power cost with a resultant slight increase in membrane (capital) costs. Furthermore, for some applications the lighter-weight, less complex membrane unit will be preferred over PSA even with a small disadvantage in energy cost.

The capital costs for NEA generation systems are compared in Figure 12 for production of 95% nitrogen. Dow system costs are plotted over a range of 1 to about 3 tons per day and are based on published information and private communications (6), (7), and (9). No information was available on the Monsanto system capital costs.

For the PSA system and the membrane systems, the slope of the capital cost versus capacity curves are similar. The Dow system is about 10% more expensive than PSA and about 50% more than the A/G Technology system on an FOB basis. For example, at a 2.6 ton per day capacity the following capital costs are estimated:

System	Capital Cost
Dow	$110,000
PSA	95,000
A/G Technology	70,000

The Dow system capital cost is from Reference (6). However it is unclear whether or not this cost includes a compressor. If it excludes the compressor, an additional charge of about $40,000 is estimated. Both the PSA and A/G Technology system capital cost projections include compressors.

The operating costs for the A/G Technology membrane system and PSA are compared in Figure 13 over a range of 3 to 20 metric tons per day of 95% NEA. Table 4 gives a detailed breakdown at the 3 ton/day capacity and includes an estimate of the Dow system operating costs at a 2.6 ton/day capacity. For all three system options, similar factors were used to generate the operating costs. The major portion of these costs stem from either the system capital costs (membrane replacement, hardware depreciation and capital charges) or from the power costs. Thus the figures given should be reasonably accurate. For the membrane systems, a 5 year membrane life was used to reflect pressurized mode operation as opposed to the 10 year life assumed for OEA vacuum operation.

The operating costs for the A/G Technology system and the PSA system are similar, about $35/ton and $37/ton, respectively. The Dow system operating costs are estimated to be about $55/ton. These operating costs translate to the following costs per 100 SCF 95% nitrogen at a 2 to 3 ton/day level:

System	Cost, $/100 SCF
Dow	0.21
PSA	0.13
A/G Technology	0.13

The costs for all three alternative systems will fall as capacity increases (see Figure 13).

By comparison, liquid nitrogen production costs are estimated to be $0.15/100 SCF with selling prices between $0.25 and $0.50/100 SCF to the end user depending on user location and market conditions.

In published literature, the Dow system operating costs are stated to be $0.12 to 0.14/100 SCF in "typical" situations (7), however the detailed calculations for this operating cost range were not presented. By maintaining the same factors for the operating costs for all systems, it is believed that the comparison presented in this paper represents an accurate relative ranking of the systems in terms of operating costs.

SUMMARY

The industrial uses for low purity oxygen (<40%) and medium purity nitrogen (90% to 98%) are well-documented. Historically, for the small volume user, on-site generation has been prohibitively expensive and delivered LOX or LN were employed if the overall economics of a given application were favorable.

Recently, two options have emerged as viable on-site air separation systems: Pressure Swing Adsorption via molecular sieves and Semi-Permeable Membrane Technology. Little technical differentiation is observed among competing PSA systems, however significant differences are found in competing membrane systems due to the intrinsic oxygen to nitrogen separation factors and oxygen permeability levels of the differing mem-brane materials.

For high purity requirements, PSA is generally preferred over membrane systems on a purely cost basis. However, for low purity oxygen and medium purity nitrogen generation, certain membrane systems offer advantages in terms of operating costs, power requirements and capital costs.

The unique performance characteristics of the A/G Technology Corporation hollow fiber membrane system makes it the system of choice for both OEA and NEA generation. The high permeability of these membranes combined with a sufficiently high selectivity for oxygen over nitrogen results in its projection as the lowest power, lowest operating cost and lowest capital cost system for air separation.

LITERATURE CITED

1. Private communications with a number of gas suppliers who requested not to be identified.

2 Ward, W. J., "Immobilized Liquid Membranes", Chapter In Recent Developments in Separation Science, Vol. 1, N. Li, editor, CRC Press Raton, FL, 1979.

3. Fluid Systems Division of UOP, Inc. Product Literature presented in Ref. (4).

4. Anderson, C. L. "Advanced On-Board Inert Gas Generator System Technology Assessment", Boeing Military Airplane Company, Prepared under US Air Force Contract F33615-84-C-2431, May, 1985.

5. Midland Ross Corporation, Surface Combustion Division, Engineering Manual.

6. CHEMENTATOR Section, Chemical Engineering, March 4, 1985.

7. CPI100, May-June, 1985.

8. Grenich, A. F., et. al., "Design of On-Board Inert Gas Generation Systems for Military Aircraft", Boeing Military Airplane Co., AIAA/AHS/ASEE Aircraft Design Systems and Operations Meeting, October/November, 1984, San Diego, CA.

9. Private communication, Dow Chemical Co., Midland, MI, July 23, 1985.

10. Haesloup, D.G., et. al., "Development and Demonstration of a Spiral Wound Thin-Film Composite Membrane System for the Economical Production of Oxygen-Enriched Air", Fluid Systems Division of UOP, Inc., Final Report, Phase IV, US DOE Contract DE-AC01-79CS-40294, January 15, 1985.

11. Johnson, R.L. and Billerman, J.B., "Aircraft Fuel Tank Inerting System", Air Research Manufacturing Co., Final Report for USAF Contract F33615-77-C-2023, Report No. AFWAL-TR-82-2115, July 1, 1983.

ACKNOWLEDGEMENT

The OEA membrane technology development and the OEA systems work described in this paper were sponsored by the US Department of Energy, Office of Industrial Programs under DOE Contract DE-AC07-83ID12429. The support of Albert Hayes and Ramesh Jain, DOE Project Monitors and John Patton, DOE Idaho Operations Office are grate-fully acknowledged. Technical guidance through-out the OEA development was provided by Joe Keller, EG&G Idaho, Inc.

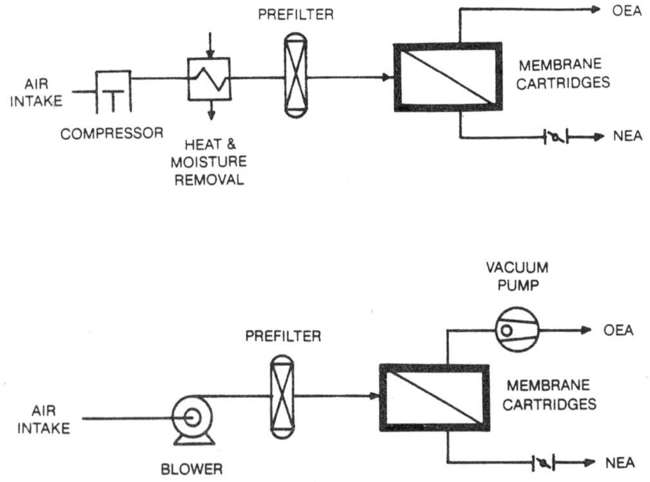

Figure 1. Simplified Diagrams of Pressurized and Vacuum Modes of Operation for Membrane Air Separation Equipment.

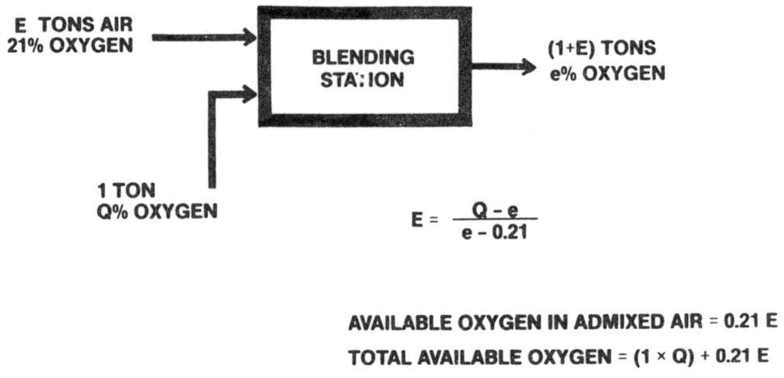

Figure 2. Mass Balance for Calculation of Available Oxygen from Admixing of Air.

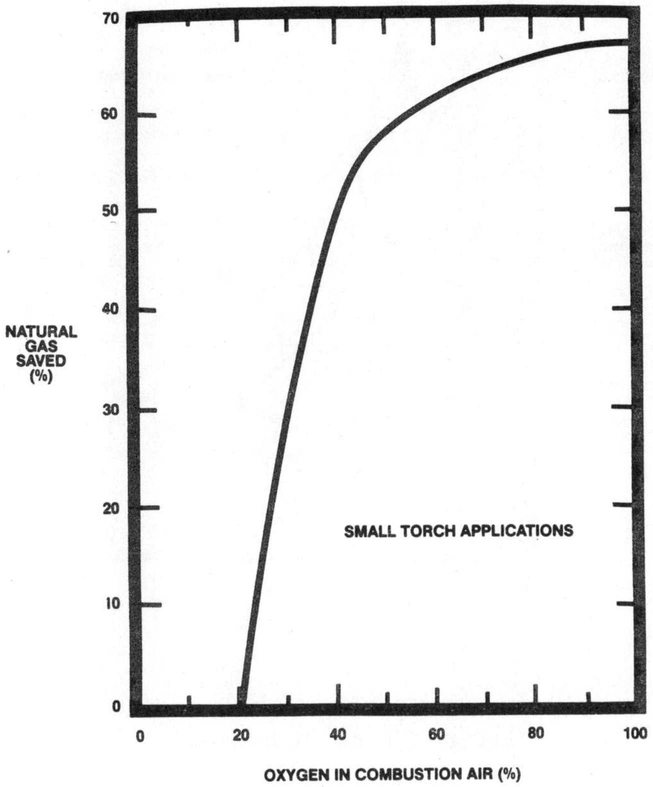

Figure 3. Natural Gas Savings versus Oxygen Concentration in Combustion Air for Small Torch Applications.

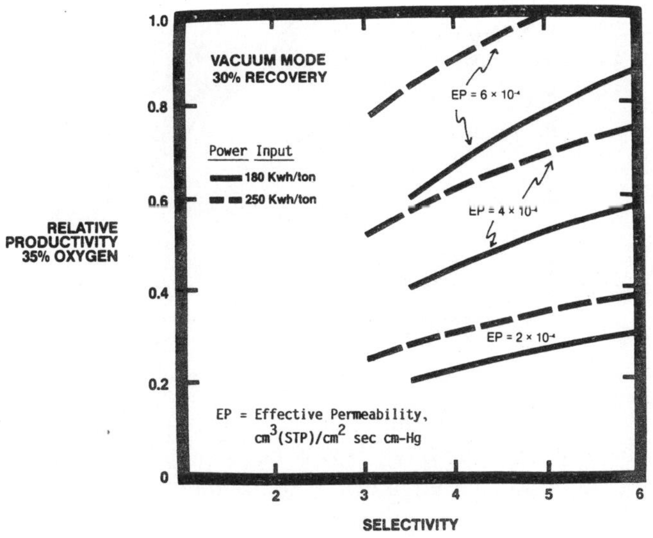

Figure 4. Relative Membrane Productivity at 35% Oxygen Purity as a Function of Membrane Performance Characteristics.

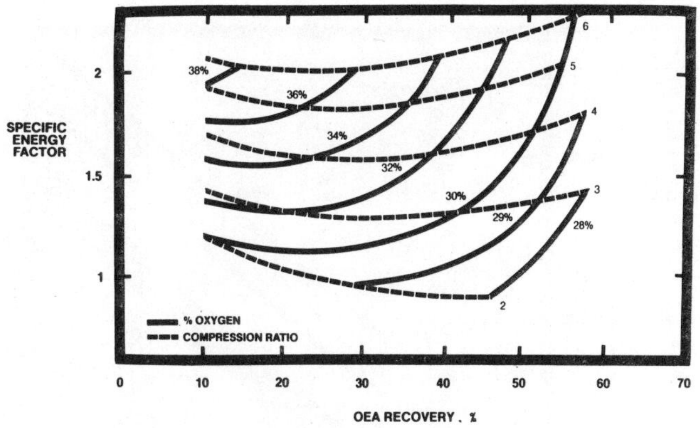

Figure 5. Typical Performance Map for A/G Technology OEA Membranes Operated in Vacuum Mode.

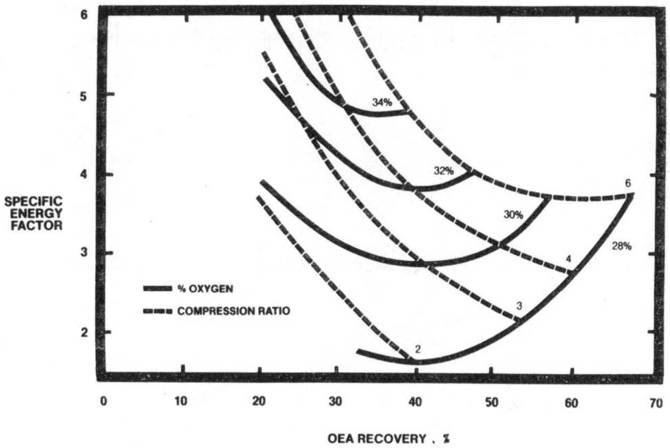

Figure 6. Typical Performance Map for A/G Technology OEA Membranes Operated in Pressurized Mode.

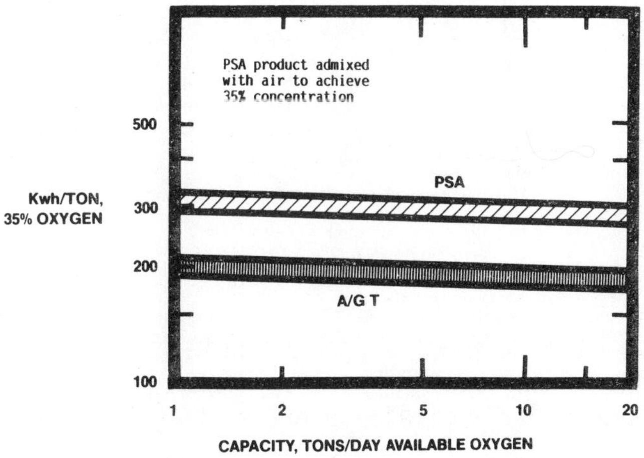

Figure 7. Comparative Power Requirements for PSA and A/G Technology Hollow Fiber Membrane System for Production of 35% OEA.

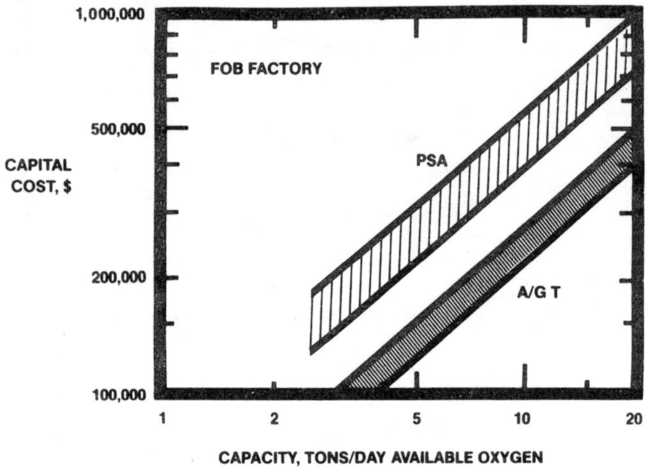

Figure 8. Comparative Capital Costs for PSA and A/G Technology Hollow Fiber Membrane System for Production of 35% OEA.

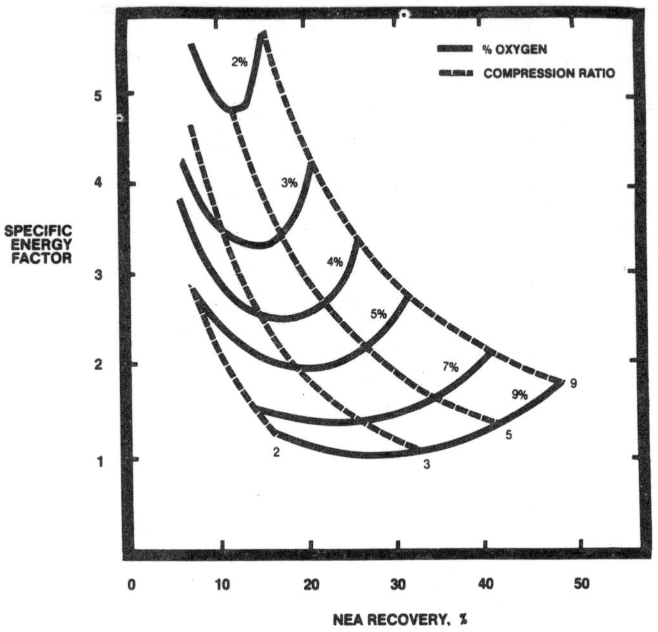

Figure 10. Typical Performance Map for A/G Technology NEA Membranes (Pressurized Mode).

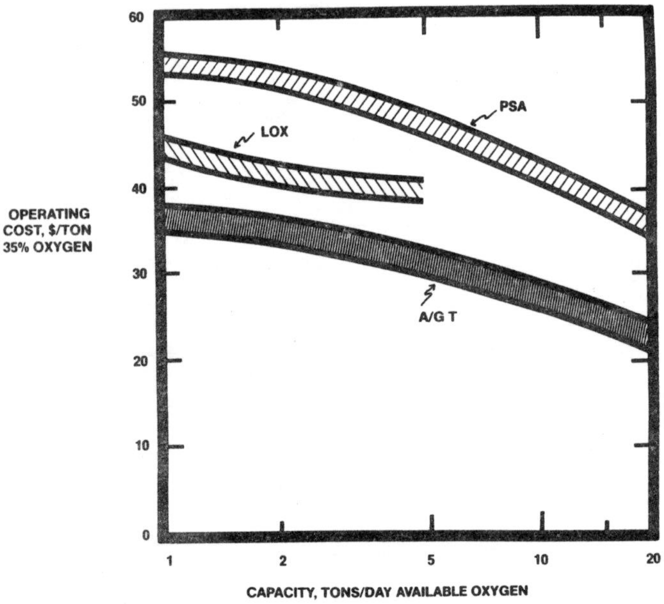

Figure 9. Comparative Operating Costs for PSA, Delivered LOX and A/G Technology Hollow Fiber Membrane System for Production of 35% OEA.

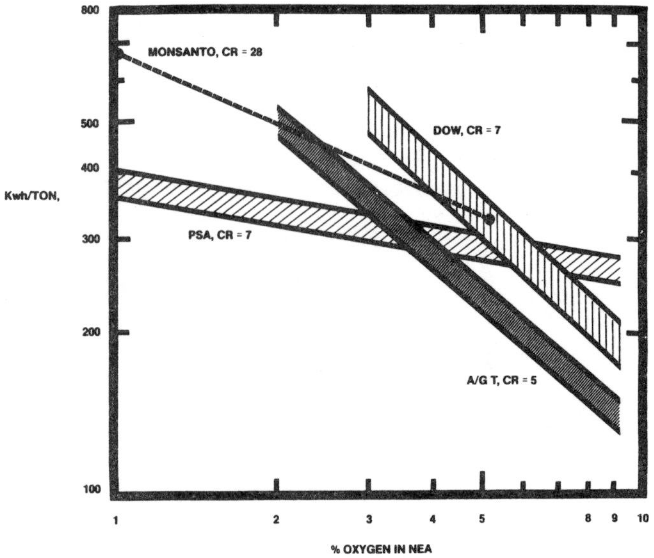

Figure 11. Comparative Power Requirements for PSA and Alternative Membrane Systems for Production of NEA.

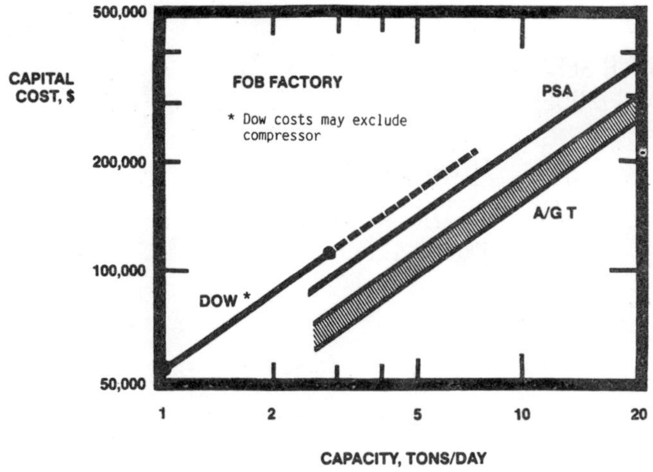

Figure 12. Comparative Capital Costs for PSA and Alternative Membrane Systems for Production of 95% NEA.

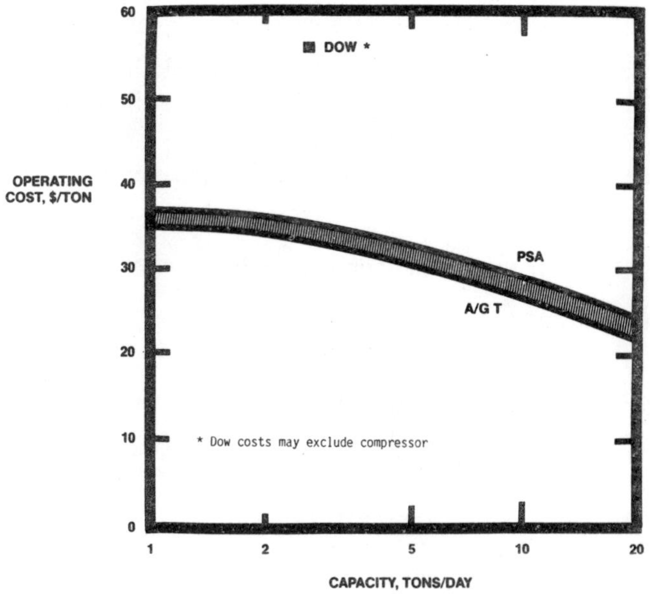

Figure 13. Comparative Operating Costs for PSA and Alternative Membrane Systems for Production of 95% NEA.

TABLE 1
BEST ESTIMATE OF COMPARATIVE MEMBRANE INTRINSIC PROPERTIES[1]

	A/G TECHNOLOGY	UOP[2]	DOW[3]	MONSANTO[4]
PRIMARY APPLICATIONS	OEA to 40% NEA to 98%	OEA to 30%	OEA to 35% NEA to 99%	NEA to 99%
MEMBRANE CONFIGURATION[5]	H F	S W	H F	H F
OXYGEN/NITROGEN SEPARATION FACTOR	3 to 4	1.8 to 2	3.5 to 4.5	4 to 5
OXYGEN EFFECTIVE PERMEABILITY[6]	1 to 4	5 to 8.5	0.03 to 0.04	0.1 to 0.2

1. All performance data at ambient temperature (25 C).
2. References 4 and 10.
3. References 11.
4. Best Estimate.
5. HF indicates hollow fiber configuration; SW indicates spiral wound configuration.
6. Units of Effective Permeability are cm^3(STP)/cm sec cm-Hg x 10^4.

TABLE 2
ASSUMPTIONS USED IN AIR SEPARATION SYSTEM ECONOMIC MODEL

PARAMETER	ASSUMED VALUE
Number of System Stages	1
Blower Energy Efficiency	65%
Vacuum Pump Energy Efficiency	60%
Compressor Energy Efficiency	70%
Hardware Life	15 years
Separation Media Life	5 years Membrane (NEA) 10 years Membrane (OEA) 15 years PSA
Power Cost	\$0.045/Kwh
Cooling Water Cost	\$0.10/1,000 gal
Operator Labor (variable with plant capacity)	2 to 5 hrs/wk
Maintenance & Taxes	2% of Hardware Cost*
Capital Charges	12%
Blower/Vacuum Pump/ Compressor Costs	Vendor Quotes
Installation Cost (Dependent on Plant Capacity)	15 to 20% of Capital Cost
Plant Utilization	90%

*Hardware Cost = Installed Cost - Separation Media Cost.

TABLE 3
OPERATING COST COMPARISON FOR
35% OEA GENERATION
(10 Ton/day Available Oxygen)

	A/G Technology Membranes	PSA
Capital Cost, FOB	$250,000	$480,000
Installation, %	15	15
Installed Cost	$287,500	$552,000
Membrane Replacement	38.11	0.00
Power Cost	85.50	130.50
Operator Labor Cost	8.60	8.60
Maintenance & Taxes	9.91	19.02
Capital Charges	105.18	201.95
Depreciation	33.03	63.41
$/day	280.33	423.49
$/ton	28.03	42.35

TABLE 4
OPERATING COST COMPARISON FOR
95% NEA GENERATION

	A/G Technology Membranes	PSA	Dow* Membranes
Capacity, ton/day	3	3	2.6
Capital Cost, FOB	$75,000	$100,000	$110,000
Installation, %	20	20	20
Installed Cost	$90,000	$120,000	$132,000
Membrane Replacement	16.01	0.00	40.24
Power Cost	35.10	40.50	38.61
Cooling Water Cost	0.43	0.43	0.39
Operator Labor Cost	4.30	4.30	4.30
Maintenance & Taxes	3.89	5.18	4.02
Capital Charges	32.93	43.90	48.29
Depreciation	12.96	17.28	13.41
$/day	105.61	111.59	149.27
$/ton	35.20	37.20	57.41
$/100 SCF	0.13	0.13	0.21

*Best estimate based on published capital cost and capacity. Capital cost may not include compressor. If compressor is excluded, about $40,000 should be added to the given FOB capital cost to provide an equal basis comparison with PSA and A/G Technology membrane systems. Such an increase would raise the Dow System operating costs to about $0.24/100 SCF.

REMOVAL OF MULTIVALENT METAL CATIONS FROM WATER USING MICELLAR-ENHANCED ULTRAFILTRATION

John F. Scamehorn, Rex T. Ellington, Sherril D. Christian,
Barry W. Penney, Robert O. Dunn and Subray N. Bhat ■ University of Oklahoma, Norman, OK 73019

In this application of micellar-enhanced ultrafiltration (MEUF), anionic surfactant is added to water at concentrations well above the critical micelle concentration. Multivalent cationic heavy metals adsorb on the resulting highly negatively charged micelles. The solution is then ultrafiltered using membranes with pore sizes small enough to block passage of micelles. Electroneutrality requires that the adsorbed heavy metals also be rejected by the membrane. The permeate contains both heavy metals and surfactant in very low concentrations. In this work, divalent copper (Cu (II)) was used as a model heavy metal cation. When it is present in dilute concentrations, rejections of 99.8% are observed, demonstrating that MEUF can be extremely effective in this separation. The permeate purity decreases as the metal concentration in the feed increases, the added surfactant concentration is reduced, the solution is progressively ultrafiltered to higher concentrations, or added monvalent salt is present in solution. Membranes with pore sizes of molecular weight cut-offs of 20,000 or below produce excellent rejections in this application of MEUF.

INTRODUCTION

Industrial wastewaters, from sources such as synfuels plants, metal plating plants, and chemical operations, frequently contain high levels of toxic heavy metal cations. These wastewaters can be hazardous to the environment and may present economic obstacles to commercialization of many processes.

Surfactant-based separation processes are an increasingly important class of separations for industrial application (1). In general, these separations require much less energy than traditional purification methods, such as distillation, or distillation followed by extraction (1,2).

Micellar-enhanced ultrafiltration (MEUF), a recently proposed surfactant-based separation technique for the removal of multivalent metal cations, is illustrated in Figure 1. As shown, anionic surfactant is added to the polluted stream. In the resulting stream, most of the surfactant present will form micelles aggregates of 60-200 surfactant molecules). In an aqueous environment, the micelle (see Figure 1) will form with the hydrophobic tail groups (made up of hydrocarbon chains) on the inside of the micelle, bunched together towards its "core," and with the hydrophilic, anionic headgroups tightly packed across the micellar surface, which forms an interface with the aqueous environment (3). The micellar surface has a high charge density and a high absolute electrical potential (4); therefore, the multivalent heavy metal cations electrostatically adsorb onto or near the micellar surface. This stream is subsequentally forced through an ultrafiltration membrane with pore sizes small enough to completely reject the micelles and associated heavy metal. The permeate which passes through the membrane will have very low concentrations of both the heavy metal and the surfactant. The retentate solution will have very high concentrations of both the heavy metal and the surfactant, and will be much smaller in volume (or flowrate) than the initial feed stream. Therefore, the ultimate disposal or subsequent treatment will be much less expensive than direct treatment or disposal of the original feed stream.

Previously, MEUF has been shown to remove dissolved organics from wastewater (1,2,5-7). Studies of this technique have been made using 4-tert-butylphenol as an organic pollutant, and cationic hexadecylpyridinium chloride as the surfactant.

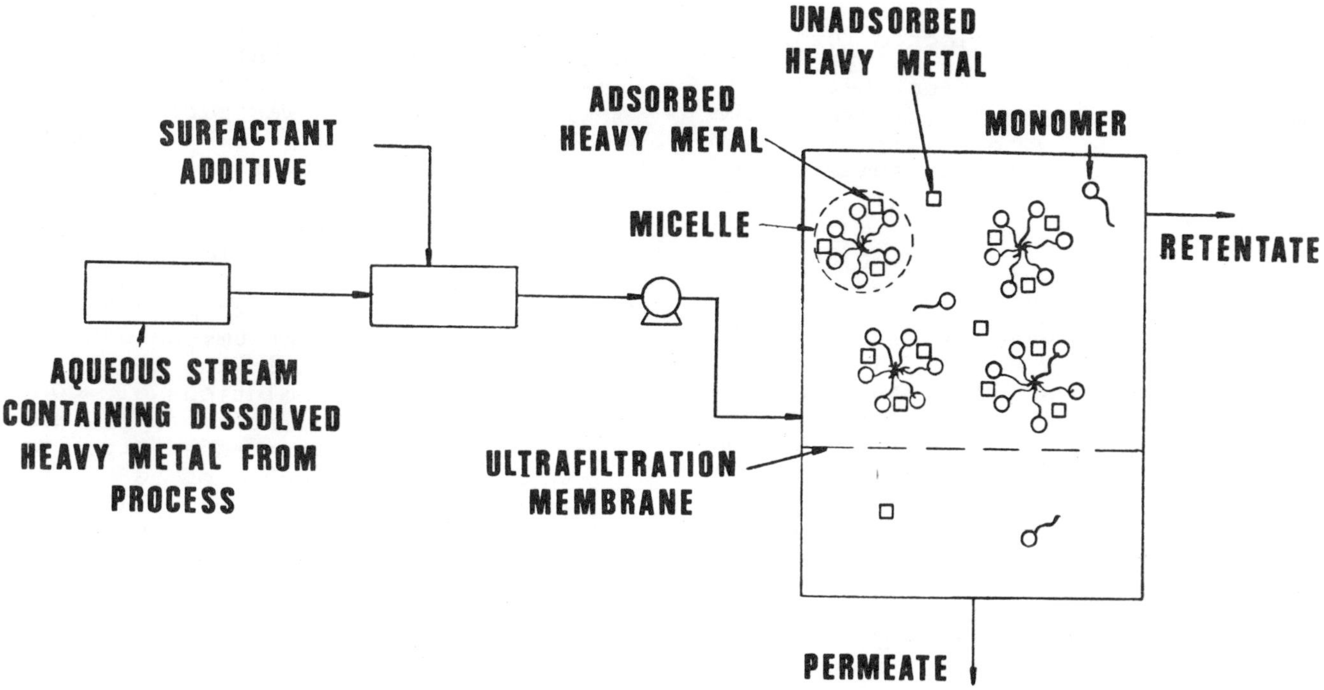

Figure 1. Schematic of micellar-enhanced ultrafiltration to remove multivalent heavy metal cations from water

Rejections of greater than 99% for both the phenolic and the surfactant studied were reported. Membranes with pore size as large as 50 000 dalton molecular weight cut-off (with a pressure drop of 414 kPa) have been shown to reject micelles efficiently (2,5). Other studies have been made applying MEUF to the removal of straight chain alcohols, reporting better than 98% rejections of alcohols using hexadecylpyridinium chloride as the surfactant (7).

Although cationic surfactants are, in general, the surfactant of choice in applying MEUF to remove nonionic organics from water (1,2); anionic surfactants must be used to effect removal of multivalent heavy metal cations. The selection of a model surfactant system for study requires several considerations. It is desired that the anionic surfactant have a long hydrocarbon tail, since this will promote micelles of higher aggregation number combined with a lower critical micelle concentration (CMC) (2). As a result, the micelles will have a higher surface charge density and a smaller fraction of surfactant will be present in non-micellar (monomer) form. Sodium dodecyl sulfate (SDS) was chosen as a surfactant for the study reported in this paper. The hydrocarbon chain length of anionic surfactants that can be used in MEUF is limited by the Krafft temperature (2), defined as the temperature below which a surfactant will precipitate as a solid hydrate (8). The Krafft temperature will increase as the hydrocarbon chain length increases (2). For example, sodium tetradecyl sulfate in aqueous solution will precipitate when cooled below 30°C, while a solution of SDS has a Krafft temperature of 16°C, well below room temperature (9). It must also be noted that some of the micellar characteristics will change upon addition of metal ions, since binding onto the micelle can affect the size and shape of the micelle, the aggregation number of the micelle, the CMC of the surfactant, and the Krafft temperature of the surfactant.

The multivalent heavy metal

cationic pollutant selected for study was the cupric ion, or Cu(II). This cation is a model divalent metal, as well as being found in effluents from many processes. The Cu(II) ions in solution were from dissolution of cupric chloride ($CuCl_2$) salt in this study.

EXPERIMENTAL

The surfactant used in this study, sodium dodecyl sulfate (SDS), was purified by recrystallization from reagent grade ethanol (plus about 50 mL distilled water per 600 mL solution) followed by freeze-vacuum drying. Dissolved cupric chloride dihydrate ($CuCl_2 \cdot 2H_2O$), the source of cupric ions for this study, was used as received. Both of these materials were obtained from Fisher Scientific Company. Water used in all experiments was distilled and deionized.

The ultrafiltration experiments (runs) were carried out in Nuclepore 400 mL batch, stirred cells. The cells were wrapped in tubing containing circulating water to maintain temperature for all runs at 30°C. Transmembrane pressure drops were maintained at 414 kPa gauge using nitrogen gas. Agitation speed was held at approximately 845 rpm for all runs, as measured by a strobe light. Ultrafiltration membranes used were 76 mm-diameter, anisotropic cellulose acetate, with an effective area of 30.2 cm^2, and were obtained from Nuclepore. Molecular weight cut-offs of 1 000, 5 000, 10 000, and 20 000 daltons were examined in this study. The membranes were pre-equilibrated overnight in a solution containing SDS at a concentration near its CMC, and Cu(II) ion at concentrations approximating those in the permeate. The cell was initially charged with 300 mL of feed solution, and the permeate solution was removed until approximately 100 mL of mother liquor remained. As samples were taken throughout a run, permeate fluxes were measured. Permeate samples of 50 mL were analyzed for Cu(II) ion content using a cupric ion-specific electrode, with a single junction reference electrode, both obtained from Orion Research Inc. The samples were adjusted by 1 mL of ionic strength adjusting solution (5 M $NaNO_3$) per 50 mL solution before analysis. In solutions containing NaCl, the analysis was performed using the reagent PAR [4-(2-pyridylazo) resorcinol monosodium salt] (10,11).

Semi-equilibrium dialysis (SED) measurements were performed under conditions similar to those in several of the MEUF experiments to determine the equilibrium, unadsorbed Cu(II) concentration in the retentate solution (12,13). Solutions of copper (II) chloride, surfactant, and NaCl (if present) were added to the retentate compartment of a 5 mL equilibrium dialysis cell (Fisher Scientific Company). The permeate compartment contained an aqueous solution having the same NaCl concentration as the retentate. After approximately 20 h, the concentration of Cu(II) in the permeate solution was determined, using the PAR reagent. Care was taken to ensure that the Cu(II) ions were completely complexed while estimating permeate Cu(II) concentrations.

THEORY

A micelle composed of ionic surfactants is highly charged. For the condition of electroneutrality to be satisfied, the concentration of counterions (ions with charge opposite to that of the surfactant) in the solution immediately adjacent to the micelle surface must be higher than the concentration in bulk solution. Thermodynamically, these counterions are present as a surface excess (14). Since adsorption is a surface excess, these counterions may also be considered as adsorbed at the micelle-solution interface. These adsorbed counterions are either bound in the Stern layer of the micelle, or present in excess in the electrical double layer surrounding the micelle (4,15,16).

The higher the valence of a counterion, the more it will tend to be adsorbed on the micelle. For example, Hirasaki and Lawson (15) have shown that the the ratio of the concentration of adsorbed calcium to that of the unbound calcium concentration in solution is much greater than that ratio for sodium around micelles composed of anionic surfactant.

Consider now the ultrafiltration of a micellar solution containing ionic surfactant, multivalent counterions, and monovalent counterions. If the micelles are completely rejected by the membrane, electroneutrality constrains the adsorbed counterions to also be rejected. If they were not, a greater number of positive charges from counterions would be permeating the membrane than negative charges from anions, and an electrical potential would be rapidly set up across the membrane which would prevent the passage of any more adsorbed counterions through the membrane (17). As a result, the adsorbed counterions are rejected by the membrane if the micelles are rejected.

In the systems studied here, only one monovalent counterion type is present, Na(I), from either dissociation of the surfactant salt upon dissolution ($NaC_{12}SO_4$), or any added salt (NaCl) which may be present. The water also contains $CuCl_2$, so the divalent counterion, Cu(II), is also present. When a micelle forms, the total adsorption of counterion charges on that micelle is dictated by electroneutrality. The effective removal of the divalent copper ion is determined by its adsorption than that of the sodium.

A thorough and rigorous calculation of the adsorption of different counterions from a mixture of counterions in a micellar solution is a formidable task. In the case of the removal of dilute heavy metals from aqueous solution using MEUF an easily described limiting case can be approximated closely.

Dilute Metal Limiting Case

Consider the situation where the multivalent ion being removed is present in much smaller concentrations than the surfactant present in micellar form. Also, assume that the concentration of surfactant present in monomer (unaggregated) form is negligible compared to that present as micelles; that is, all of the surfactant is assumed to be present as micelles. For this case, the adsorption of the multivalent metal ions on the anionic micelle will be negligible compared to the adsorption of the monovalent counterion from the dissociation of the surfactant. If added monovalent salt is also present in solution, the adsorption of the multivalent cations, compared with that of the monovalent counterion, is even less significant. This approximation would be valid even if the bulk of multivalent counterion present were bound to the micelle. Also, the nature of the Stern layer and the electrical double layer surrounding the micelles will not be affected by the presence of the heavy metal if the heavy metal is dilute enough. In other words, the heavy metal is assumed to be so dilute in solution as to not affect micelle formation in any way. This special case will be referred to as the dilute metal limiting case. This corresponds to the actual situation in many practical applications, where heavy metal must be removed from water at very low concentrations because of the extreme toxicity.

Effect of metal concentration. If the surfactant and added monovalent salt concentrations remain constant and only the dilute metal concentration changes, the volume immediately surrounding the micelle can be viewed as a separate phase in which the metal can distribute itself. If the metal concentration is low enough, the effective distribution coefficient for multivalent metal counterions between the water-micelle interface and bulk solution will be constant. This means that, for a given surfactant concentration, the adsorption of a multivalent metal counterion will be proportional to the unadsorbed metal counterion concentration in bulk solution, and also to the total metal counterion concentration in solution. This is analogous, in liquid-liquid extraction, to stating that the concentration of solute in one liquid phase is proportional to that in the other liquid phase. The metal concentration in the permeate will be equal to the unadsorbed metal concentration in the retentate since it is the adsorbed counterion which is rejected by the membrane; therefore, with all other variables being held constant, the permeate metal concentration should be proportional to the total metal concentration in the retentate for the dilute metal limiting case.

Effect of surfactant concentration. Consider a surfactant solution with no added multivalent ions. The fractional counterion binding of monovalent counterions onto charged micelles is essentially independent of the counterion concentration (4). This counterintuitive effect is the result of the interaction between the electrical double layer and the micelle surface. As the counterion concentration in solution increases, the electrical double layer becomes more compressed (18). The result of this is that the absolute magnitude of the electrical potential at the micelle surface is reduced. This results in a reduced driving force for adsorption due to the electrostatic attraction. Yet, the mass action effect of increased counterion binding due to higher counterion concentration tends to offset this electrostatic effect. These two effects tend to approximately cancel, and the counterion binding is almost independent of the monovalent counterion concentration.

Now consider the effect of surfactant concentration on adsorption of multivalent metal counterions at constant total metal concentration. As more surfactant is added to solution, the total unbound monovalent counterion concentration increases from dissolution of the surfactant salt. This results in a decrease in electrical potential on the micelle surface, which tends to cause a decrease in the adsorption of metal ions due to a decrease in the electrostatic attraction to a given micelle. On the other hand, there are more micelles in a given volume of solution, tending to increase the number of adsorbed metal counterions per unit volume of solution. Although an exact solution of the problem would be complex, the Donnan model, proposed by Hirasaki and Lawson (15), can be used to show that these effects approximately cancel, and that the adsorption of the metal counterion per unit volume of solution should be independent of surfactant concentration for the dilute metal limiting case. As a result, the permeate metal counterion concentration should be independent of surfactant concentration for the dilute metal limiting case.

Effect of added monovalent salt. As just discussed, an increase of monovalent counterion concentration in solution will decrease the absolute magnitude of the electrical potential at the micelle surface, and reduce the driving force which would cause adsorption of the metal ion on the micelle surface. Unlike the effect of increasing surfactant concentration, there are no additional micelles being formed to counteract this effect. Consequently, added monovalent salt should tend to decrease adsorption of the multivalent metal ion and increase permeate metal concentration.

Effect of permeate/feed concentration ratio. As a retentate is further concentrated in a batch or a continuous unit, at the high rejections generally observed in this work, the [metal]/[surfactant] ratio in MEUF would remain constant in the retentate as both concentrations increase. From previous discussion of the dilute metal limiting case, the permeate metal concentration should be independent of surfactant concentration and proportional to metal concentration in the retentate. Therefore, as the metal concentration increases in the retentate along a path, at constant [metal]/[surfactant] ratio, the permeate metal concentration should increase in direct proportion to the retentate metal concentration.

RESULTS

The effect of the Cu(II) concentration in the retentate on the Cu(II) concentration in the permeate is shown in Figures 2 and 3. Figure 2 shows the data over a wide range of retentate concentrations and Figure 3 shows the data in the low concentration region. The data are tabulated with the corresponding rejections in Table 1. Rejection is defined as:

$$\text{Rejection (\%)} = 100 \cdot \left[1 - \frac{(\text{permeate [Cu(II)]})}{(\text{retentate [Cu(II)]})} \right]$$

The effect of the surfactant concentration in the retentate on the Cu(II) concentration in the permeate is shown in Figure 4, and in Table 1.

As the feed solution is treated, the retentate becomes increasingly concentrated in rejected species. At the conditions used in this work

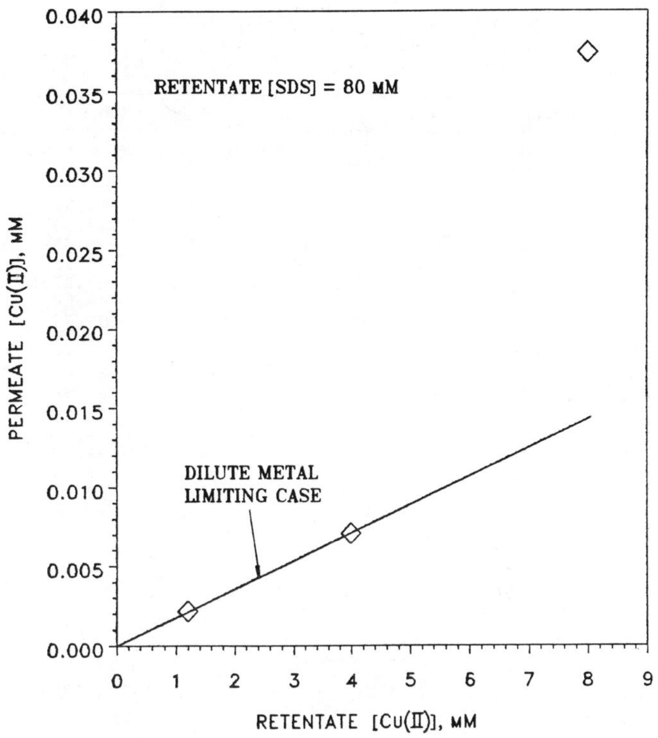

Figure 3. Effect of retentate [Cu (II)] on permeate [Cu (II)] at low concentrations.

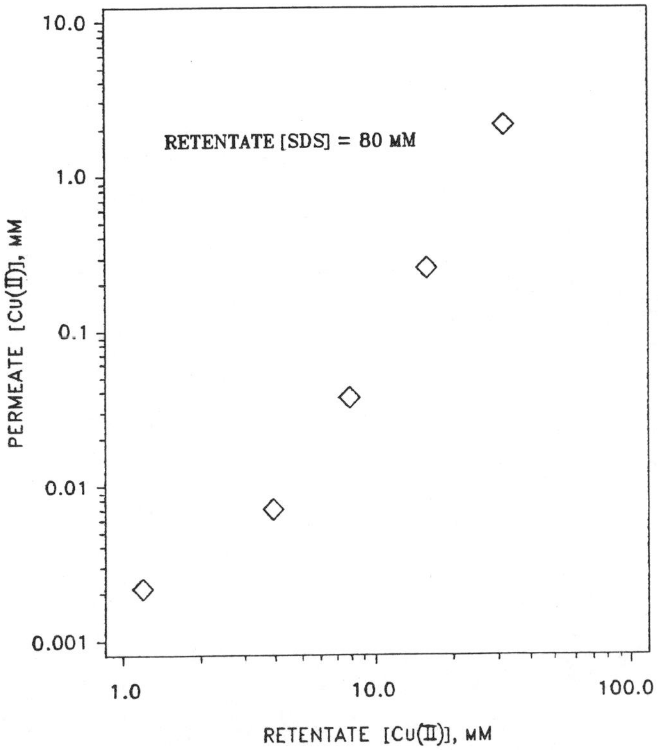

Figure 2. Effect of retentate [Cu (II)] on permeate [Cu (II)].

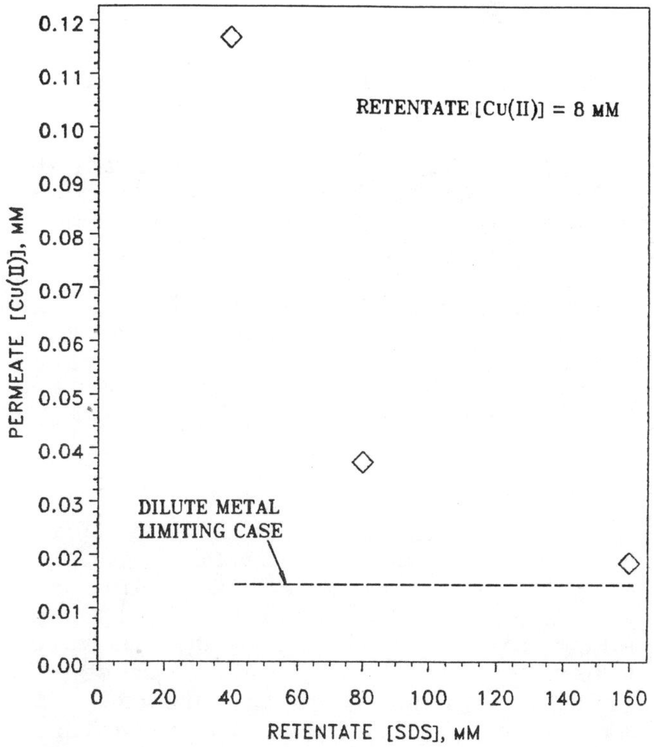

Figure 4. Effect on retentate [surfactant] on permeate [Cu (II)].

Table 1. Results from MEUF runs.

Retentate			Membrane	Permeate	
[Cu(II)] (mM)	[SDS] (mM)	[NaCl] (mM)	Pore Size (MWCO)	[Cu(II)] (mM)	Rejection (%)
1.2	80	0	1 000	0.00220	99.82
4	80	0	1 000	0.00710	99.82
8	80	0	1 000	0.0374	99.53
16	80	0	1 000	0.257	98.39
32	80	0	1 000	2.16	93.25
8	40	0	1 000	0.117	98.54
8	160	0	1 000	0.0186	99.77
16	160	0	1 000	0.0840	99.48
32	320	0	1 000	0.215	99.33
64	640	0	1 000	0.316	99.51
8	80	100	1 000	1.22	84.75
8	80	200	1 000	2.32	71.00
8	80	400	1 000	3.07	61.62
8	80	0	5 000	0.0184	99.77
8	80	0	10 000	0.0247	99.69
8	80	0	20 000	0.0389	99.51

(except at high added salt levels), rejection of surfactant and copper is so high that the ratio of concentration of copper to that of surfactant in the retentate essentially remains constant as the solution is treated. In order to examine the effect of retentate composition as a solution is treated, the effect of the Cu(II) concentration in the retentate on Cu(II) concentration in the permeate at a constant [Cu(II)]/[surfactant] ratio in the retentate is shown in Figure 5 and Table 1.

The effect of added monovalent salt concentration on the Cu(II) concentration in the permeate is shown in Figure 6, and in Table 1.

The effect of membrane pore size on the Cu(II) concentration in the permeate is shown in Figure 7, where membrane pore size is quantified in terms of molecular weight cut-off (MWCO), a common way of characterizing ultrafiltration membranes.

In order to determine the equilibrium, unadsorbed Cu(II) concentration in the retentate solution, semi-equilibrium dialysis (SED) was employed (12,13). In SED, osmotic pressures are the only pressure gradients present, so the predominant mechanism of transport is diffusion. Using appropriate analysis techniques (12,13), the resultant concentration differences of heavy metals across a

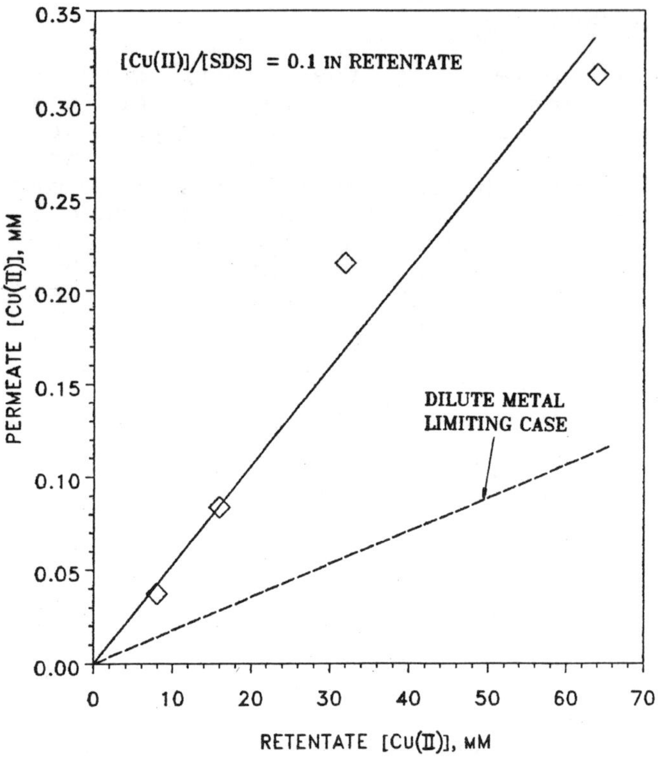

Figure 5. Effect on retentate [Cu (II)] on permeate [Cu (II)] at constant retentate [Cu (II)]/[SDS].

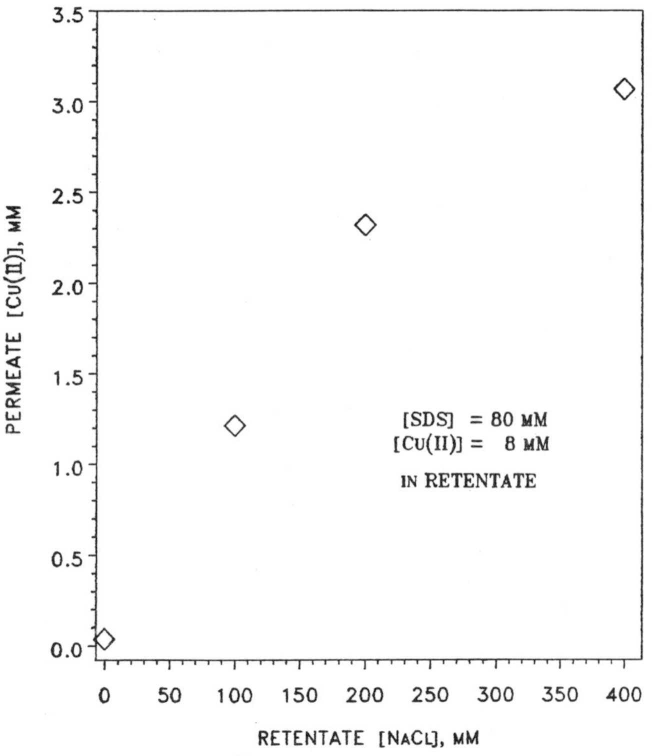

Figure 6. Effect of added NaCl on permeate [Cu (II)].

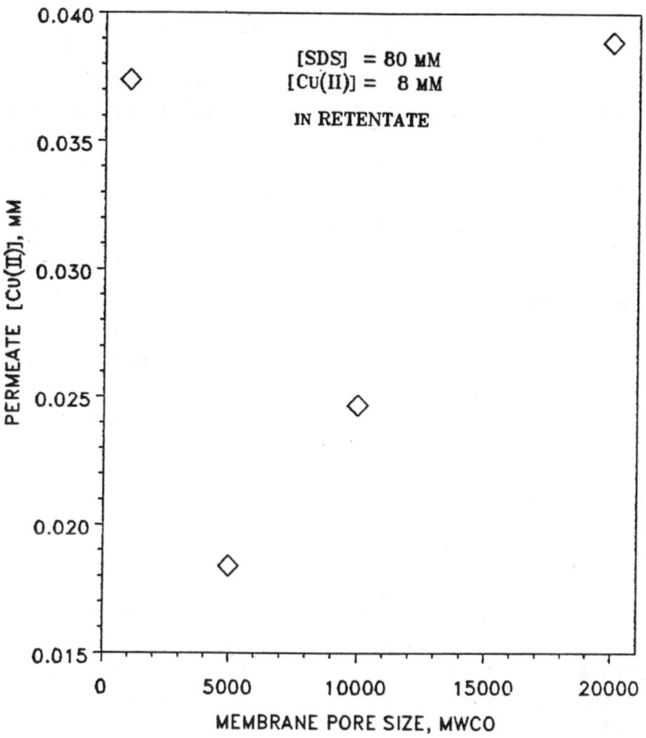

Figure 7. Effect of membrane pore size on permeate [Cu (II)].

dialysis membrane can be used to deduce the unadsorbed and adsorbed concentration of the metals in the retentate surfactant solution. At retentate concentrations of 80 mM SDS, 8 mM Cu(II) and 100 mM NaCl, the unadsorbed Cu(II) concentration was measured and averaged to be 1.220 mM, using SED.

DISCUSSION

In Figure 3, the permeate [Cu(II)] increases in proportion to the retentate [Cu(II)] at low metal concentrations, as predicted by the dilute metal limiting case. At these low concentrations, the rejection is 99.82%, as seen in Table 1. This means that the permeate has a 500-fold lower concentration than the retentate. MEUF can be an extremely effective separation technique for heavy metals in this region.

As the retentate [Cu(II)] increases further, the permeate contains a higher [Cu(II)] than predicted by the dilute metal limiting case (Figure 2). This is because the divalent copper is now present in high enough concentrations to constitute a significant fraction of the bound counterions on the micelle, and to cause a reduction in micelle surface electrical potential by compressing the electrical double layer. The copper is now causing the same deleterious effects that added monovalent salt does when at high enough concentrations; however, even when the copper concentration in solution (in Normality) is comparable to that of the surfactant, an extremely good separation of the heavy metal is possible. For example, at 32 mM Cu(II) concentration (64 mN) in the retentate (compared to 80 mM surfactant concentration), the rejection is still greater than 93%, a respectable separation factor. In this region, the separation could be improved by increasing the surfactant concentration in the retentate.

When the surfactant becomes concentrated enough, the dilute metal limiting case is approached. The resulting permeate [Cu(II)]/retentate [Cu(II)] ratio should be independent of both surfactant and copper concentration. The value of this ratio in this region from Figure 3 is 0.0018 for this system. In Figure 4, the permeate [Cu(II)] is shown that corresponds to this state, under the conditions stated. At high surfactant concentrations, this value is appoached. At lower surfactant concentrations, the permeate is less pure. This is due to two effects. The effect that the copper has on counterion binding and the electrical double layer at high retentate [Cu(II)]/[surfactant] ratios has already been discussed. In addition, the fraction of the surfactant which is in micellar form instead of monomeric form decreases as the surfactant concentration decreases. Using the Shinoda equation (19), with appropriate parameters measured for this surfactant (20), leads to the following estimates of the fraction of surfactant in the retentate present in micellar form, for each point in Figure 4:

Total [SDS] in Retentate (mM)	Fraction of SDS Present as Micelles (%)
40	88.1
80	95.5
160	98.6

This effect limits the minimum amount of surfactant that can be added to solution to effect a separation, no matter how low in concentration the metal ion in the solution being treated for a given surfactant. The surfactant concentration must be well above the critical micelle concentration (CMC). For SDS, the CMC is approximately 8 mM.

As the retentate concentration of both copper and surfactant simultaneously increase, the permeate [Cu(II)] increases approximately linearly with the retentate [Cu(II)], as shown by Figure 5. The permeate is less pure than it would be in the dilute metal limiting case, whose value is shown in Figure 5 (calculated from the dilute data in Figure 3). It is interesting that the rejection remains fairly constant as the retentate becomes more concentrated (see Table 1), while the permeate becomes less pure. This is an anomaly due to the manner in which rejection is defined. In removal of organics by MEUF, it has been pointed out that the permeate concentration remains constant as retentate becomes more concentrated, but rejection decreases for this same reason (2,5,7).

As expected, from the theoretical discussion, added monovalent salt decreases permeate purity (Figure 6) and decreases rejection (Table 1). This reduction is substantial, with only 0.1 M NaCl reducing rejection from 99.82% to 84.75%. At 0.4 M NaCl, rejection is 61.62%. Using MEUF to remove heavy metals would not be nearly as effective when high concentrations of dissolved salts are present.

It has been shown that in the removal of dissolved nonionic organics, the micelles formed by the cationic surfactant used (hexadecylpyridinium chloride) could be effectively rejected by membranes with pore sizes as large as 10 000 dalton MWCO, with only slight leakage with pore sizes up to 50 000 MWCO (5). SDS has a lower molecular weight than that cationic surfactant, and forms micelles of lower aggregation number (21,22) in the absence of solubilized organics or bound heavy metals. Consequently, SDS micelles may be considered likely to require a smaller membrane pore size, for high rejections, than the cationic surfactant micelles previously studied. The effect of membrane pore size is shown in Figure 7 for membrane pore sizes of up to 20 000 daltons. The permeate [Cu(II)] is less for a 5 000 MWCO membrane than for a 1 000 MWCO membrane. This is probably due to the presieving effect of the gel layer set up by concentration polarization more prevalent for higher pore size membranes, as observed for MEUF of organics by Dunn, et. al. (5). The increased permeate [Cu(II)] at high pore size membranes indicates that some micellar leakage is probably occurring; however, even for the most porous membrane studied (20 000 MWCO), this effect is almost negligible. Therefore, high MWCO membranes can produce extremely pure permeate in MEUF.

At the retentate conditions used in the SED experiments (80 mM SDS, 8 mM Cu(II), and 100 mM NaCl), the permeate [Cu(II)] from MEUF from table 1 is 1.22 mM. This may be compared to an equilibrium, unadsorbed [Cu(II)] of 1.220 mM from SED. These values show that, for the conditions studied, the permeate metal concentration is essentially equal to the equilibrium unadsorbed metal concentration in the retentate. Therefore, the MEUF process, under normal operating conditions, is occurring slowly enough that the metal counterions are at equilibrium between the permeate and retentate. This means that SED, a rapid and inexpensive analysis technique, can estimate the effectiveness of removal of an arbitrary multivalent heavy metal.

Comparison to Removal of Nonionic Organics

In previous applications of MEUF to treat aqueous streams, we have studied removal of such dissolved organics as phenolics and alcohols in some detail (1,2,5,7). These organics solubilize or dissolve in the micelle interior. Although dipole-ion effects are important in the palisade layer of the micelle, electrostatic effects are secondary in nonionic organic removal by MEUF. It is interesting to compare the effects of operating variables on performance for multivalent heavy metal removal to those for nonionic organic removal using MEUF.

The organic concentration in the permeate is nearly a linear function of the retentate organic concentration under normal conditions. This is only generally true for heavy metal removal at very dilute concentrations.

The organic concentration in the permeate is approximately inversely proportional to surfactant concentration in the retentate. For heavy metals, the surfactant concentration has a large effect on permeate metal concentration at low surfactant concentration (see Figure 4). At high surfactant concentrations, the surfactant concentration has no effect on permeate purity for heavy metal removal.

As the retentate becomes more concentrated during an operation (constant organic/surfactant concentration), the permeate organic content remains nearly constant. For heavy metals, the permeate metal concentration is roughly proportional to the retentate metal concentration as the retentate is processed.

Because electrostatic effects are secondary in nonionic organic removal by MEUF, added salt effects are expected to be small (mainly salting out or salting in effects on the organic). In contrast, added salt has a substantial negative effect on heavy metal removal using MEUF.

The dynamics (flux rates, gel points, mass transfer coefficients) of the ultrafiltration process have been studied in detail for organic removal (1,2,5,7). Preliminary data obtained in this work indicates that these parameters are very similar when heavy metals are being removed using MEUF.

ACKNOWLEDGEMENTS

Financial support for this work was provided by the following organizations: the Office of Basic Energy Sciences of the Department of Energy - Contract DE-AS05-84ER13175, the Oklahoma Mining and Minerals Resources Research Institute, and the University of Oklahoma Energy Resources Institute.

LITERATURE CITED

1. Scamehorn, J.F. and Harwell, J. H., "Surfactant-based Treatment of Aqueous Process Streams," in Surfactants and Chemical Engineering, Wasan, D.T., Shah, D.O., and Ginn, M.E. (Eds.), Marcel Dekker, New York (In Press).

2. Dunn, R.O., Scamehorn, J.F., and Christian, S.D., Sep. Sci. Technol., 20, 257 (1985).

3. Rosen, M.J., in Surfactants and Interfacial Phenomena, Chap. 3, Wiley, New York (1978)

4. Rathman, J.F. and Scamehorn, J. F., J. Phys. Chem., 88, 5807 (1984).

5. Dunn, R.O., Scamehorn, J.F., and Christian, S.D., Sep. Sci. Technol., In Press.

6. Leung, P.S., in Ultrafiltration Membranes and Applications, Cooper, A.R., (Ed.), p. 415, Plenum, New York (1979).

7. Gibbs, L.L., Scamehorn, J.F., and S.D. Christian, J. Membrane Sci., Submitted.

8. Rosen, M.J., in Surfactants and Interfacial Phenomena, p 160, Wiley, New York (1978).

9. Weil, J.K., Smith, F.S., Stirton, A.J., and Bristline, R.G. J. Amer. Oil Chem. Soc., 40, 538 (1963).

10. Anderson, R.G. and Nickless, G., Analyst, 92, 207 (1969).

11. Iwamoto, T., Bull. Chem. Soc. Japan, 34, 605 (1969).

12. Christian, S.D., Tucker, E.E., Smith, G.A., and Scamehorn, J.F. Langmuir, 1, 564 (1985).

13. Smith, G.A., Christian, S.D., Tucker, E.E., and Scamehorn, J. F., J. Soln. Chem., In Press.

14. Adamson, A.W., in Physical Chemistry of Surfaces, 4th Ed., Chap. 3, Wiley, New York (1982).

15. Hirasaki, G.J. and Lawson, J.B., SPE Reservoir Eng., 1, 119 (1986).

16. Beuen, J.A. and Ruckenstein, E., J. Colloid Interface Sci., 96, 469 (1983).

17. Helfferich, F., in Ion Exchange, p. 134-5, McGraw-Hill, New York (1962).

18. Hiemenz, P.C., in Principles of Colloid and Surface Chemistry, Chap. 12, Marcel Dekker, New York (1986).

19. Shinoda, K., in Colloidal Surfactants, Shinoda, K., Nakagawa, T., Tamamushi, B., Isemura, T. (Eds.), Chap. 1,Academic Press, New York (1963).

20. Rathman, J.F. and Scamehorn, J. F., Langmuir, submitted.

21. Rosen, M.J., in Surfactants and Interfacial Phenomena, p. 90, Wiley, New York (1978).

22. Bushong, D.S., "Osmometry Studies of Surfactant Aggregation and Solubilization by Micellar Systems," PhD Dissertation, p. 69, University of Oklahoma, Norman (1985).

FRUIT JUICE ULTRAFILTRATION

Robert G. Blanck and William Eykamp ■ Koch Membrane Systems, Inc.
850 Main Street, Wilmington, MA 01887

Since the 1970's the fruit juice industry has enjoyed an increase in consumption of clear fruit juices and natural fruit juice products. Parallel to this recent popularity of fruit juices is a drive to develop efficient processing techniques to produce highly clarified versions of these juices. The membrane process of ultrafiltration is providing a cost-effective alternative to traditional fining and clarification techniques where the objective is production of stable juices free of nonbiological haze or sediment. The clarified juice is normally bottled for direct consumption, concentrated to 70°Brix, or is blended with other juices for specialty juice products. Over the past five years, apple juice has been the major fruit processed with membrane technology, although new applications with grape, pear, pineapple, cranberry, and citrus juices are rapidly developing. In related food applications, ultrafiltration provides a viable alternative to conventional techniques in wine and vinegar processing.

Ultrafiltration of fruit juice is a simple unit operation that allows for clarification and fining in one step. Figure 1 illustrates a representative juice ultrafiltration process where a minimal amount of upstream equipment is required for clarified juice production. Fresh pressed juice from any one of a number of extraction processes is pasteurized at 72°C for 16 seconds to inhibit yeast growth and prevent fermentation. After cooling to 55°C by thermal regeneration, the juice is partially depectinized using enzymes. At the ultrafiltration system, the semi-permeable membrane separates juice constituents on the basis of their molecular size, thereby retaining all haze components and suspended solids in the retentate. Brilliant, clear juice permeate is collected in a balance tank. At the end of the process run, an optional diafiltration step maximizes system yield by leaching residual sugar from the retentate. The clarified product can be hot packed, sterile filtered, or evaporated to 70°Brix.

The single unit operation of the ultrafilter for clarification and fining is in contrast with the multi-step conventional processing technique. Traditional juice processing requires the use of fining agents, enzymes, centrifugation, and diatomeaceous earth filtration to provide a clarified product and is illustrated in Figure 2. After juice extraction, centrifugation or vibrating screen filtration is required for removal of the coarse suspended solids from pasteurized juice. Removal of colloidal particles and haze-inducing components requires the use of surface active fining agents such as bentonite, gelatin, silica sols, and fining proteins in a time- and concentration-dependent batch step. Similarly, pectin and starch removal is accomplished by a batch enzymatic hydrolysis with pectinase and amylase, respectively. The flocculant generated by fining requires treatment in a series of steps including centrifugation, settling and racking, or diatomaceous earth filtration. Additional polish filtration is required for production of a brilliant, clear juice. The success of any fining or clarification step in producing immediate and long-term juice stability is influenced by apple type, ripeness, juice extraction procedures, fining agent concentrations, and filtration efficiency.

In the juice plant, the primary function of the ultrafilter is to remove suspended solids, colloidal haze particles, haze proteins, and condensed polyphenolics from the juice stream. Studies on the chemical nature of juice haze components have been most accurately described for apple juice by Heatherbell et al. (1, 2), Johnson et al. (3), and VanBuren and Robinson (4), although the complexity of differential haze development prohibits adequate description. In general, fruit juice hazes are summarized in Figure 3. Biological haze results from micro-organism growth in the nutrient-rich juice, while non-biological haze has its origins in high molecular weight components and their associated complexes.

The acceptance of ultrafiltration by the fruit juice industry has been led by improved processing efficiency, enhanced juice yields, and superior product quality. In Figure 4, a comparison of the two process modes is presented, and the advantages of the membrane process are highlighted. Juice extraction for either process can be identical, although the process time for the ultrafiltration method is on the order of two to four hours compared to 12 to 36 hours for conventional techniques. Suspended solids removal before ultrafiltration is not a prerequisite for processing with tubular ultrafiltration, although thin-channel ultrafiltration systems such as spiral wound modules and capillary tubes require extensive prefiltration. Enzyme treatment of the juice prior to membrane processing is optional from a product quality point of view, as pectin and starch retention is generally on the order of 99%. Pectin causes reduced flux rates due to concentration polarization at the membrane surface and fouling; therefore, partial or full removal of pectin from the process stream enhances ultrafiltration performance. For most apple juice applications the amount of pectinase added will be between 25% and 50% of the amount used in traditional clarification. This enzyme concentration will vary according to fruit type, fruit age, extraction method, and temperature. The elimination of conventional clarification and fining steps enhances the yield of the overall process. In practice, yields of 96 to 99% are realized from an ultrafilter, in contrast to 80 to 94% yields for established techniques. The capital expenses and operating costs of the ultrafilter are rapidly offset by savings illustrated in Figure 5. Ultrafiltration allows for increased product yield, elimination of rough and polish filtration, elimination of diatomaceous earth disposal, a guarantee of product quality, and a reduction in enzymes and labor costs.

To realize these advantages of membrane processing in commercial applications, a synthesis of membrane and process development is necessary. Membrane system design must be optimized by considering all aspects of juice pretreatment, process operating parameters, and careful selection of the membrane type and module configuration. A typical juice ultrafiltration system consists of a feed tank, a feed pump, and the ultrafilter as shown in Figure 6. To insure maximum juice yield, most juice ultrafiltration systems are operated in a batch or modified batch method. Over the course of a 20-hour process period, juice is fed to the ultrafilter at a rate equal to the rate of continuous clarified juice collected as permeate. The retentate is recirculated through the ultrafiltration modules until the concentration of retained species begins to limit the flow and flux.

The relationship between feed suspended solids and membrane configuration is an important consideration when processing fruit juices with ultrafiltration technology. The high suspended solids and fiber content of retentate when achieving maximum product yield inhibits the use of thin-channel devices such as spiral wound technology and capillary tube systems. The fiber-laden retentate acts as a Bingham plastic and forms pockets of suspended solids in the thin channel feed spacers which are not reliably removable with standard or aggressive cleaning cycles. Because residual product in the system leads to sanitation problems, most fruit juice processors favor tubular ultrafiltration systems with minimal or no prefiltration.

Koch Membrane Systems has developed and optimized a tubular ultrafiltration configuration to enhance system performance (Figure 7). The 19-tube Supercor product is capable of processing fluids with suspended solids concentrations up to 90% when measured on a volume basis. The membrane, rated as a 18,000 nominal molecular weight cutoff, has a retention capability designed for fruit processing. The module is resistant to temperatures up to 60°C, cleaning solutions at pH 2.0 to 11.0, and chlorine levels of 200 ppm at pH 10. The ultrafiltration module is incorporated into a system (Figure 8) designed for continuous processing and maximum product yields. The tubular ultrafiltration system shown is capable of producing 70,000 gallons of clarified apple juice per day.

Productivity of membrane systems on juice streams is dependent on juice type, pretreatment conditions, temperature, and operating pressures. Characteristic flux rates for some fruit juice streams are given in Figure 9. Highest flux rates are found in warm, depectinized streams such as apple and pineapple juice, where average flux values are on the order of 80 to 110 liters/meter^2-hour, respectively. Reduced flux rates are found in non-depectinized apple and pear trials. Grape juice processed at 7.0°C typifies low-temperature processing of fruit juices where temperature-dependent viscosity effects limit flux.

The volumetric yield of juice during the concentration stage of ultrafiltration is dependent on the maximum achievable concentration of suspended solids, colloids, and other retained species in the retentate. Low suspended solids juices are concentrated approximately 50 to 100 fold corresponding to a 98 to 99% yield. High suspended solids effects limit concentration factors to about 25 because the viscosity increase with concentration results in a lower flow rate, lower Reynolds number, and therefore lower flux. Viscosity increases rapidly during the last 0.5 hours of processing. Figure 10 describes an apple juice process run where flux is plotted against concentration factor. Flux stability is excellent at low concentration factors, but flux declines dramatically at high concentration factors.

The ability to process juice to its maximum suspended solids level represents a substantial yield increase over traditional process methods. Figure 11 illustrates the projected gross revenue increase per day for one juice plant producing 70,000 gallons per day. A one percentage point increase in yield has the potential of an $875/day revenue increase. This translates to $131,250 per year assuming a conservative 150-day process year. A five percentage point increase in yield over conventional techniques results in a $4,375/day increase or $656,250 over the course of a year.

Additional savings can be reached when diafiltration is used to remove residual sugar from the retentate (Figure 12). This allows for the greatest economic payback by increasing sugar yields to levels of 99% regardless of the juice yield during the concentration phase. For this process, diafiltration water is added to the final retentate at a rate equal to the permeate removed. Maximum performance of the ultrafilter is achieved when diafiltration is initiated at the highest possible concentration factor. The diafiltration step results in a diluted permeate that requires concentration via evaporation or reverse osmosis to achieve the standard permeate concentration. The water from the concentration process can be recycled as the diafiltration water.

The yield incentives for use of ultrafiltration in the juice plant are paralleled by product quality improvements. The membrane filter allows for a predictable, consistent-quality juice that is insensitive to variable concentrations of colloidal particles and haze-inducing components, eliminating significant laboratory and quality control expense inherent in conventional juice processes. Figure 13 describes permeate clarity from an apple juice ultrafiltration process. Average ultrafiltered permeate turbidity values are 0.3 nephlometric turbidity units (NTU) where all samples are below the conventional process standard of 1.5 NTU. Protein removal from the juice stream results in a permeate resistant to protein-induced haze. Concentration of true protein in the permeate is consistent with a 98 to 99% true protein retention by the UF membrane. As seen in Figure 14, true protein concentrations in the product are reduced to trace levels during the entire UF process, while feed true protein increases according to volumetric concentration factor. Non-protein nitrogen is poorly retained by the membrane, and is not concentrated in the feed.

Independent of yield improvement and product quality enhancement, the labor and operating cost alone justifies ultrafiltration in the juice plant. Figure 15 illustrates a comparison of ultrafiltration with conventional processing techniques with respect to major production costs. Typical reductions in enzyme treatment are on the order of 75%. The reusable ultrafilter membrane diminishes the cost of disposable filtration media by another 75%. As labor costs of conventional filtration and its associated quality control is a major portion of the production cost, its reduction by 92% is a significant element in the economic analysis of an ultrafiltration plant. Other associated costs are reduced to a lesser degree. The installed cost of an ultrafilter is comparable to the annual savings. The return on investment is high. For a case recently computed, assuming 1985 tax rates, depreciation schedules, and investment tax credit, the internal rate of return exceeded 100%. Cash payback occurred in the 16th month of ownership after only eight months of actual operation. With performance now demonstrated for six seasons, ultrafiltration is becoming firmly established as the unit operation for juice clarification.

SUMMARY

Ultrafiltration has proven itself as a cost effective commercial clarification and fining method for processing apple, grape, pineapple, and other fruit juices. The single unit operation of the ultrafilter increases juice yield, eliminates diatomaceous earth filtration, and reduces operating costs when compared to conventional juice processing techniques. Ultrafiltered juice has a clarity of 0.3 NTU, excellent flavor characteristics, and can be processed in continuous mode over a 24 hour day.

Literature Cited

1. Heatherbell, D. A., J. L. Short, and P. Strubi. Confructa. 22:157-169 (1977).

2. Heatherbell, D. A. Alimenta. 15:151-154 (1976).

3. Johnson, G., B. J. Donnelly, and D. K. Johnson. J. Food Sci. 33:254-257 (1986).

4. VanBuren, J. P. and W. B. Rominson. J. Agr. Food Chem. 17:772-777 (1969).

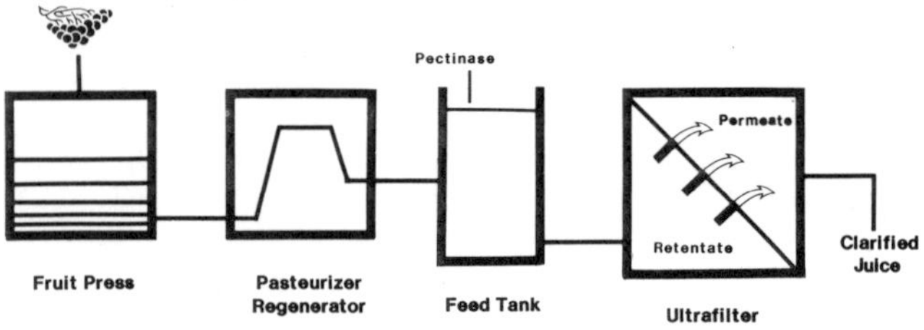

Figure 1. Process schematic for producing clarified juice with ultrafiltration.

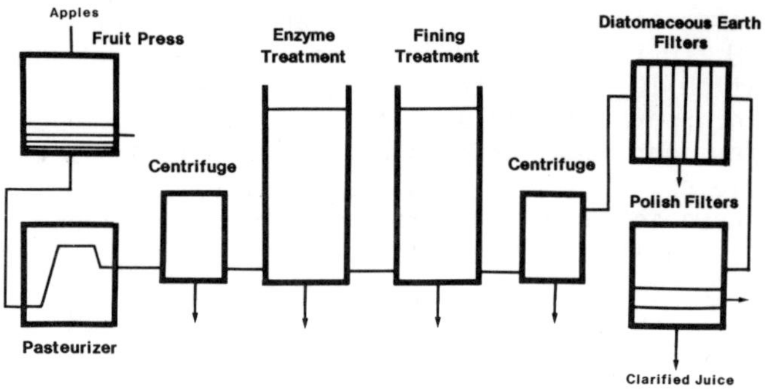

Figure 2. Process schematic producing clarified juice using conventional procedures.

Biological Haze	Non-Biological Haze
Yeast	Suspended Solids
Bacteria	Starch and Carbohydrates
Molds	Starch-Phenolic Complexes
	Protein
	Protein-Phenolic Complexes
	Protein-Metal Ion Complexes
	Condensed Polyphenolics
	Pectin

Figure 3. Haze components of fruit juice.

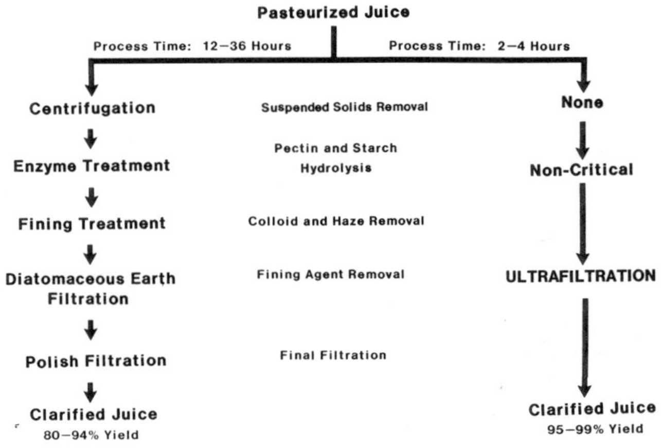

Figure 4. A comparison of ultrafiltration and conventional clarification techniques.

Increased Product Yield

Elimination of Rough and Polish Filtration

Elimination of Diatomaceous Earth Disposal

Consistency in Product Quality

Reduction in Operating Costs
- **-Reduced Labor Costs**
- **-Reduced Enzyme Costs**

Figure 5. The major advantages of ultrafiltration for fruit juice clarification.

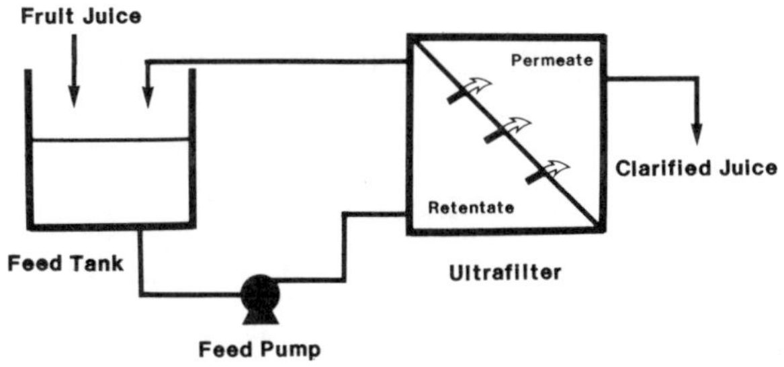

Figure 6. Unit operation of an ultrafilter.

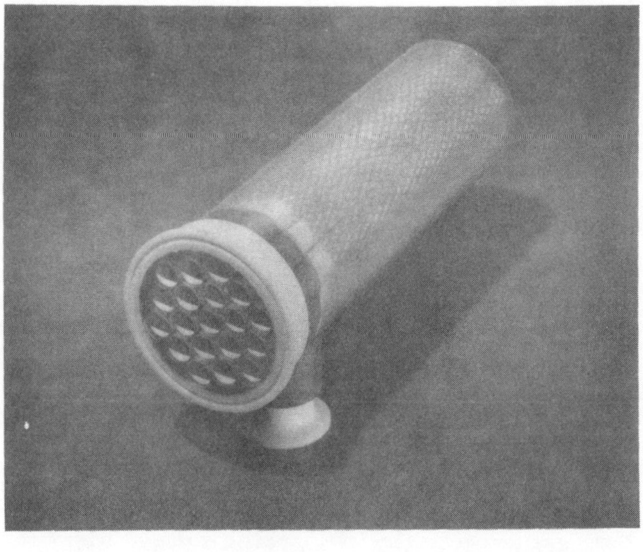

Figure 7. Cutaway of a Koch membrane system SUPER-COR ultrafiltration module.

Figure 8. A commercial apple juice ultrafiltration system using the Koch Tubular Design.

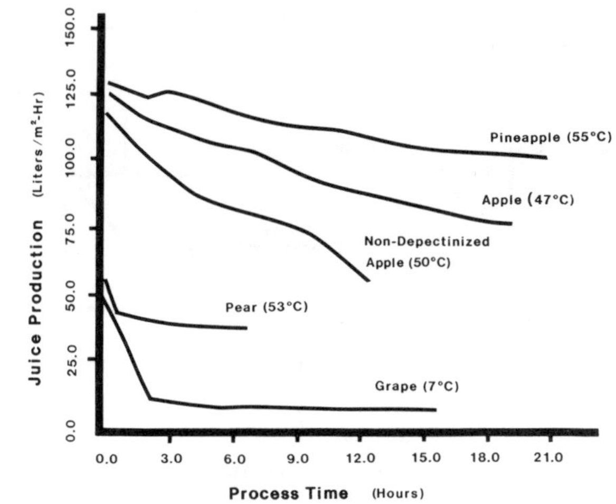

Figure 9. Ultrafiltration process flux versus time for selected fruit juices.

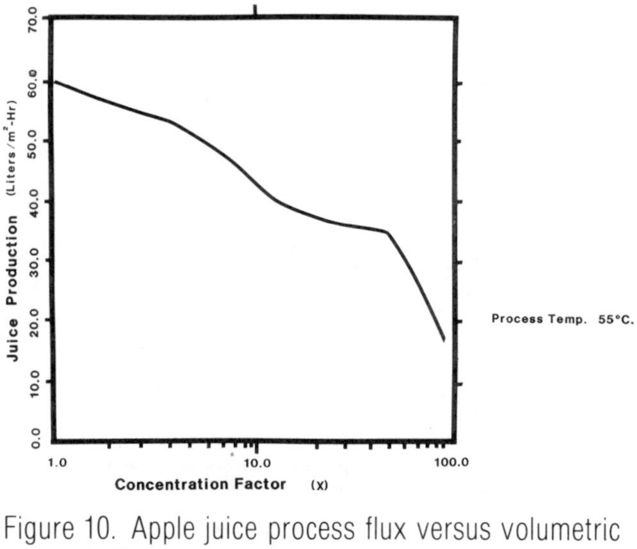

Figure 10. Apple juice process flux versus volumetric concentration.

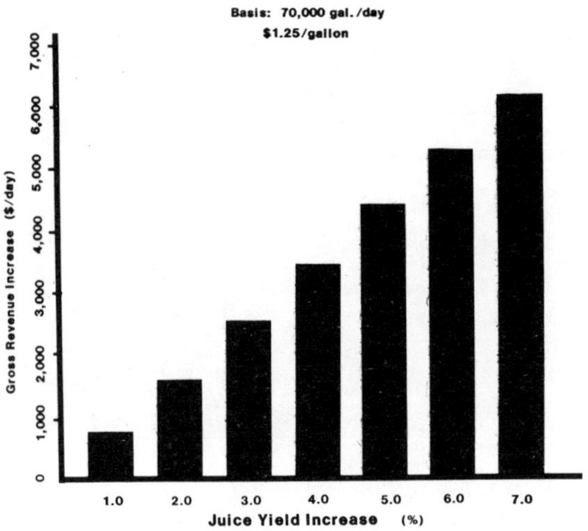

Figure 11. Potential gross revenue increase versus juice yield increase using the ultrafiltration process.

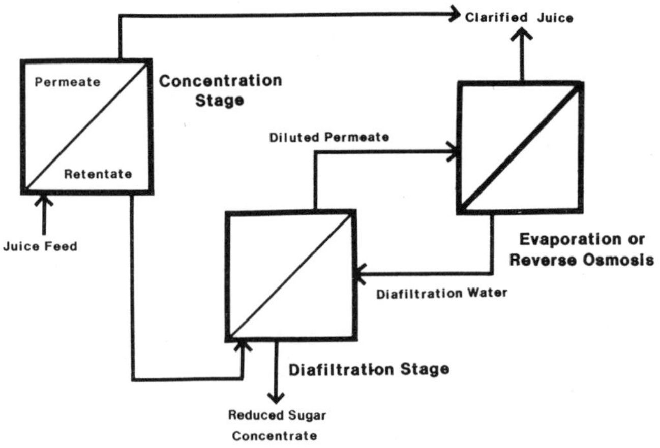

Figure 12. Process schematic for enhanced juice sugar yield using diafiltration with a subsequent concentration step.

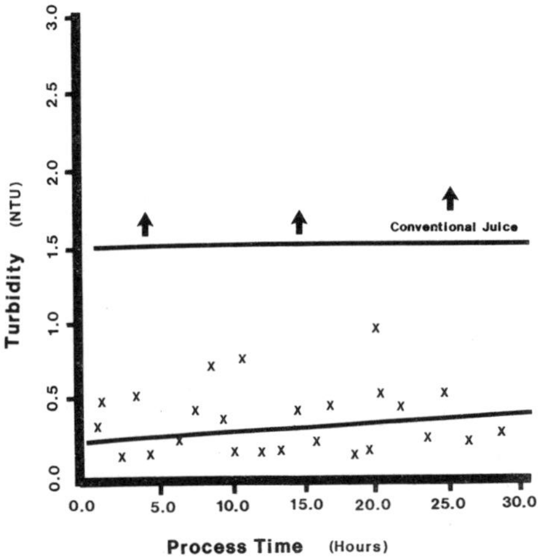

Figure 13. Clarity of ultrafiltered apple juice.

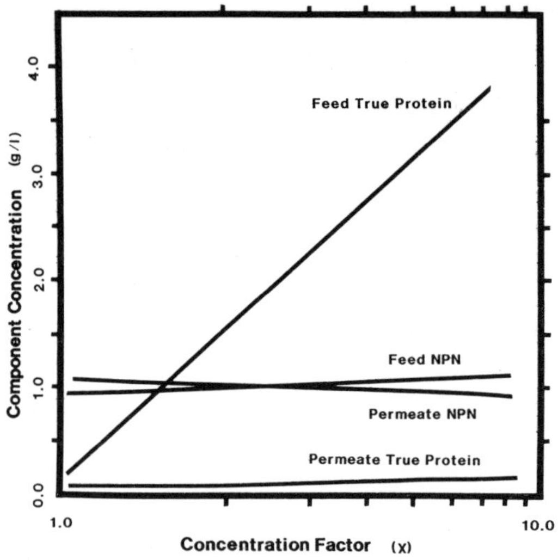

Figure 14. Feed and permeate true protein and non-protein nitrogen (NPN) concentration versus concentration factor.

Basis: 70,000 gallons/20 hour day
150 days/year

Ultrafiltration		Conventional Clarification	
Pretreatment Enzymes	$150.00	Pretreatment Enzymes	$600.00
Membrane Replacement (2 yrs)	207.00	Diatomaceous Earth	700.00
Cleaning Chemicals	16.00	Filter Pads	160.00
Maintenance	12.00	Maintenance	25.00
Power (32 hp)	32.00	Power (75 hp)	75.00
Labor (4 man-hours/day @ $12.50/hr)	50.00	Labor (48 man-hours/day @ $12.50/hr)	600.00
Total/Day	$467.00	Total/Day	$2 160.00
Total/Year	$70,050.00	Total/Year	$324,000.00

Total Yearly Savings = $253,950.00

Figure 15. A cost comparison of ultrafiltration and conventional filtration techniques.

ULTRAFILTRATION AS A COMPETITIVE UNIT PROCESS

Ivars Bemberis and Karen Neely ■ Dorr-Oliver, Incorporated, Stamford, CT

Membrane separations have become commonplace in many industrial processes. Ultrafiltration (UF) is used extensively in the food, pharmaceutical and electrocoating industries. A recent survey cited over 100,000 m^2 of installed membrane capacity in the dairy industry alone (1).

The unique ability to modify the selectivity of membranes and to choose the appropriate geometry of the membrane module and system configuration makes UF a very special unit operation.

In this paper a number of examples will be examined of the use of UF in commercial processes and of the competing unit operations it displaces. The comparisons drawn will illustrate the economic benefits of UF separations. These evaluations, however, cannot show the process improvements that result from the 'much finer' separation membranes provide. Often the characteristics of the retentate and/or permeate differ from those of the competing unit operation. These differences are an intrinsic property of the UF process. Such differences often enhance the performance of downstream operation and/or the quality of the product produced. These 'pluses' don't usually appear on the balance sheet because they are difficult to measure or value.

Experience over the last 20 years has shown that the capabilities of UF separations are not fully appreciated when UF is first used in the process train. After UF is in place in the flowsheet and operating experience has been gained, the greater potential of this technique is understood, and this leads to a revision of the process. In its final form, the process train is usually simplified and the point of use of the UF equipment has moved further upstream.

The following examples include the use of UF to replace precoat vacuum filters, plate and frame presses, and centrifuges and to reduce the load on thermal dewatering equipment. The data presented are based on actual plant operating experience to ensure that the economic comparisons are valid.

The benefits of a given project to the user may be measured in a variety of ways, including discounted cash flow analysis, annual cost comparison, present-worth comparison, and rate of return analysis. We have chosen to use rate of return analysis because it is the most straightforward method of communicating the advantages of making an extra investment. "Rate of return" means the interest which makes costs equivalent to income. (2) Looking at capital expenditures in this way focuses on the benefit of that investment in terms of how that expenditure compares with other uses for the same capital.

AN OVERVIEW OF ULTRAFILTRATION (3)

Ultrafiltration is a pressure-driven membrane separation process with selectivity occurring on a molecular scale. The membrane discriminates on the basis of molecular size and shape, retaining feed constituents larger than the membrane pores and passing materials of smaller dimension. The solution containing the membrane-retained species is referred to as the concentrate or retentate; the solution flowing through the membrane is referred to as the permeate.

In an operating system the feed solution is pumped across the membrane surface to provide the crossflow velocity needed to maintain high throughputs. The pressure created by the pump also provides the driving force to transport solvent and solutes through the membrane.

Figure 1 depicts the fundamental concept of a crossflow ultrafilter.

Figure 1. Ultrafiltration Process

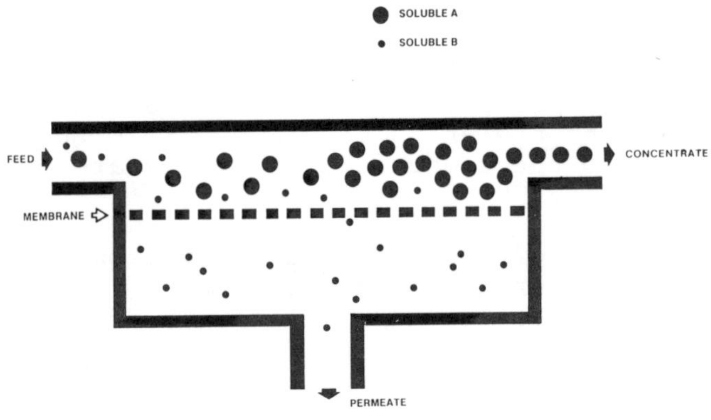

The concentration of membrane permeable components is the same on either side of the membrane, while the concentration of retained species increases because they are excluded from passing through the membrane. UF membranes are commonly used to separate materials in the range of 100 to 1,000,000 molecular weight, including soluble macromolecules such as proteins and dispersions such as fats and colloids. Figure 2 shows a membrane cut-off selection chart which indicates that the cut-offs of membranes are not precise. The normal variation in membrane pore diameters must be considered when choosing the appropriate membrane. Typically, the pore size should be several times smaller than the species to be retained or several times bigger than the material to be permeated to ensure the maximum retention and/or permeation.

Figure 2. Membrane Selection Guide

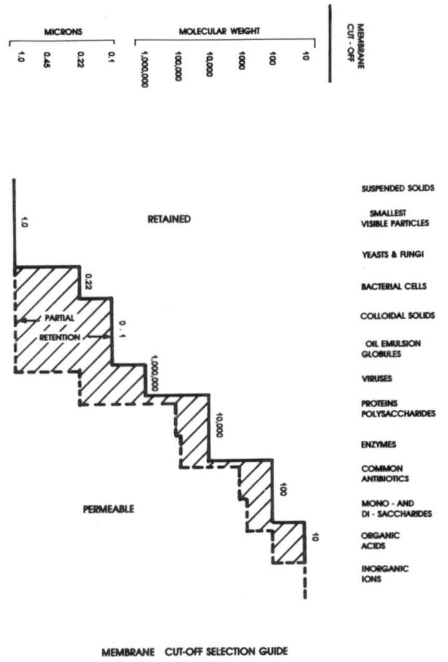

Today, membranes are made of a wide variety of polymers which provide the required mechanical strength and chemical and thermal stability needed for industrial processing. Membranes can be made with various structures and separation properties. The salient feature of most membranes is their anisotropic or asymmetric structure which provides very high fluxes. The dense, thin surface layer (0.2 um thick) permits these membranes to act as true surface filters (4). Rejected components are retained at the membrane surface and can be removed by the surface crossflow velocity. The membrane substructure (about 100 um thick) acts only to support the surface skin which fully determines the properties of the membrane.

FACTORS AFFECTING FLUX

Flux is defined as the permeate rate per unit of membrane area. Thus, maximizing flux will reduce system size and cost. A number of parameters influence flux rate. Flux data can be correlated by the following expression:

$$J \propto \frac{(V)^x \quad (D)^y}{(\nu)^z} \left[\log (C_w/C_b)\right] \quad \text{[Equation 1]}$$

where J is the process flux, V is the linear crossflow velocity, D is the diffusivity, ν is the kinematic viscosity, C_w is the solute concentration at the membrane surface, C_b is the solute concentration in the retentate, x, y, and z are empirical constants.

As permeate is removed, the concentration of retained components builds up at the membrane surface forming a dynamic boundary layer (gel layer). The resulting concentration gradient becomes the driving force for moving retained materials from the boundary layer back to the bulk solution. This phenomenon is illustrated in Figure 3. The accumulation of retained solids at the membrane surface creates a hydraulic resistance which may lower the permeability of the membrane (5).

The relationship of flux and retained solute concentration is typically presented as a flux vs. log concentration plot, as depicted in Figure 4. Also shown in the figure is the influence of crossflow velocity. As predicted by equation 1, increasing the crossflow velocity will increase the flux due to changes in the dynamic

Figure 3. Concentration Polarization

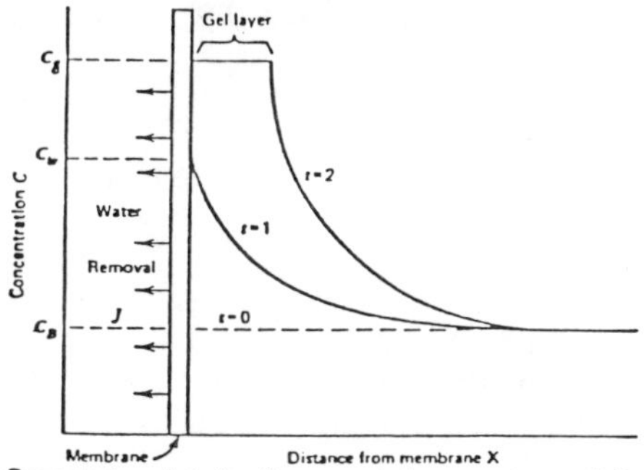

Concentration polarization. C_w = concentration at membrane wall, C_B = bulk concentration, C_{B_i} = bulk concentration of species i, J = flux, and C_g = gel concentration.

Figure 4. Flux-Concentration -Relationship

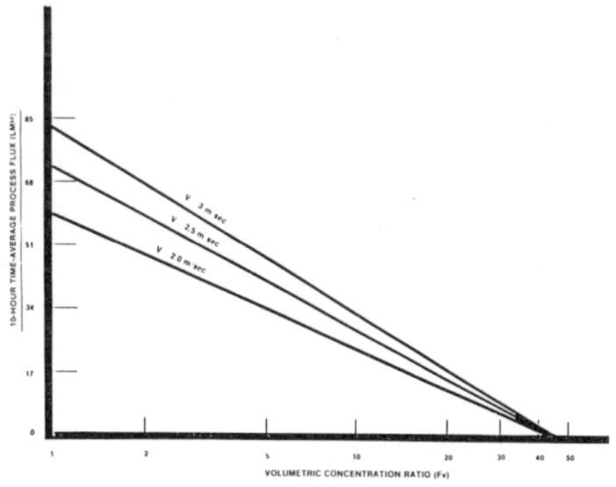

boundary layer. Clearly, the rate of flow past the membrane surface will influence the thickness of the boundary layer. The greater the surface velocity, the thinner the flux-retarding boundary layer, and hence, the higher the process flux at any given concentration.

Temperature is also a significant flux performance parameter. The kinematic viscosity of the solution and the diffusivity are strongly temperature dependent. Increasing the processing temperature will lower the viscosity and increase the diffusivity; both effects will increase the flux as predicted by Equation 1.

The flux is relatively independent of the transmembrane pressure difference. (The permeate is usually at atmospheric pressure, and the feed is pressurized.) The flux/pressure relationship is described by a Darcy's law type expression:

$$J = k (P)^{M} \quad \text{[Equation 2]}$$

where k is a constant, P is the transmembrane pressure difference, and M is an empirically-derived exponent. The shape of the flux vs. pressure curve is highly foulant dependent.

Finally, flux is also time dependent. As new components are introduced into the gel layer and interactions occur, the flux may show a time-related decrease characteristic of membrane fouling. The effect of fouling is most pronounced in the beginning of a process run and at low non-permeable solids concentrations. In these cases, there is little gel layer resistance, and any added resistance is very significant. Figure 5 demonstrates the effect of time and pressure on flux. Time-dependent fouling, if severe, may lower the flux sufficiently to require more frequent cleaning of the membranes.

Figure 5. Effect of Transmembrane Pressure on Process Flux

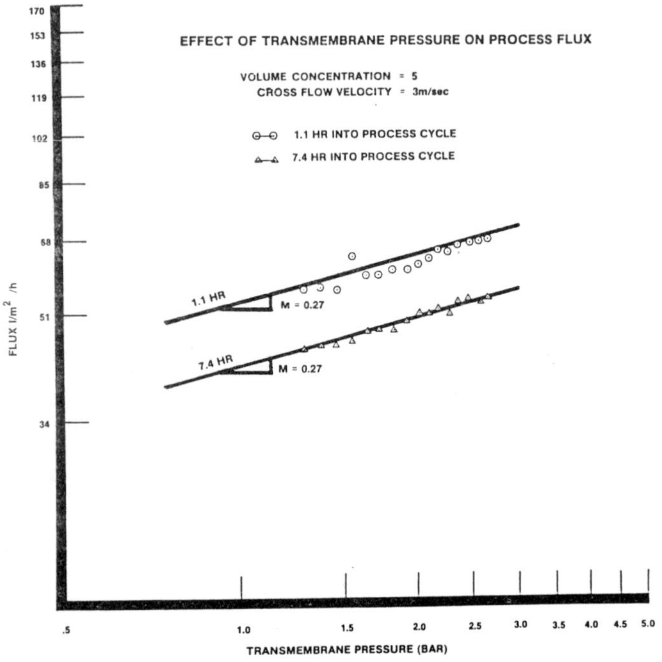

MEMBRANE CONFIGURATIONS

The versatility of phase inversion methods used to manufacture the polymeric structures we call membranes allows them to be 'cast' into numerous shapes: self-supported fibers (hollow fibers, capillary membranes); larger bore supported tubes; and flat sheets for parallel leaf, plate and frame and spiral wound elements. The various types are differentiated below:

Flat Sheets: Sheet stock is used as the replaceable element. The sheet is held in place in a device which provides for the proper distribution of the feed and permeate flows.

Flat Leaf/Plate: Membrane is directly cast onto or attached to a support which is then assembled into a cartridge or module. The replaceable element is the entire leaf.

Spiral Wound: Sheet stock is arranged to form porous sheet, sandwich-type structures which are then rolled around a central (permeate) tube. The laminate is sealed at the three sides and the porous central tube. These wraps are typically placed in a round (pipe) pressure vessel.

Wide-bore Tubes (10 to 25 mm I.D.): Membranes are cast in place or inserted into individual tubes. Tubes can be arranged in a variety of configurations of single or multiple tubes in parallel inserted into a larger housing for permeate/feed containment.

Narrow-bore Tubes (2 to 10 mm I.D.): Membranes are cast in place in smaller diameter tubes. Packaging is achieved by assembling tubes in bundles surrounded by a housing.

Hollow Fibers (1 to 2 mm I.D.): Spun hollow fine fibers are bundled and potted into cartridge housings. The feed channel can be either the inside or the outside of the fiber.

Inorganic Membranes are also entering the market. Both ultrafiltration and microfiltration capabilities are available with membranes made from zirconium oxide and aluminum oxide. The configuration is currently limited to narrow bore tubes of 3 to 6 mm diameter.

TABLE 1. COMPARISON OF UF DEVICE DESIGNS

Device Type		Established Suppliers	Advantages	Disadvantages
Hollow Fiber (1-2 mm bore)		Romicon Amicon	Low Capital Cost Backflush Low Hold Up	High Membrane Replacement Cost Energy Input Limited Pressure Limited Easily Plugged
Narrow-Bore Tube (2-10 mm)	Polymer Membrane	Koch		
	Ceramic Membrane	SFEC (6 mm) Ceraver (4 mm) Norton (3 mm)	Microbiological Control (Feed/Perm.) Long Life High Energy Inputs (Flux) Backflush	(UF Limited mwco) High Membrane Replacement Cost
Wide-Bore Tube (10-25 mm)	Inserted Tube	PCI (12.5 mm) Watilin (14.4 mm)	Low Membrane Replacement Cost Resistance to Plugging Hygienic Permeate Channel	Difficult Membrane Replacement High Hold Up Volume Energy Inefficient
	In-Situ Cast Tube	Koch (25 mm) Koch (12.5 mm)	Resistance to Blockage Long Membrane Life	High Space Required/Hold Up High membrane Replacement Cost Open Permeate Flow Energy Inefficient
Spiral-Wound		Koch Osmonics	Low Membrane Replacement Cost Low Capital Cost Low Hold Up	Easily Plugged Unhygienic Permeate Channel
Flat-Sheet		DDS Rhone-Poulenc	Low Membrane Replacement Cost High/Efficient Energy Input (DDS) User Independent of Supplier Relative Resistance to Blockage	Flow Channeling Difficult Membrane Replacement
Flat Leaf/Plate		Dorr-Oliver Millipore	Ease of Membrane Replacement Relative Resistance to Blockage	Moderate Membrane Replacement Cost Pressure/Energy Limited

These configurations provide a high degree of flexibility for matching membrane capabilities to process needs. The choice of appropriate membranes and module configurations can have a significant impact on process performance in a given application. Module design has a measurable effect on the hydrodynamic performance of crossflow membrane devices. A comparison of the merits of various membrane devices is shown in Table 1 (6) which highlights the strengths and weaknesses of the devices.

ULTRAFILTRATION AS A REPLACEMENT FOR ROTARY VACUUM FILTRATION

Kaolin Dewatering

Kaolin is manufactured by both dry and wet processes. In the wet process, crude clay is blunged to disintegrate it into fine particles. Turbulent agitation of the clay in the presence of suitable dispersants produces the desired disintegration. The fluid properties of the slurry are determined by the dispersant which alters the slurry viscosity to create a fluid or watery state.

Next, the kaolin is fractionated by hydroseparators or centrifuges. The various fractions are then bleached to the desired brightness with zinc or sodium salts of hydrosulfurous acid. Chemical reduction of the insoluble oxides creates the brightening effect. The clay is acidified during bleaching to cause it to agglomerate and make it easier to filter.

Continuous rotary drum vacuum filters are generally used to filter the kaolin. During filtration the clay is also washed to remove residual chemicals and dispersants. Drum filter cakes are typically 50 to 60% solids by weight. This cake can be reslurried by the addition of a dispersant and sold in slurry form, or it can be dried (by drum or spray dryers) and sold as a dry powder. Occasionally, plate and frame pressure filters are used as an alternative for dewatering.

Ultrafiltration in the Kaolin Flowsheet

Three basic applications can be identified for UF in the wet process:

1. After bleaching, the clay slurry is at 15 to 17% solids and must be concentrated to about 30% to be fed to the drum filters. Centrifugation is normally used for this concentration. UF can replace this centrifugation and increase the yield because there would be no clay loss to the centrate. For centrifugation, dispersants are added to disperse the particles for improved seperation.
2. Losses of kaolin to the filtrate from the drum or plate and frame filters average from 1 to 3% of the feed solids. With very fine grades, the losses can be even greater requiring additional filter area. To keep filtrate losses low, the filter cloth is changed frequently. UF can capture the clay that would be lost and permit less frequent filter cloth changes.
3. The kaolin is filtered and washed to achieve cake solids of 50 to 60%. If sold in slurry form, the clay must be thermally dewatered to about 70% solids. UF can be used to concentrate kaolin to the 60 to 70% range. Even at the low flux rates when operated at high solids, operating cost benefits are substantial.

Ultrafiltration cannot replace the classification unit operations performed by hydroseparators and centrifuges, but it is appropriate for any of the dewatering operations whether the desired final solids are 15, 30, 60 or 70%. As long as the feed is highly dispersed, UF can perform the dewatering steps without any production process changes. The productivity of the UF system will be determined by the desired final product concentration since the flux varies according to the concentration. Membrane abrasion has not been detected during protracted test periods.

Tables 2 and 3 list the parameters used in the economic analyses for two types of clay - a kaolin clay of normal size distribution and a very fine submicron kaolin clay, respectively. Vacuum filters are used to increase the solids concentration from approximately 30 to 50% solids before spray drying. A UF system can perform precisely the same function.

TABLE 2 PARAMETERS FOR COST COMPARSION OF ULTRAFILTRATION (UF) AND ROTARY DRUM VACUUM FILTRATION (VF) OF KAOLIN CLAY

PROCESS DUTY: 13.61 Metric ton/hr.*, 24 hr./day, 260 days/hr. 34% - 53% TS

	UF	VF
ENERGY (.061 $/Kwhr)	37.3 Kw	73 Kw/filter
LABOR ($10/hr. - operator) ($15/hr. - supervisor)	4 man hr./24 hr. day	2 operations/hr. + 1 supervisor/hr.
MAINTENANCE	2% of capital cost/yr.	5% of capital cost/yr. (includes filter cloth replacement)
THROUGHPUT	14.6 Kg clay/m² hr.	34.17 Kg clay/m² hr.
REQUIRED FILTRATION AREA	1156 m²	426 m² (3 - 13.5′ x 36′ drums)
MEMBRANE REPLACEMENT	Every 2 years $179/m²	
CAPITAL COST	$969/m²	$3521/m² ($500,00/filter excluding buildings)

* (15 ton/hr.)

TABLE 3 PARAMETERS FOR COST COMPARSION OF ULTRAFILTRATION (UF) AND ROTARY DRUM VACUUM FILTRATION (VF) OF SUBMICRON KAOLIN CLAY

PROCESS DUTY: 13.61 Metric ton/hr.*, 24 hr./day, 260 days/hr. 30% - 50% TS

	UF	VF
ENERGY (.061 $/Kwhr)	37.3 Kw	73 Kw/filter
LABOR ($10/hr. - operator) ($15/hr. - supervisor)	4 man hr./24 hr. day	2 operations/hr. + 1 supervisor/hr.
MAINTENANCE	2% of capital cost/yr.	5% of capital cost/yr. (includes filter cloth replacement)
THROUGHPUT	14.6 Kg clay/m² hr.	12.94 Kg clay/m² hr.
REQUIRED FILTRATION AREA	1156 m²	1135 m² (8 - 13.5' x 36' drums)
MEMBRANE REPLACEMENT	Every 2 years @ $179/m²	
CAPITAL COST	$969/m²	$3521/m² ($500,00/filter excluding buildings)

* (15 ton/hr.)

TABLE 4 COST COMPARSION OF ULTRAFILTRATION (UF), ROTARY DRUM VACUUM FILTRATION (VF) AND SPRAY DRYING (SD) KAOLIN CLAY FROM 34% - 53% TS

	UF	VF	SD
ENERGY	0.96	1.00	13.12
CHEMICALS	0.32	2.65	
MAINTENANCE	0.26	0.88	1.39
LABOR	0.12	1.84	
MEMBRANES	1.23	-	-
TOTAL OPERATING COST ($/Metric Ton)	2.89	6.37	14.51
CAPITAL COST ($) (13.61 metric tons/hr.)	1,120,000	1,500,000	
SAVING ($/yr.) *	986,684	691,189	
% IRR	59.2	33	

* Compared to spray drying.

For both cases, the membrane area is the same even though the flux is higher for the submicron clay. For the finer clay, the higher flux is offset by the greater water removal requirement. In the submicron clay example, the UF area exceeds the filter area, even though the UF has a higher throughput rate. Selection of optimal system configuration resulted in a UF system with slightly more membrane area than required from flux considerations alone.

The operating costs and capital costs for UF and rotary drum vacuum filters for kaolin clay dewatering are detailed in Table 4. The major differences in operating cost are high chemical and labor requirements for the vacuum filter. The high cost component for the UF system is membrane replacement.

While capital costs for the UF and vacuum filter are similar, the difference in operating costs compared to spray drying is quite striking. The internal rate-of-return (IRR) for the rotary drum vacuum filter is 33% (payback period of 3.5 years). The lower capital and operating costs of the UF system increases the IRR to 59.2%.

When the characteristics of the clay are changed such as with submicron clay the economic picture shifts dramatically. Finer clays command a higher market price but are more difficult to dewater by vacuum filtration. In this case, detailed in Table 5, the high capital costs for the vacuum filters reduces the IRR to 8.5%. However, the UF system is extremely attractive with an IRR of 74.2%.

TABLE 5 COST COMPARSION OF ULTRAFILTRATION (UF), ROTARY DRUM VACUUM FILTRATION (VF) AND SPRAY DRYING (SD) SUBMICRON KAOLIN CLAY FROM 30% - 50% TS

	UF	VF	SD
ENERGY	0.96	2.67	16.59
CHEMICALS	0.32	2.98	
MAINTENANCE	0.26	2.36	1.39
LABOR	0.12	2.57	
MEMBRANES	1.23	-	-
TOTAL OPERATING COST ($/Metric Ton)	2.89	10.58	17.98
CAPITAL COST ($) (13.61 metric tons/hr.)	1,120,000	4,000,000	
SAVING ($/yr.) * (2 yr. life)	1,281,331	628,353	
% IRR	74.2	8.5	

* Compared to spray drying.

TABLE 6 OPERATING COST COMPARSION FOR FERMENTATION BROTH CLARIFICATION USING ULTRAFILTRATION (UF) AND PRECOAT ROTARY DRUM VACUUM FILTRATION (VF)

COST ITEMS	UF	VF
FILTER AID	-	6.34
MEMBRANE REPLACEMENT	2.22	-
ENERGY	0.25	0.08
LABOR	0.08	0.32
MAINTENANCE	0.04	0.14
CLEANING CHEMICALS	0.43	-
TOTAL OPERATING COST ($/1000 liters)	3.02	6.88
CAPITAL COST ($/1000 liters/day)	2,940	695
% IRR	31	

TABLE 7 PARAMETERS FOR COMPARATIVE PROCESS ECONOMICS OF BROTH CLARIFICATION FOR ULTRAFILTRATION (UF) AND PRECOAT ROTARY VACUUM FILTRATION (VF)

ANNUAL PROCESS DUTY: 16 hrs./day, 250 days/yr.

	UF	VF
THROUGHPUT	17 ℓ/m²h	410 ℓ/m²h
MEMBRANE LIFE	1 year	-
MEMBRANE REPLACEMENT COST	$150 m²	-
FILTER AID CONSUMPTION	-	24 kg/1000 liters filtrate
FILTER AID COST	-	$0.26/kg
POWER CONSUMPTION	0.14 kw/m²	1.07 kw/m²
ENERGY COST	$0.03/kwhr.	$0.03/kwhr.
LABOR REQUIREMENTS	1 hr./day/560 m²	4 hrs./day/23 m²
LABOR COST	$12/hr.	$12/hr.
ANNUAL MAINTENANCE COST	0.5 hr./day/560 m²	$226/m²
CLEANING COST	$0.1/m²	-
CAPITAL COST	$800/m²	$4,500/m²

result is an IRR of 31% when comparing the difference in operating savings to the capital cost of the UF system. These types of savings are typical for fermentation broths which are easily filterable. The additional cost of the precoat waste disposal was not included in this analysis because such costs are highly site specific.

Antibiotic Recovery - Case II

A UF system can significantly outperform precoat filters on slimy fermentation broths which produce poor filtrate rates. For example, a UF system of 1070 m^2 membrane area has been operating for six years in a fermentation broth clarification. The broth is concentrated to 3X and then diafiltered with three turnovers to produce an actual reported yield of 98% (7). At this plant, both precoat filters and centrifuges were considered before the UF system was installed.

In tests of precoat filters, the broth formed a very compressible, slimy filtercake which gave very low filtrate rates. Addition of filter aid decreased the compressibility, but the filtrate rates remained poor. Furthermore, washing the cake to increase recovery was impractical since it required re-suspension of the cake in water and re-filtering. At best, 96% yield was achieved. The use of a precoat filter also increased costs and created a waste disposal problem.

In tests of centrifugation, the settling rates were so slow, even with flocculent added, that only a dilute underflow sludge formed, and it had to be re-diluted and centrifuged to recover the substantial amount of product it contained. The resulting high capital costs eliminated the centrifuge for this application.

The cost advantages of UF were very attractive in this case. The membrane cost is one fourth that of the filter aid. The labor cost is one third that of the precoat filter. Furthermore, since in this case the filtrate rates are so poor, the capital cost of the UF is 20% less than that of an equivalent precoat filter.

Finally, it should also be noted that a precoat filter frequently passes some filter aid and broth solids which affect downstream recovery operations. The UF always produces an exceptionally clear permeate eliminating concerns about downstream problems.

Both cases demonstrate an advantage for UF due primarily to lower energy, labor and capital costs. Additional savings are also obtained with the UF system since there will be no loss of product in the filtrate (permeate). These savings were not included in the IRR analysis. There is considerable potential for product enhancement with UF resulting from removal of salts which affect final product viscosity. Wash out of these salts lowers the viscosity and produces a higher value end product.

ULTRAFILTRATION AS A REPLACEMENT FOR PRECOAT FILTRATION

Fermentation Broth Clarification

Many commercial products are made by fermentation. Examples of products produced by fermentation include citric acid, acetic acid, antibiotics, amino acids and animal vaccines. In the recovery process, the first step usually requires the separation of the fermentation broth into a cell-free fraction and a broth solids mixture. Often this clarification step is carried out by precoat vacuum filters. Since extra-cellular fermentation products are small, having molecular weights of 500 to 2000 daltons, UF membranes with a 10,000 to 30,000 molecular weight cut-off may be used for clarification. The extraneous materials such as suspended solids, cells, cell debris, and other macromolecules which are larger than the membrane pores are retained while the clarified product passes through.

Most fermentations are batch operations and, therefore, the UF is also run in a batch mode. However, if the stability of the product is such that short residence times or low operating temperatures are required, a continuous mode can be employed. Since flux is temperature dependent, processing should be at the highest allowable temperature, usually determined by the thermal stability of the product.

As the clarification proceeds, the retained solids concentration builds up and lowers the flux.

Economic factors usually dictate how far a concentration may proceed before the flux rate becomes too low. Some permeable solids (the product) remain in the concentrate. Further recovery of the product can be achieved by diafiltration, a washing procedure where water is added to the system at the same rate as permeate is removed. The retained or non-permeable solids concentration remain unchanged while the concentration of permeable solids decreases. In this manner, more product can be permeated, and the UF system can operate at an economically reasonable flux.

As an example, a volumetric concentration to five fold (5X) implies that 80% of the feed has permeated while 20% has been retained; thus the recovery is only 80%. If diafiltration is then performed using four turnovers (four volumes of water per volume of retentate), 99.6% product recovery can be achieved. Furthermore, diafiltration can be performed anywhere in the process before, during or after concentration depending on the volume of water required and the flux performance desired.

Antibiotic Recovery - Case I

An economic comparison (3) of a UF system using diafiltration and a precoat filter with cake washing can be seen in Table 6. The assumptions used to generate this comparison are shown in Table 7. Equal product recoveries, volumes of wash water and total permeate (filtrate) have been assumed.

As can be seen in Table 6, the capital cost of the UF system is four times greater than that of the precoat filter, but the operating costs are only 44% of the precoat operating costs. Major differences in running costs are that the filter aid cost is three times the membrane replacement cost, precoat maintenance cost is three and one half times greater than maintenace for UF, and labor costs are four times greater for the precoat filters. These higher costs far outweigh the higher energy and chemical costs of the UF. The

Antibiotic Recovery - Case III

In another case, a Japanese pharmaceutical manufacturer experienced a low process yield (90%) and high operating costs ($264,000/year for filter aid and power alone) with precoat filters which severely reduced the profitability of the product. A UF system performed the same separation at a fraction of the cost. While the slimy, hard-to-dewater nature of the broth required much precoat area, the comparable membrane system was considerably smaller.

UF and diafiltration of the broth achieved a 99% product recovery at a moderate capital investment. The IRR and cash flow analyses are shown in Table 8. The high rate of return clearly establishes the advantage of UF processing for this manufacturer.

TABLE 8 OPERATING COST COMPARSION FOR FERMENTATION BROTH CLARIFICATION USING ULTRAFILTRATION (UF) AND PRECOAT ROTARY DRUM VACUUM FILTRATION (VF)

COST ITEMS	UF	VF
FILTER AID	-	88.16
MEMBRANE REPLACEMENT	6.22	-
ENERGY	4.52	52.80
LABOR	5.33	-
MAINTENANCE	0.04	-
CLEANING CHEMICALS	.36	-
TOTAL OPERATING COST ($/1000 liters)	16.47	140.96
SAVINGS ($/1000 liters)	124.99	-
CAPITAL COST ($/1000 liters/day)	8333.	-
% IRR	114.6	

ULTRAFILTRATION AS A REPLACEMENT FOR FILTER PRESSES

Apple Juice Clarification (8)

Of the four common types of apple juice produced, natural, crushed, clarified, and clear, clear juice is a candidate for membrane processing. Clear juice is typically produced by enzyme treatment of the press juice to hydrolyze the pectin (which also makes the juice hard to filter) and improve filtration characteristics. The enzyme treatment is followed by precoat filtration.

Process modifications are possible because membrane systems permit a more continuous operation. The key is eliminating the holding/decanting steps to increase production capability through expanded utility of existing tankage. Also, greater certainty in production scheduling is achieved since processing times are much more predictable. Finally, complete automation of the process is possible.

Many apple processors are currently using or testing UF equipment. Process economics comparing juice clarification by plate and frame presses and UF are shown in Table 9. (The analysis is very sensitive to the process yield improvement credited to the UF step.)

TABLE 9 COST COMPARISON OF ULTRAFILTRATION (UF) AND PLATE & FRAME PRESSES (P&F) FOR APPLE JUICE CLARIFICATION

	UF	VF
ENERGY	0.26	-
CHEMICALS	0.26	2.90
MAINTENANCE	0.65	-
LABOR	0.82	2.64
MEMBRANE/FILTER AID	0.79	6.08
TOTAL OPERATING COST ($/1000 liters)	2.78	11.62
SAVINGS (per 100 day season)	$301,444	
CAPITAL COST	$430,000	
% IRR	81.3	

Based on operating cost savings of $\$8.84/m^3$, the internal rate of return is 81%. This analysis does not take into account increased revenues made possible by productivity improvements made by converting a laborious batch process into a continuous operation.

ULTRAFILTRATION AS A REPLACEMENT FOR CENTRIFUGES

Substitution of UF for centrifugation is more complicated. The rheological properties of the feed as well as shear and thermal effects must be considered when using centrifuges. Typically, centrifuges are selected when the retained solids are very forgiving of the mechanical forces they are exposed to in the centrifuge.

Centrifuges are usually used for dewatering. Some UF systems are not capable of achieving the high solids concentrations, for example, in harvesting cells from fermentation broths. If a final cell concentration of 200 gm/l is acceptable, then the following analysis is valid. A $45,000 hollow fiber filtration system and a $150,000 centrifuge were compared by R. J. Tutunjian (9) for a system processing 3,000 l/hr. The analysis showed a marginal savings in operating cost, $2900 per year, and a rate of return of -1.58%. Based on these savings, the purchase of a UF system to replace an existing centrifuge would not be justified. However, based on savings in capital when a UF system is purchased instead of a centrifuge, the internal rate of return is very attractive, 127.2%.

ULTRAFILTRATION AS A SUPPLEMENT FOR SPRAY DRYING

Removal of water by non-thermal means provides a cost advantage to spray dryer operators. The benefit can be two-fold. First, it reduces the operating costs for an existing dryer because the energy required to remove water by UF is only one to two percent of that required to evaporate water. Second, an advantage can be gained when planned production increases require added spray dryer capacity. Pre-concentrating the feed to the spray dryer permits increasing the solids throughout without requiring additional water removal capacity.

Egg White Concentration

UF has been applied in the food industry since its commercial introduction, and it has been employed extensively in the dairy industry for milk and whey concentration. It is also now used by the egg industry as a supplement to spray drying. In the traditional process, eggs are broken and separated into yolk and the liquid egg white fractions. The whites are strained to remove shells and the chalazae. The liquid egg whites are pasteurized and fermented to lower the glucose level to prevent browning during heating. Finally, they are spray dried to a powder. This powder is used in a variety of products including cake mixes.

Since the egg whites are only 11.2 to 11.7% solids, spray drying to three percent moisture requires removal of a large quantity of water. Clearly, removal of part of the water by non-thermal means will greatly reduce the energy costs of drying the egg whites. As a result, UF is employed to provide supplemental dewatering prior to spray drying. Usually UF removes half the volume of water of the incoming whites, increasing the solids level to 20%. A proportional loss of glucose and other minerals which readily pass through the 10,000 molecular weight membrane occurs during the process. There is virtually no passage of protein since the proteins all have molecular weights of 14,600 or greater, exceeding the pore size of the membrane. Protein recoveries typically exceed 99.5%.

After UF, the concentrate is fermented to reduce the glucose level nearly to zero, and the whites are spray dried to a powder. The characteristics of egg white powder which has been ultrafiltered and spray dried are comparable to those which are spray dried directly. The cost comparison in Table 10 shows that there are substantial economic advantages to utilizing UF before spray drying. The operating costs to concentrate to 20% total solids using

UF are one fourth those of a spray dryer removing the equivalent amount of water. This difference results from the savings in energy since UF dewatering does not require a phase change to remove the water. The economic comparison shows a 61.5% IRR with 1.42 year payback.

Further diafiltration to remove salts from whites which have been chemically treated is also possible. As stated earlier, once UF is in place, product modifications are easily made.

When UF is used prior to spray drying, it may also be possible to de-sugar during the UF process to take advantage of the elevated temperature and operate at higher process fluxes.

CONCLUSION

Ultrafiltration is a unit operation which is finding increased application in the process industries. The ability to choose membrane selectivity and equipment configuration provides to the process engineer a unique capability to fully optimize separations processes and, often, increase the value of the end product. As new membrane materials become available with a wider range of molecular weight cut-offs, the use of UF will continue to expand.

The potential uses of UF are quite diverse as is evident from the applications discussed: kaolin dewatering, antibiotic recovery, apple juice clarification, and egg white concentration. Economic advantages for employing UF instead of conventional dewatering methods have been clearly demonstrated. The operating cost savings and simplified process operations provided by UF are strong incentives for using this competitive separations technology.

LITERATURE CITED.

1. Hanemaaijer, J.H. and Hiddink, J., "The Expansion of Membrane Filtration in the Dairy Industry," North European Dairy Journal. No. 2 (85).

2. Taylor, G.A., Managerial and Engineering Economy, D. Van Nostrand Co., New York (1964).

3. O'Sullivan, T.J., et al, "Applications of Ultrafiltration in Biotechnology," Chemical Engineering Progress (January, 1984).

4. Strathman, H., "Membranes and Membrane Processes in Biotechnology," Trends in Biotechnology, Vol. 3, No. 5 (1983).

5. Klinkowski, P.R., "Ultrafiltration," Kirk-Othmer Encyclopedia of Chemical Technology, Vol. 23, John Wiley & Sons, New York (1983).

6. Beaton, N.C., Private communication (1985).

7. Gravatt, D.P. and Molnar T.E., "Recovery of an Extracellular Antibiotic by Ultrafiltration," Engineering Foundation Conference (February, 1984).

8. Bemberis, I., "Membrane Processes for Biotechnology and Food Processing," 1st Annual Membrane Technology/Planning Conference, Danvers, Massachusetts (1983).

9. Tutunjian, R.S., "Scale-Up Considerations for Membrane Processes," Biotechnology, Vol. 3 (July, 1985).

TABLE 10 **COST SUMMARY AND ECONOMIC ANALYSIS OF ULTRAFILTRATION (UF) AND SPRAY DRYING (SD) OF EGG WHITES**

AUNNUAL UTILIZATION: 16 hrs./day, 250 days/yr., 10,000 1/day of 11.2-11.7% concentrating to 2X

COST ITEMS	UF	VF
ENERGY	2,223	85,360
LABOR	10,000 (4 hrs./day, $10/hr.)	40,000 (16 hrs./day, $10/hr.)
MEMBRANE REPLACEMENT (Every 2 years, $179/m^2)	5,457	-
CLEANING COST	2,002	-
TOTAL OPERATING COST ($/year)	19,419	125,360
SAVINGS ($/yr.)	105,941	-
CAPITAL COST (61.5 m^2 x $1869/m^2)	$115,000	-
% IRR	61.5	

HYDROCARBON SEPARATIONS WITH POLYMERIC MEMBRANES

S.S. Kulkarni, E.W. Funk and Norman N. Li ■ Signal Research Center Inc.
50 East Algonquin Road, Box 5016, Des Plaines, IL 60017-5016

I. INTRODUCTION

This paper presents a potentially new use of polymeric membranes for the separation of hydrocarbon mixtures.(1) The focus is on the separation of hydrocarbons which are chemically similar but which differ significantly in molecular weight. This type of separation by membranes has large potential applications in fields such as petrochemicals, synthetic fuels and environmental systems.

Our research addresses a particular application in the petrochemical field. As shown in Figure 1, crude oil is typically first fractionated in an atmospheric distillation column; non-volatile material is often then separated in a vacuum distillation tower. As feedstocks become heavier and synthetic fuels are introduced, a significant fraction of the feedstock cannot be fractionated by either atmospheric or vacuum distillation. For example, the oil from the Athabasca Tar Sands(2) contains 50% non-volatile hydrocarbons.

Hydrocarbons which are non-volatile can be further separated using solvent extraction. This is typically done in a refinery which seeks to maximize its output of light hydrocarbons for fuels or a refinery involved in the manufacturing of lubricating oils and greases.(3) These solvent extraction processes generally are very high solvent/oil ratios (4) and recover the solvent by energy-intensive distillation. We see significant energy savings in the use of membrane technology to provide a more efficient solvent recovery by avoiding the vaporization of the solvent.

This paper presents membrane permeation data applied to solvent deasphalting; the new processing concept using membranes(1) is illustrated in Figure 2. Paraffinic solvents (propane through hexane) are used to divide the heavy crude into a high-quality oil which is low in metals (Ni, V, Fe) and hetero-atoms (N, O, S) and a low-quality asphaltene fraction which may serve as process fuel. Figure 3 presents basic system performance for solvent deasphalting with pentane.(5) It can be used to highlight the potential advantages of membrane technology. First, deasphalting processes use high solvent/oil weight ratios (4-6 pentane/oil) to give a good separation of oil and asphaltenes and thus a large amount of energy is consumed in vaporizing solvent. Second, the viscosity of the pentane/oil remains essentially constant down to a pentane/oil ratio of about 1; then the viscosity rises rapidly. Membrane modules can operate efficiently with feeds up to 3-5 cp and thus a membrane system can be used to reduce the solvent/oil ration down to near unity; steam stripping would then be used to recover residual solvent.

An important advantage of this processing concept is that it does not require highly selective membranes. Oil which permeates the membrane is simply recycled back to the primary separation unit and may have a small effect on the separator size.

II. EXPERIMENTAL PROCEDURE

This experimental membrane permeation unit is shown schematically in Figure 4. It can be used to study both flat-sheet membranes and spiral-wound elements. In this test unit, the liquid hydrocarbon feed is supplied from a three-liter reservoir. The reservoir rests atop a top-loading balance which serves as a fail-safe measure that allows for constant monitoring of feed supply. A diaphragm pump is used to obtain flowrates from 100-500 mL/min. Most data was obtained at a flowrate of 400 mL/min. The range of operation is from 25-75°C. Thermocouples monitor temperature fluctuations in the membrane modules and feed reservoir.

Under normal operation the permeate streams are continuously recycled. The liquid hydrocarbon feed is pressurized in the range of 50-250 psig, while the pressure regulators on the permeate side are maintained at 30-35 psig to avoid vaporization of the light solvent and subsequent errors in material balances. The recovery per pass ratios of the spiral-wound modules are typically 0.1. Small permeate and feed samples are collected 1 to 3 times a day. The unit is designed to run unattended 24 hours a day.

Figure 5 shows a schematic diagram illustrating the construction of a spiral-wound element and Figure 6 shows the engineering design of the pressure housing used to evaluate these membrane modules in the laboratory permeation unit. The spiral-wound modules were made in a single-leaf design and contained approximately 1 ft^2 of active membrane surface Vexar screen was used as the feed-side spacer and a melamine-impregnated Tricot was used for the permeate spacer. The glues used were of the polyurethane and epoxy types. The permeate draw-off tubes were 0.75" in diameter and 15" long. These tubes had either a row of 12 holes each of 0.1 diameter and spaced 0.50" apart of a narrow slit of 6" in length. The slit design was used particularly in cases of high flowrate (>20 gal/ft^2-D).

The heavy oil used for the majority of the studies was Boscan Vacuum Resid; the molecular weight distribution of the deasphalted oil determined by gel permeation chromatography is shown in Figure 7. N-Pentane in a weight ratio of 4:1 solvent/oil was selected as the paraffinic solvent for deasphalting. In later studies an Arabian Long Resid was used as the heavy oiul; after deasphalting this has a $\bar{M}w$ of ∿2100 compared to 1600 for the Boscan crude.

III. MEMBRANE SCOPING

A number of different types of commercial ultrafiltration membranes were evaluated to select potentially effective membranes. The results of these scoping studies are summarized in Table 1. The membrane selectivity, S, is defined as (6)

$$S = (\text{Solvent/Oil})_{\text{permeate}}/(\text{Solvent/Oil})_{\text{feed}} \quad (1)$$

where the solvent and oil fractions are on a weight basis. The data in Table 1 show that the commercially available membranes tested fell into two classes: those with negligible flux and those with a high flux (greater than 20 GSFD) but with no selectivity as defined by equation (1).

While it is possible to develop a membrane which possesses high stability in hydrocarbon medium and attractive selectivity and flux, we have developed a composite membrane which meets the stability, selectivity and flux requirements.(1) The composite membrane is composed of a polymeric membrane and a gel layer, illustrated in Figure 8. The polymeric membrane serves as a backing material for the gel layer. It does not, therefore, need to meet the selectivity requirement. It is however required to be stable in a hydrocarbon medium and should have the right porosity to give suitable flux. Several ultrafiltration membranes, especially those of hydrophobic type, may be used. Although the selectivity of the composite membranes resides on the gel layer, increase in its thickness usually leads to no increase of selectivity but only decline in flux. Details of the separation mechanism in terms of pore size and surface chemistry of the membranes will be given in a forthcoming publication.

IV. PERMEATION STUDIES WITH UF MEMBRANES

In fabricating the composite membrane, 12 key variables were sleected for detail studies and statistically-designed experiments (7 to 9) were used to optimize the membrane performance in terms of membrane flux and selectivity.

Figure 9 presents typical permeation results for a feed of pentane/Boscan Vacuum resid having a weight ratio of 4. The system temperature was 50°C and the upstream pressure was maintained at 220 psig (30 psig on the downstream pressure). Those data show that membranes gave moderate pentane/oil selectivities (3 to 5) and with fluxes in the range of 10 gal/ft^2-D (typical of the range of flux obtained in commercial desalination operations.) The flux did show a decline with time; subsequent examination of the membrane showed build-up of a gel layer on the membrane surface. This gel-layer formation is typical of ultrafiltration processes (6 to 10).

The flux and selectivity shown in Figure 8 are of commercial interest if cleaning techniques can be developed to maintain the membrane flux. Three techniques were explored for controlling the growth of the surface gel layer: 1) reducing the system pressure for 10 minutes while continuing normal flow, 2) increasing the temperature by 10°C for 10 minutes and 3) washing the membrane surface with clean pentane for 10 minutes. Figure 10 shows that all three techniques are effective in restoring the membrane flux. Process considerations will dictate the most desirable one to use. (1)

An important aspect in development of this type of new membrane separation is the performance of the membrane in the configuration which would be used in commercial application. Spiral-wound modules were selected since this is a well-developed module configuration (6) and is easily applied to particulate free feed materials such as used in the present study.

Figures 11 and 12 present membrane permeation data obtained with spiral-wound elements. The flux and selectivity show very good agreement with the data obtained for the flat-sheet membranes. The permeation test with the Boscan Resid Feed shown in Figure 11 was run continuously for 2 weeks; the gel layer was controlled by the pressure reduction technique; this cleaning technique was used for 5 minutes every 48 hours and was sufficient to maintain the flux about 15 gal/ft^2-D.

The Arabian Long Crude showed a greater tendency towards fouling; however, as shown in Figure 12, increasing the frequency of cleaning (every hour) maintained the flux around 10 gal/ft^2-D for 30 days. (1) The membrane selectivity with the Arabian Crude was higher than that with the Boscan Resid. This increased selectivity is a consequence mainly of the difference in feed characteristics, not of the cleaning frequency.

Figure 13 shows the effect of system temperature on the flux; these data are for a spiral-wound element. Over this temperature range of 25 to 50°C, there is very little effect of temperature. There is, however, an effect on the selectivity as shown in Figure 14. The higher selectivities obtained at 25°C are expected to be due to the structure of the surface gel layer rather than changes in the polymeric membrane itself.

Figure 15 presents the flux as a function of the upstream pressure. Extrapolation of the flux/pressure curve to lower pressures leads to an estimate of zero flux at 30 psig--the downstream pressure. This type of behavior demonstrates that this is a basic ultrafiltration membrane process and that osmotic pressure does not have a significant influence on the separation system.

Figure 16 shows that the flux increases with the rate of liquid flow across the membrane surface (cross-flow mode). This is typical of ultrafiltration separations (6, 12, and 13) and shows that concentration polarization and gel-layer formation are important factors in controlling the separation.

V. CONCLUSIONS

New permeation data are presented on the use of ultrafiltration membranes for the energy-efficient recovery of low-molecular weight paraffinic solvents from high-molecular-weight oils. The focus of this work has been on solvent deasphalting processes as used in the petrochemical industry. This study has shown that UF membranes can be stable in the hydrocarbon environment at least up to 50°C and can give fluxes of 10 to 15 gal/ft^2-D with solvent/oil selectivities of 3 to 10. Three cleaning techniques were effective in maintaining a stable flux for both flat sheets and spiral-wound modules and for continuous permeation runs as long as 30 days.

VI. ACKNOWLEDGEMENTS

We would like to acknowledge the financial support of DOE and the advice and encouragement of W. M. Sonnett, DOE Office of Industrial Programs and S. A. Leeper, EG&G Idaho, Inc.

VII. LITERATURE CITED

1. U.S. Patent Applications pending.

2. Bowman, C.W., "Molecular and Interfacial Properties of Athabasca Tar Sands," Proc. 7th World Petroleum Congress, Vol. 3, 583 (1967).

3. Nelson, W.L., "Petroleum Refining," 4th Edition, McGraw-Hill, New York, N.Y. (1958).

4. Hengstebeck, R.J., "Petroleum Processing" McGraw-Hill, New York, N.Y. (1959).

5. Funk, E.W., Pirkle, J.C. and May, W.G., "Processing Approach for the Solvent Extraction of Athabasca Tar Sands," Energy Progress, Vol. 4, No. 1, 12 (1984).

6. Li, N.N., section on "Membrane Processes" p. 14-17, Perry's Chemical Engineers' Handbook, McGraw-Hill, 6th Edition(1984).

7. Cabasso, I., Practical Aspects in the Development of a Polymer Matrix for Ultrafiltration," Ultrafiltration Membranes and Applications, A.R. Cooper, Editor, Plenum Press, New York, N.Y. (1980).

8. Peterson, R.J., Cadotte, J.E., and Buettner, J.M., "Development of FT-30 Membranes in Spiral Wound Modules," OWRTC-80318-S(8547), Office of Water Research and Technology, Washington, D.C., (1982).

9. Box, G.E.P., Hunter, W.G., and Hunter,J.S. "Statistics for Experimenters," John Wiley New York, New York (1978).

10. Blatt, W.F., Dravid, A., Michaels, A.S. and Nelson, L., "Solute Polarization and Cake Formation in Membrane Ultrafiltration -Causes, Consequences and Control Techniques," in Membrane Science and Technology, J.E. Flinn, Editor, Plenum Press, New York, N.Y. (1970).

11. Leeper, S.A., Membrane Technology and Applications: An Assessment, EG&G Idaho, Inc., (1984).

12. Scott, J., Editor, Membrane and Ultrafiltration Technology, Noyes Data Corporation, Park Ridge, N.J., (1980).

13. Blatt, W.F., "Principles and Practice of Ultrafiltration," Membrane Separation Processes, P. Meares, Editor, Chapter 3, Elsevier Publishing Co., Amsterdam, (1976).

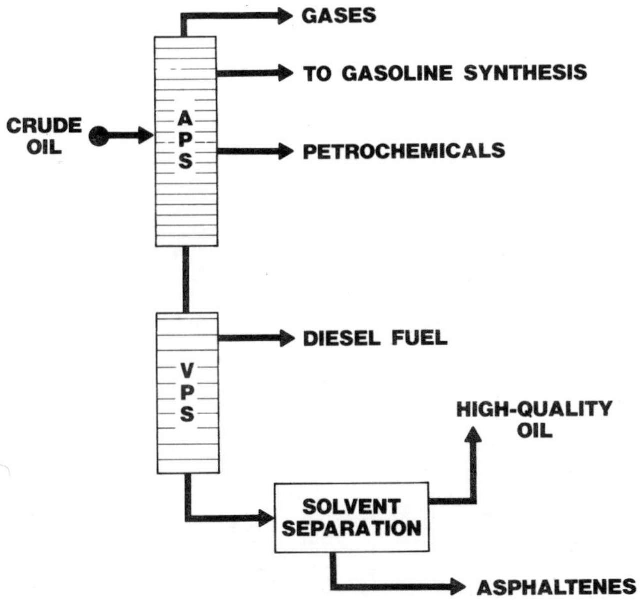

Figure 1. Overview of petroleum refining. APS and VPS refer to atmospheric and vacuum pipe stills, respectively.

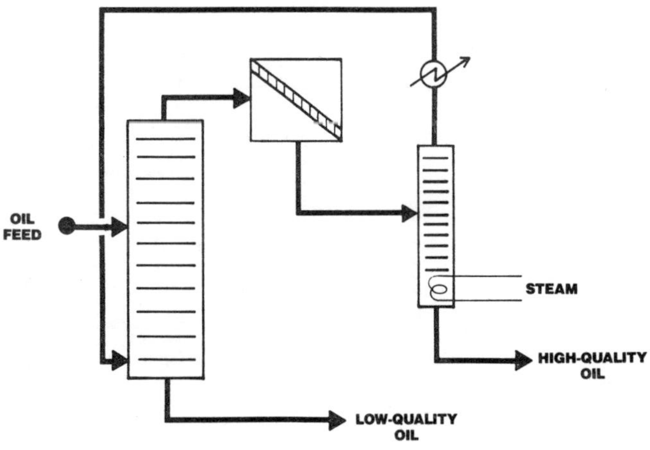

Figure 2. Conceptual use of membranes in solvent deasphalting.

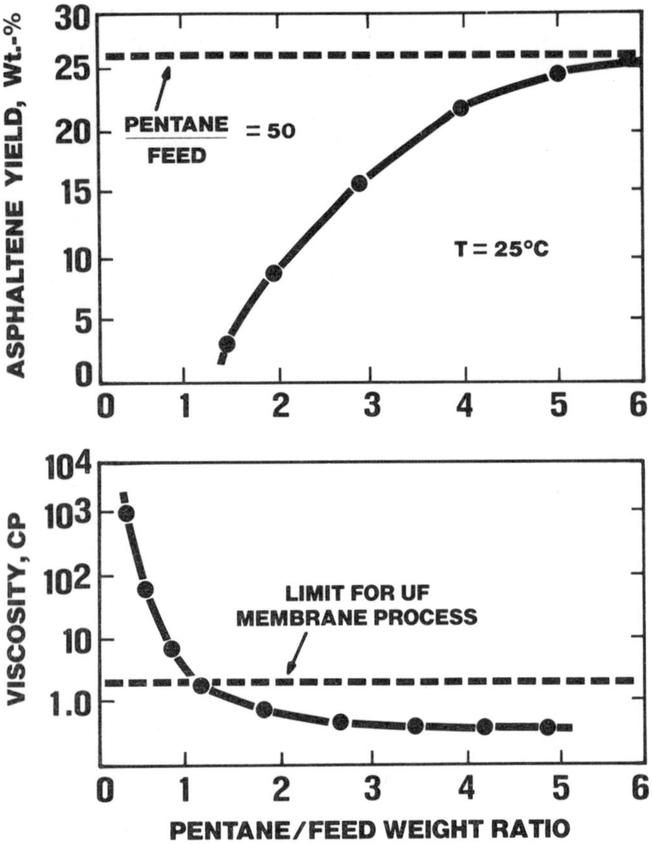

Figure 3. Effect of solvent/feed ratio on viscosity and asphaltene yield for basic deasphalting system.

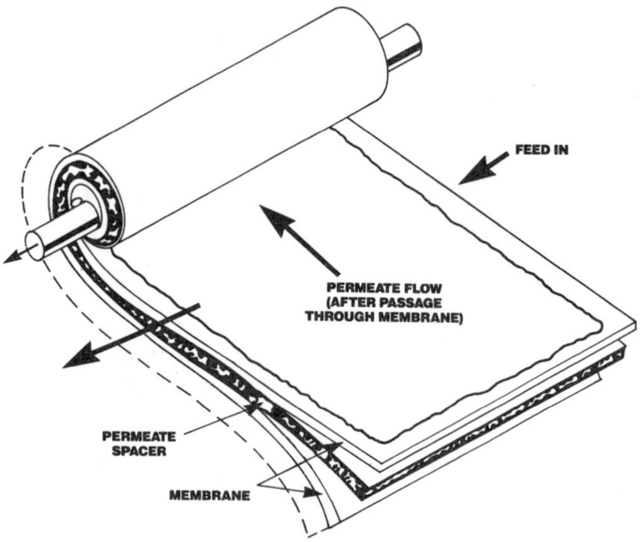

Figure 5. Schematic of spiral-wound membrane module.

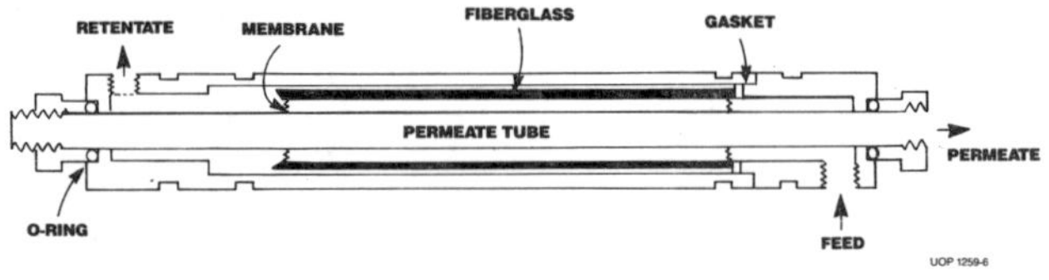

Figure 6. Pressure housing for spiral-wound module.

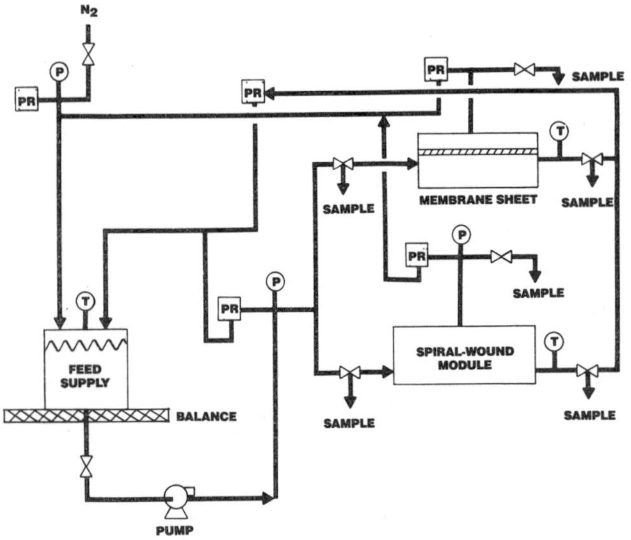

Figure 4. Experimental membrane permeation unit. PR = pressure regulator.

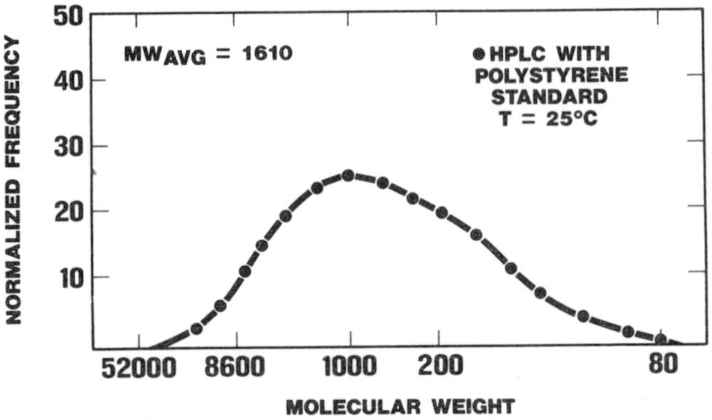

Figure 7. Molecular weight distribution of n-Pentane deasphalted Boscan Vacuum Resid. This represents the feed to the membrane unit on a solvent-free basis.

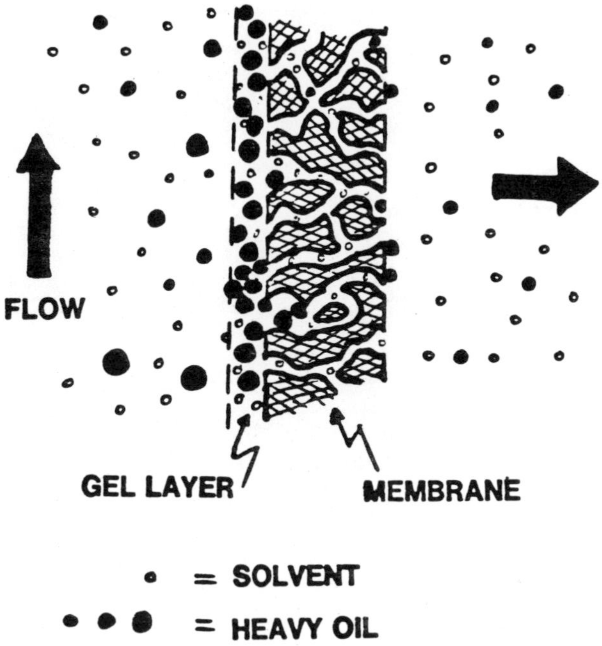

Figure 8. Proposed mechanism of solvent/oil separation on the basis of molecular size.

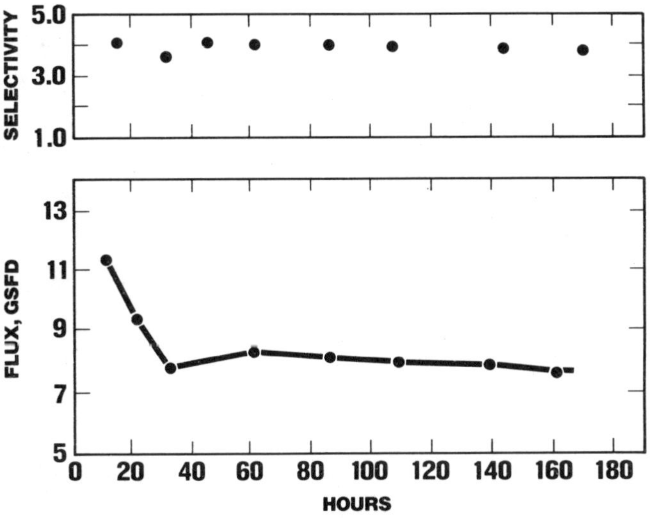

Figure 9. Typical flat-sheet permeation performance. Feed was 4:1 n-Pentane/Boscan Vacuum Resid at 50°C and upstream/downstream pressures of 220/30 psig.

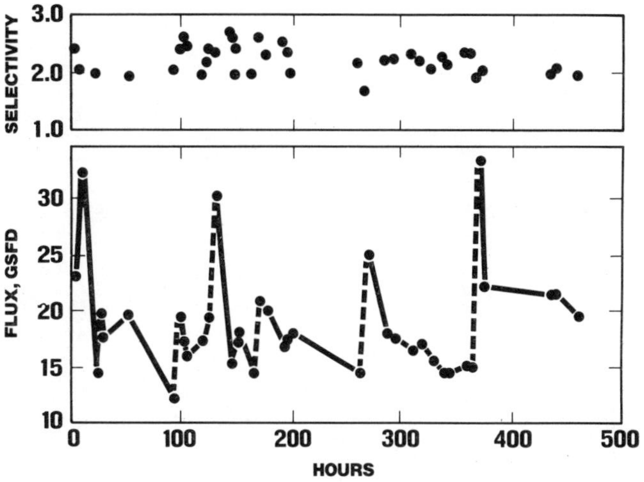

Figure 10. Effect of membrane cleaning on flux and selectivity of flat-sheet membranes with the n-Pentane/Boscan Vacuum Resid system. SP refers to a ten-minute pressure reduction, ST refers to a ten-minute temperature increase and SF refers to cleaning with n-Pentane.

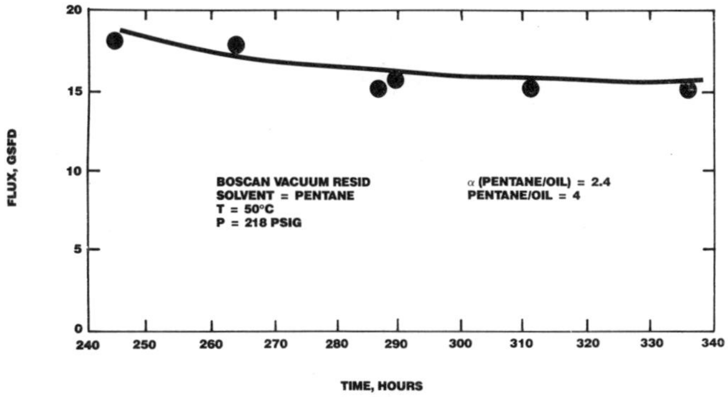

Figure 11. Spiral-wound module performance with Boscan Vacuum Resid. Flux was maintained by pressure reduction cleaning every 48 hours.

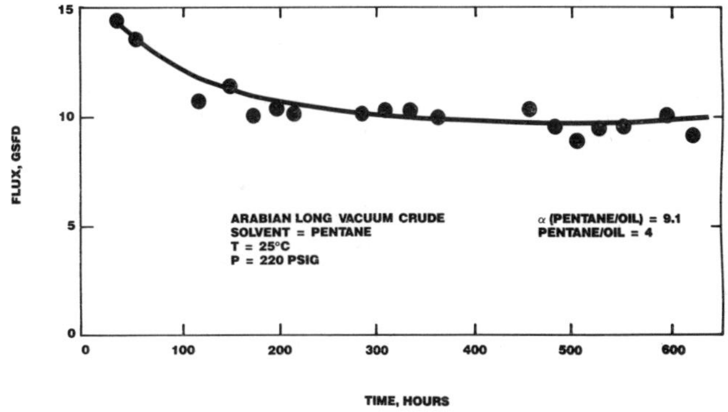

Figure 12. Spiral-wound module performance with Arabian Long crude. Flux was maintained by pressure reduction cleaning every hour.

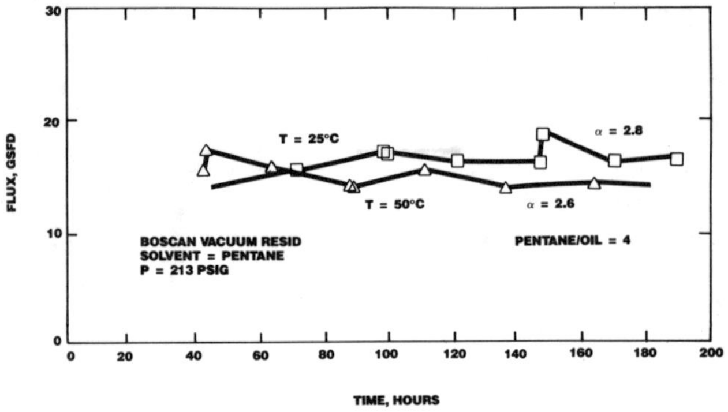

Figure 13. Effect of system temperature on flux of spiral-wound module.

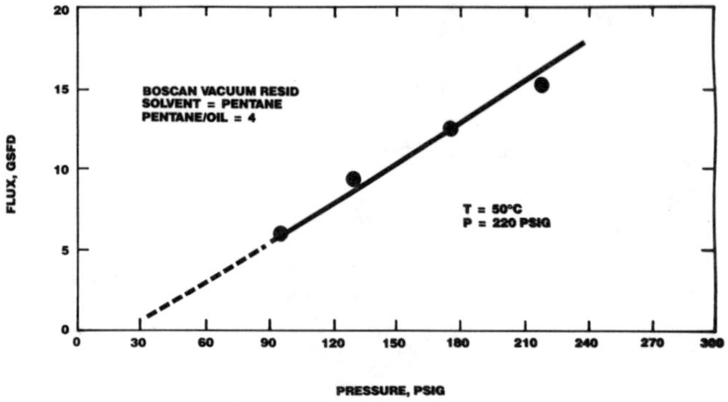

Figure 15. Effect of upstream pressure on flux of spiral-wound module. Intercept on X-axis corresponds to downstream pressure indicating negligible osmotic effect.

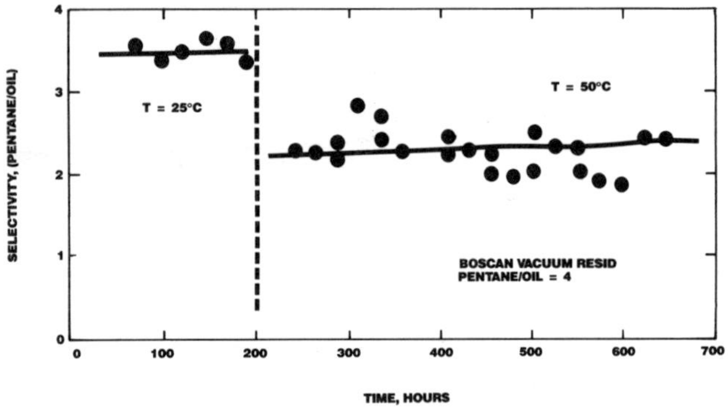

Figure 14. Effect of system temperature on selectivity of spiral-wound module.

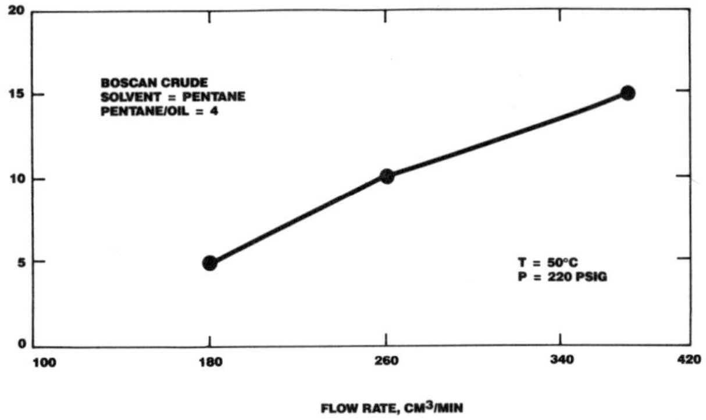

Figure 16. Effect of feed flow rate on flux of spiral-wound module.

CHEMICALLY AND ELECTRICALLY CONTROLLED MEMBRANES: SIZE SPECIFIC TRANSPORT OF FLUORESCENT SOLUTES THROUGH PMAA MEMBRANES

Aryeh M. Weiss, Alan J. Grodzinsky and Martin L. Yarmush ■ Department of Chemical Engineering
Massachusetts Institute of Technology
Cambridge, MA 02139

This paper focuses on the use of crosslinked, polyelectrolyte hydrogel membranes to achieve real time control of the flux of charged or neutral solutes by chemical or electrical means. For this purpose, a sensitive system capable of measuring the transmembrane flux of fluorescent solutes was developed, together with a family of fluorescent solutes of various sizes and electrical charge. Our experimental results show that large, reversible, size-specific permeability changes can be induced in crosslinked poly(methacrylic acid) (PMAA) membranes in response to changes in bath composition and to applied electric currents. The data show that permeability changes as high as a factor of 50 can be induced by altering the bath pH surrounding crosslinked PMAA membranes. The magnitudes of the changes in permeability depended on the relation between solute size and membrane equivalent "pore size" (membrane hydration) for solutes up to 6,000 MW. Devices based on these phenomena have potentially important applications in the area of actively controlled time variable membrane separation and drug delivery systems, as well as low energy analytical and preparative systems.

This paper focuses on the use of crosslinked, polyelectrolyte hydrogel membranes to achieve real time control of the flux of charged or neutral solutes by chemical or electrical means. For this purpose, a sensitive system capable of measuring the transmembrane flux of fluorescent solutes was developed, together with a family of fluorescent solutes of various sizes and electrical charge. Our experimental results show that large, reversible, size-specific permeability changes can be induced in crosslinked poly(methacrylic acid) (PMAA) membranes in response to changes in bath composition and to applied electric currents. Devices based on these phenomena have potentially important applications in the area of actively controlled time variable membrane separation and drug delivery systems, as well as low energy analytical and preparative systems.

Membrane systems whose permeability can be controlled electrically or chemically were reviewed by Grodzinsky and Weiss (1). Several previous studies have examined the effects of changes in bath pH on the permeability of hydrogel membranes. Kost et al (2) measured the permeability of modified poly(HEMA) hydrogels to ethylene glycol at pH 4, 7.4, and 10, and found that the permeability was about 40% higher at pH 4 than at pH 10. Kirstein et al (3) reported that diffusivity of ^3H_2O, ^{14}C-sucrose, and other small solutes though HEMA-MAA copolymer membranes increased by factors of about five when the bath pH was raised from 5 to 8. The changes in permeability observed in both the above studies were attributed to a pH-induced change in ionization state of the membrane's fixed charge groups, and a concomitant swelling of the membrane. Changes in permeability to larger solutes (MW > 350) were not measured by these investigators. Furthermore, the kinetics of the change in permeability in response to imposed pH changes were not reported. While PMAA membranes have received attentión with respect to mechanochemical transductive coupling (4) and protein ultrafiltration (5), the effect of chemical and electrical control of membrane hydration on the permeability of PMAA membranes has not been quantified.

The use of transmembrane electric fields to alter permeability has been less studied until recently. Pasechnik et al. (6) modified the permeability of an ultrafiltration membrane to protein by application of a transmembrane electric field. They ascribed their result to an unknown electrohydrodynamic effect which increased the pore size of the

Continuum Electromechanics Group, Laboratory for Electromagnetic and Electronic Systems, Department of Electrical Engineering and Computer Science, and *Department of Chemical Engineering, Massachusetts Institute of Technology, Cambridge, MA 02139.

membrane. Burgmeyer and Murray (7) studied an "ion gate membrane" in which a voltage-controlled electrochemical reaction within a poly(pyrrole) redox membrane cast on a gold grid changed the permeability of the membrane to certain ions. More recently, Eisenberg and Grodzinsky (8) reported that an applied transmembrane electric field was able to alter intramembrane ionic profiles in collagen membranes via an electrodiffusion mechanism. This caused a selective change in the permeability of those membranes to ^{14}C-sucrose relative to $^{3}H_2O$. They showed that this change resulted in part from a change in the restricted diffusion of sucrose within the membrane.

In this paper, data are presented which show that permeability changes as high as a factor of 50 can be induced by altering the bath pH surrounding crosslinked PMAA membranes, and hence the ionization state of the membrane's fixed charge groups. Permeability changes were induced both chemically (by direct addition of HCl or KOH to the baths) and electrically (by current-controlled electrode reactions). The magnitudes of the changes in permeability depended greatly on the relation between solute size and membrane equivalent "pore size" (membrane hydration) for solutes up to 6,000 MW.

We have focused particularly on the concept of membranes as dynamic structures whose transport properties can be actively controlled during use. Therefore, our system for measuring solute flux was designed to enable high temporal resolution of the kinetics of solute flux in response to changes in membrane properties. The real time nature of our transport experiments provided insight into the mechanisms responsible for changes in solute flux. These experiments, taken together with a continuum theoretical model that relates membrane hydration to bath pH, ionic strength, and applied electric field (9), are expected to provide the basis for a rational approach to membrane design.

METHODS AND PROCEDURES

The transmembrane flux of fluorescent solutes was measured using the transport cell shown schematically in Figure 1. Membrane sections 4 cm by 3.5 cm were clamped on all sides by rubber gaskets between two 100 ml half cells, leaving about 3.8 cm² of active area. The half cells were made from 1/4 inch black poly(methyl methacrylate), and were filled with electrolyte having the desired pH and ionic strength. The baths were recirculated by a dual channel peristaltic pump via inlet and outlet ports which were drilled into each half cell. The inlets were such that the fluid streamed adjacent to the membrane surface, in order to minimize the effect of stagnant fluid layers which form at the membrane/electrolyte interfaces. Magnetic stirring bars assured uniform mixing of solutes within each half cell. Fluorescent solute probes were placed in the "upstream" chamber at a known concentration. "Downstream" solute concentration was measured using techniques to be described. The pH of the baths was monitored via Orion 81-55 pH electrodes. For experiments involving electric fields applied across the membrane, the pH electrodes were electrically isolated from each other and from ground.

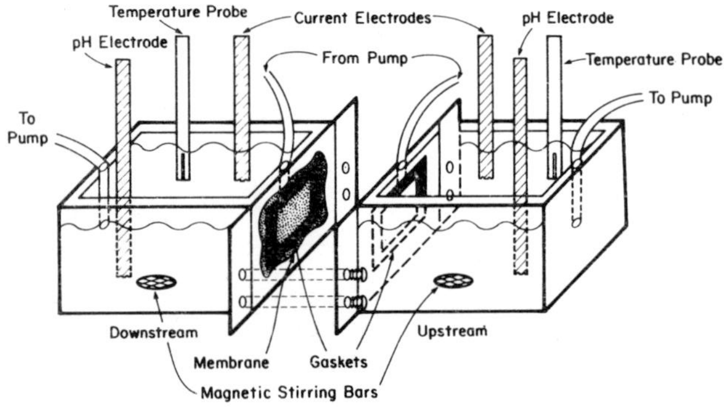

Figure 1. Schematic of transport cell used in flux experiments. For applied current, electrodes consisted of platinum strips in polyacrylamide salt bridges, or bare platinum

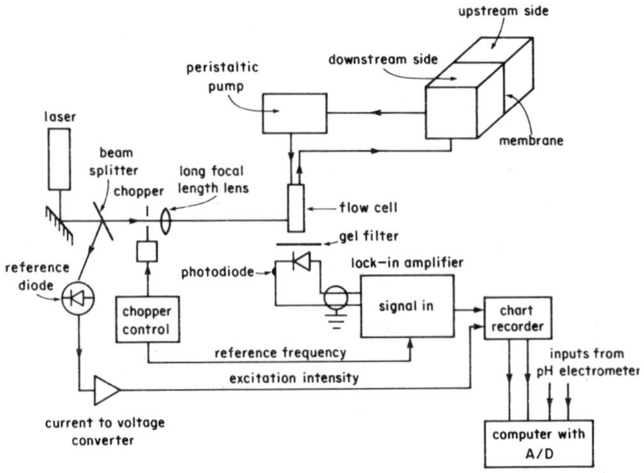

Figure 2. Schematic of optical apparatus for measuring fluorescent solute concentration on the downstream side in real time.

The apparatus for measurement of fluorescent solute concentration on the downstream side is shown schematically in Figure 2. The fluid on the downstream side was circulated by the peristaltic pump through a 10 mm x 10 mm x 30 mm flow cuvette made of special optical glass (Hellma #171 green). The fluorescence of the solution in this cell was excited by a laser, which was chosen to correspond with the spectral properties of the dye in use. A photographic gel filter was used to attenuate the Raleigh and Tyndall scattering while passing the fluorescence signal. Table 1 lists the excitation wavelength/filter combinations which were used with various dyes. A photodiode (UDT PIN-10-SBI) was used as the detector. The photodiode output was fed through a shielded cable directly to a PAR Model 124A lock-in amplifier which was equipped with a Model 184 current sensitive preamplifier. The excitation beam was chopped with a 400 Hz tuning fork chopper to allow lock-in detection of the fluorescence with respect to the 400 Hz reference frequency. A beam splitter provided an optical reference signal, which was used to normalize the output of the lock-in amplifier to any changes in excitation beam intensity. A multichannel chart recorder (Linseis LS-44) recorded the fluorescence output, the downstream pH, and, for most of the experiments, the excitation reference signal. These signals, as well as the upstream pH, were then fed to a COMPAQ portable computer equipped with a Metrabyte DASH-16 A/D card.

The permeability P was computed from the rate of change of downstream dye concentration c_d by (see Appendix A)

$$P = \frac{\Delta c_d}{\Delta t} \frac{V_d}{A c_u} \qquad (1)$$

Table 1. Fluorophore, excitation wavelengths, and filters used in experiments.

Fluorophore*	MW	Charge	λ_{ex}(nm)**	Filter (Kodak Wratten)
SRB	558	negative	514	#16
LIS-MH	1711	neutral	514	#16
LIS-DEX	6000	neutral	514	#16
NSPA	301	neutral	350-360	#4

*SRB = Sulforhodamine B
LIS-MH = Lissamine-maltoheptaose
LIS-DEX = Lissamine-dextran
NSPA = N-(3-Sulfopropyl) - acridinium

**Excitation wavelength

where c_u is the upstream dye concentration, V_d is the downstream bath volume, and A is taken to be the area defined by the membrane gasket (Figure 1). The apparatus of Figure 2 was capable of resolving changes in the downstream dye concentration of 10^{-10} M. The measurement of solute flux was accurate only to within ~15%, due primarily to the uncertainty in the downstream dye concentration.

Solutes

Dextran of average molecular weight 6000 was conjugated to lissamine-rhodamine B to produce LIS-DEX. Maltoheptaose was conjugated to lissamine-rhodamine B to produce LIS-MH (MW 1711). These were custom made and purified by Molecular Probes, Inc. (Eugene, OR) using a two step synthesis developed at MIT (10). First, the reducing end of the polysaccharide was aminated by reaction with ammonium acetate and sodium cyanoborohydride. Then, a sulfonamide was produced by reacting the sulfonyl chloride group on lissamine-rhodamine B with the aminated dextran. Sulforhodamine B was produced from lissamine-rhodamine B by hydrolysis of the sulfonyl chloride in water. N-(3-sulfopropyl)-acridinium (NSPA) was purchased from Molecular Probes, Inc.

The fluorescence intensity of these solutes was found to be unaffected by changes in pH and ionic strength within the range used in our transport experiments.

Membrane Preparation and Characterization

Crosslinked PMAA membranes were prepared by modifying a formulation published by Sefton (11) as described by Nussbaum (9). The formulation included:

- 10 ml methacrylic acid (MAA)
- 0.4 ml ethylene glycol dimethacrylate (EGDMA)
- 2 ml deionized water
- 3 ml ethylene glycol (EG)
- 0.3 ml 40% ammonium persulfate (AP) (40g AP per 100ml H_2O)
- 0.3 ml 15% sodium metabisulfite (SMBS) (15g SMBS per 100 ml H_2O)

Membranes were cast between two optical windows (obtained from Edmund Scientific, Barrington, New Jersey, USA), separated by 75 µm thick teflon spacers. While polymerization usually occurs at room temperature over a two to three day period, most membranes were heated in a water bath in order to speed polymerization. The membranes were then stored hydrated in preservative (0.1% sorbic acid in 10% ethanol) to inhibit growth of bacteria or mold. Membranes are labeled here as S.X/Y, where 0.X is the volume of EGDMA per 10 ml of MAA, and Y is half the volume of water added at the time of casting in ml. For example, the above formulation corresponds to an S.4/1 membrane. Had only one ml of water been used, the membrane would be an S.4/(1/2) membrane.

A curve of membrane fixed charge density per unit dry volume versus bath pH, computed from titration measurements, is shown in Figure 3 (9). Ionization of the membrane's fixed carboxyl groups with increasing pH (Figure 3) produced significant swelling of the membrane, and the swelling was observed to occur isotropically. Figure 4 shows two cylindrical disk specimens after two days of equilibration at different pH values (specimen (a), pH ~ 3; specimen (b), pH ~ 8). These specimens were cut from adjacent regions of the same S.2/2 PMAA sample, and were initially the same size after preequilibration at pH 3. After two days at pH 8, specimen (b) showed a clear increase swelling.

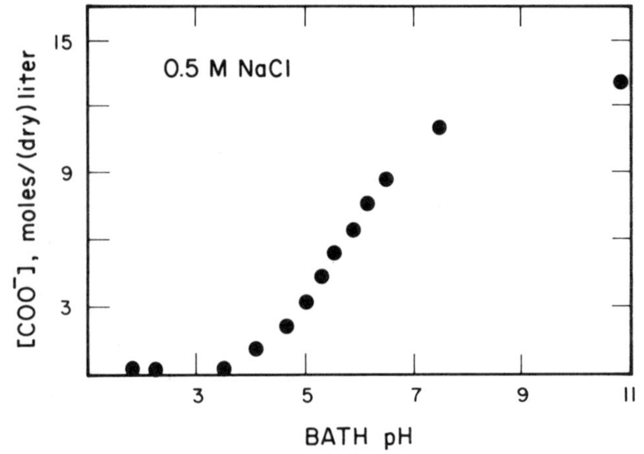

Figure 3. Charge density of PMAA membrane v.s bath pH, computed from titration measurement. Titration behavior was observed to be independent of crosslink density [based on data from Nussbaum (9)].

Membrane swelling is reported in terms of the hydration H, defined as the ratio of fluid volume V_w to solid volume V_s of the membrane:

$$H = \frac{V_w}{V_s} \quad (2)$$

H is related to the total membrane volume V by the relation:

$$V = V_s + V_w = V_s\,(1 + H) \quad (3)$$

Measured values of the wet weight and the dry weight of membrane samples were used to compute the hydration of membranes equilibrated at specified values of bath pH. For this purpose, the density of the solid was approximated to be the same as water. Table 2 compares the measured hydration of S.4/2 and S.2/(1/4) membranes equilibrated at pH 3 and pH 5.4.

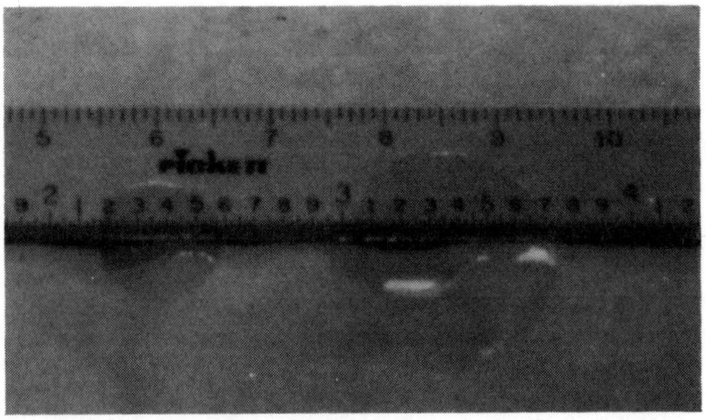

Figure 4. Two identical samples of S.2/2 PMAA membrane which were cored from the same membrane are shown in the above photo. The sample on the left was equilibrated for 2 days at pH 3 (uncharged membrane), while the sample on the right was equilibrated for 2 days at pH 8. Bath salt concentration was 0.001 M (KCL).

Table 2. Effect of pH on membrane hydration.

Membrane Composition (PMAA)	Hydration*	
	pH 3**	pH 5.4**
S.4/2	1.19	4.06
S.2/¼	1.12	3.00

* Hydration = $\frac{\text{Volume of Fluid}}{\text{Volume of Solid}}$

** pH 5.4 solutions were buffered in 0.1M sodium acetate. pH 3 solutions were unbuffered.

Applied Electric Fields

Transmembrane electric fields were applied by means of electrodes mounted in the baths or mounted in polyacrylamide gel salt bridges that were placed in the baths (Figure 1). Polyacrylamide gel salt bridges were prepared by casting 10% gels in 1/2 inch diameter glass rods to a depth of 2 to 3 inches. The gels were then equilibrated in KCl solutions having an ionic strength as close to that of the electrolyte baths as possible in order to minimize changes in bath salt concentration due to diffusion of salt out of the gel. The resulting salt bridges were mounted close to the membrane to minimize ohmic heating in the baths.

Platinum electrodes were cut from 25 μm thick platinum sheet. These electrodes were used directly in the electrolyte baths with no salt bridge when a current-induced change in bath pH was desired. Anodic current from a bare platinum electrode was found to alter the fluorescence properties of the SRB fluorophore under the conditions of our experiments. This may have resulted from the production of oxidizing agents at the platinum anode (12). NSPA was not affected by the platinum anode; cathodic currents at platinum electrodes did not alter the fluorescence of any of the solutes which we used.

Protocol for Transport Experiments

PMAA membranes were first rinsed and then soaked in deionized water for at least 1/2 hour, and then equilibrated at a desired pH (ionization state). The membrane was then mounted between the two half cells (Figure 1), and bath solutions of the desired pH and ionic strength were added to each chamber. Small volumes of 1 molar KOH, HCl, and/or KCl were added to the bath solutions to set bath pH and ionic strength.

The entire system, as shown in Figure 2, was operated for one to two hours prior to the addition of dye, in order to establish a baseline and allow the equipment and environment (temperature and pH) to stabilize.

Dye was then added from a concentrated stock solution to the upstream side, bringing the upstream dye concentration to approximately 10^{-5} M. The change in fluorescence intensity with time on the downstream side was monitored for 1 to 2 hours, by which time a steady flux (or no flux) was observed. In one series of experiments, the bath pH or ionic strength was then changed by addition of small volumes of acid (HCl), base (KOH), or salt (KCl) to one or both sides as desired. In another series, electric currents were applied via electrodes with or without salt bridges. When bare platinum electrodes were used, the pH of the bath in which that electrode was placed was observed to vary in a known manner which could be controlled by the magnitude of the applied current.

The fluorescence intensity on the downstream side was periodically calibrated relative to the upstream concentration by addition of 10 μl of upstream solution to the downstream bath. For 100 ml bath volumes, addition of 10 μl of upstream solution to the downstream side increased the downstream dye concentration (c_d) by 10^{-4} (10μl/100ml) times the upstream dye concentration, c_u. Therefore, the change in fluorescence that was observed after such addition could be used to relate the fluorescence intensity to the change in downstream dye concentration $\Delta c_d/\Delta t$ in terms of the upstream dye concentration. This information was used to compute the permeability P in Equation (1), since only $\Delta c_d/c_u$ is needed. This calibration was done during periods of steady state transport, so that the resulting increase in fluorescence would not be confused with fluorescence changes resulting from a change in membrane permeability.

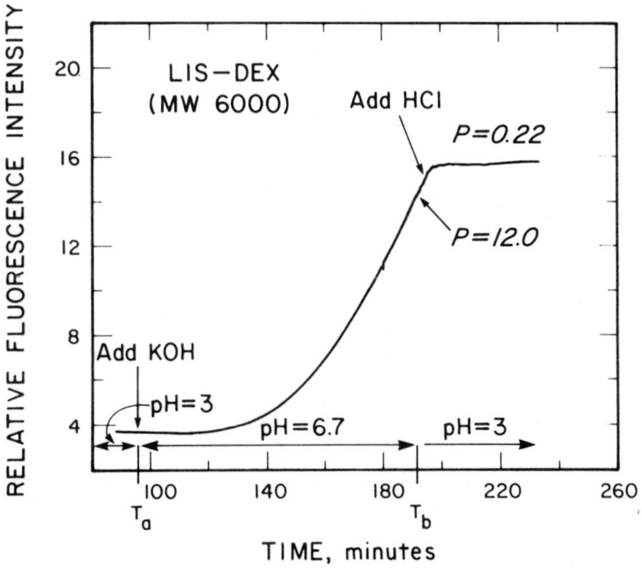

Figure 5. Changes in relative fluorescence of downstream solute versus time in response to changes in bath pH. P = permeability X 10^6 cm/s; Membrane: PMAA S.4/2 -100 μm thick at pH 3; Solute: LIS-DEX (MW 6000, neutral); electrolyte baths: 0.02 M KCl; *(bath pH decreased from 7.3 to 6.7 during this period).

RESULTS

Transport experiments in which the bath pH was abruptly changed by direct addition of acid or base were carried out using fluorescent solutes of varying size which were either electrically neutral or negatively charged. In Figure 5, the flux of LIS-DEX (average MW 6000, charge neutral) across an S.4/2 PMAA membrane is shown. For the first 100 minutes of the experiment, while the baths were at pH 3, no transport of dye was observed. The bath pH was then raised by the addition of KOH to each bath at time T_a (Figure 5) following which the solute flux increased. At time T_b the bath pH was lowered to pH 3 by addition of HCl. The ensuing result demonstrated not only that dye flux changed dramatically (a decrease of a factor of 54), but also that the change effected by raising the bath pH was reversible. The time required for the solute flux to increase following addition of KOH is seen in Figure 5 to be much longer than the time required to decrease the solute flux following addition of HCl to the baths. This difference was repeatable and probably represents the kinetics of diffusion-reaction processes involving the entire thickness of the membrane (for example, at time T_a) as opposed to a thin surface layer (T_b), as will be discussed.

Figure 6 shows the result of a similar experiment which was carried out using NSPA (MW 301, charge neutral) as the solute, with an S.4/1 PMAA membrane. The solute flux increased after the bath pH was raised at time T_a. The permeability decreased to its initial value after the bath pH was lowered at time T_b. The ratio of the maximum to minimum permeability for NSPA was 1.64, which is much lower than the permeability changes obtained for larger solutes such as LIS-DEX in Figure 5. The permeability of PMAA to NSPA was found to be much larger than the permeability PMAA membranes to LIS-DEX at all pH values (Figures 5 and 6).

Transport experiments were also carried out using charged solutes. Figure 7 shows the result of an experiment using SRB (MW 558, charge negative) and an S.4/1 PMAA membrane. The solute flux increased by a factor of 9.6 upon addition of KOH to the baths at time T_a. An additional factor of 1.7 increase in permeability was induced by further addition of KOH at time T_b. The permeability decreased by a factor of 11.1 following addition of HCl to the baths at time T_c. This result is

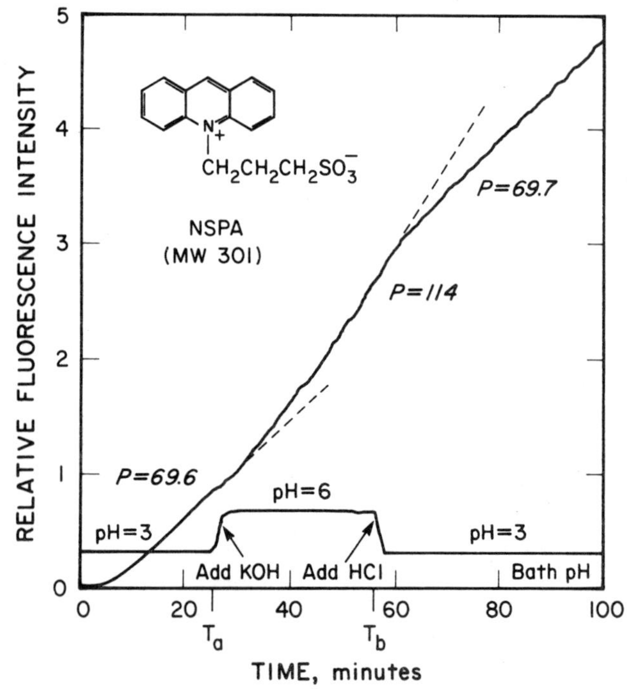

Figure 6. Changes in relative fluorescence of downstream solute versus time in response to changes in bath pH. P = permeability X 10^6 cm/s; Membrane: PMAA S.4/1 -100 μm thick at pH 3; Solute: NSPA (MW 301, neutral); electrolyte baths: 0.02 M KCl.

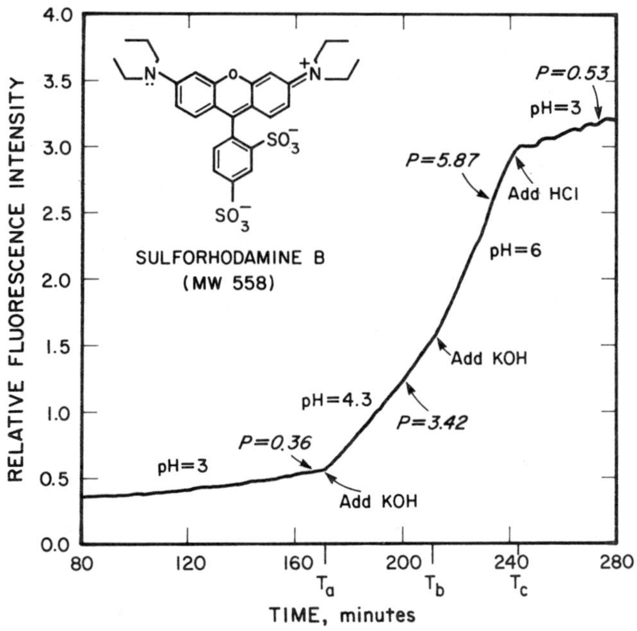

Figure 7. Changes in relative fluorescence of downstream solute versus time in response to changes in bath pH. P = permeability X 10^6 cm/s; Membrane: PMAA S.4/1-100 μm thick at pH 3; Solute: SRF (MW 558, negative electrolyte baths: 0.01 M KCl.

especially important because the membrane fixed charge is negative, so that the negatively charged solute would tend to be excluded from the membrane due to electrostatic (Donnan) interactions. Such exclusion alone would tend to decrease membrane permeability. The increase in permeability that is observed in Figure 7 occurred in spite of this exclusion.

In the next series of experiments, electric currents were used to modulate membrane permeability by changing bath pH through current-controlled electrode reactions. Figure 8 shows the result of a transport experiment using an S.4/1 PMAA membrane and LIS-MH (MW 1711, charge neutral). In this experiment, the baths on both sides of the membrane were at neutral pH at the start of the experiment. Addition of HCl to the upstream side of the membrane at time T_a reduced the permeability by a factor of 20. At time T_b, a 10 mA current (2.6 mA/cm^2 membrane area) was passed via a bare platinum cathode which had been placed in the upstream bath. This caused an increase in the upstream bath pH to 8.5. When the current was turned off, a subsequent drop in bath pH to 6 occurred due to dissociation of the membrane's ionizable carboxyl groups. This is seen in the lower trace of Figure 8, which is a record of the upstream bath pH. A 16-fold increase in solute flux resulted from the current-controlled increase in upstream bath pH. Addition of HCl to the upstream bath at time T_c reduced the solute flux by a factor of 11. (Electrode reaction was not employed to lower bath pH in this instance, because the fluorophore (SRB) was degraded in the presence of a platinum anode.) Note that the decreases in permeability that follow times T_a and T_c were caused by lowering the pH only on the upstream side of the membrane, while the downstream pH was maintained above 5 by addition of KOH.

The permeability of an S.4/1 PMAA membrane to NSPA (MW 301, charge neutral) was first increased and then decreased by using current-controlled reactions at a platinum electrode to change upstream bath pH (Figure 9). The bath pH on both sides of the membrane was in the neutral region at the start of the experiment. The permeability was reduced by a factor of 1.5 by addition of HCl to the upstream bath at time T_a. A 10 mA current was passed through a platinum cathode at time T_b, raising the upstream pH, as in the experiment described above. The permeability increased by a factor of 1.09. At time T_c, a 10 mA current was passed through a platinum anode in order to lower the upstream bath pH. The permeability then decreased by a factor of

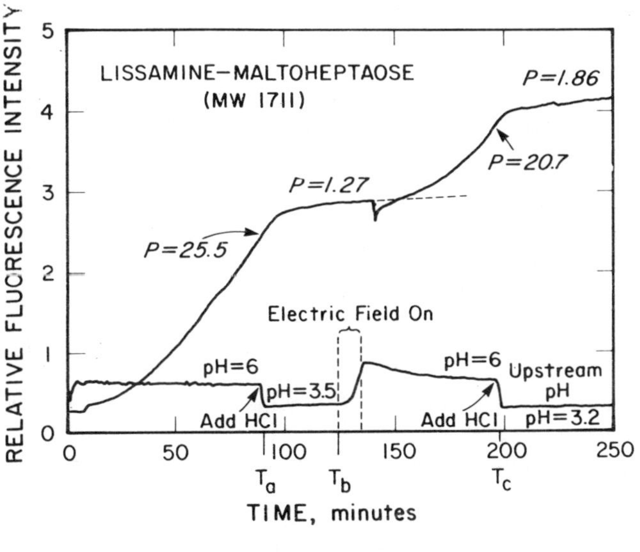

Figure 8. Changes in relative fluorescence of downstream solute versus time in response to changes in bath pH induced by current-controlled electrode reaction and by direct addition of HCl; P = permeability X 10^6 cm/s; Membrane: PMAA S.4/1 –100 μm thick at pH 3; Solute: LIS-MH (MW 1711, neutral); electrolyte baths: 0.05 M KCl.

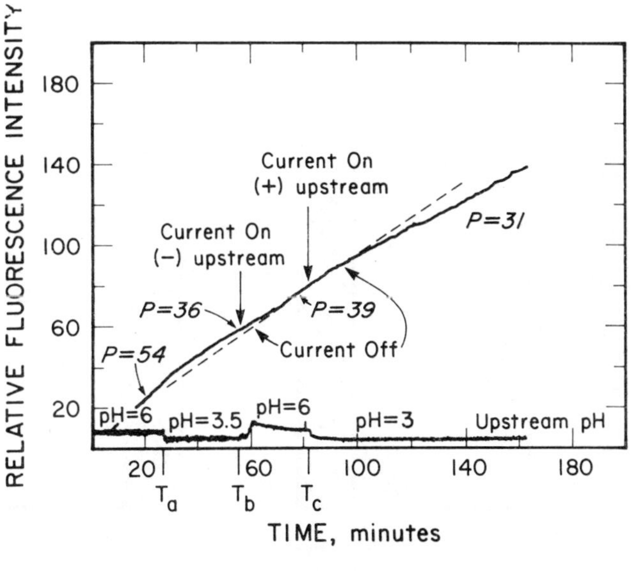

Figure 9. Changes in relative fluorescence of downstream solute versus time in response to changes in bath pH induced by current-controlled electrode reaction and by direct addition of HCl; P = permeability X 10^6 cm/s; Membrane: PMAA S.2/(1/4) –100 μm thick at pH 3; Solute: NSPA (MW 301, neutral); electrolyte baths: 0.02 M KCl.

1.22. As in the data of Figure 6, the electrode-induced permeability changes for NSPA were much smaller than the permeability changes observed for larger solutes such as LIS-MH and LIS-DEX.

Specific alterations in membrane composition were found to have significant effects on membrane transport properties. Figure 10 shows the maximum and minimum permeabilities that were observed when two different membranes were subjected to changes in bath pH, using the methods of Figure 5. The solute used was LIS-DEX. The membranes differed only by the amount of crosslinker (EGDMA) per monomer used at the time of polymerization. The maximum permeabilities of the two membranes were similar, but the more heavily crosslinked membrane had a much lower minimum permeability.

DISCUSSION

We have observed that changes in bath pH can lead to large changes in the permeability of a polyelectrolyte gel membrane to neutral or negatively charged solutes of various sizes. In the experiments presented here, the pH was changed by direct addition of acid or

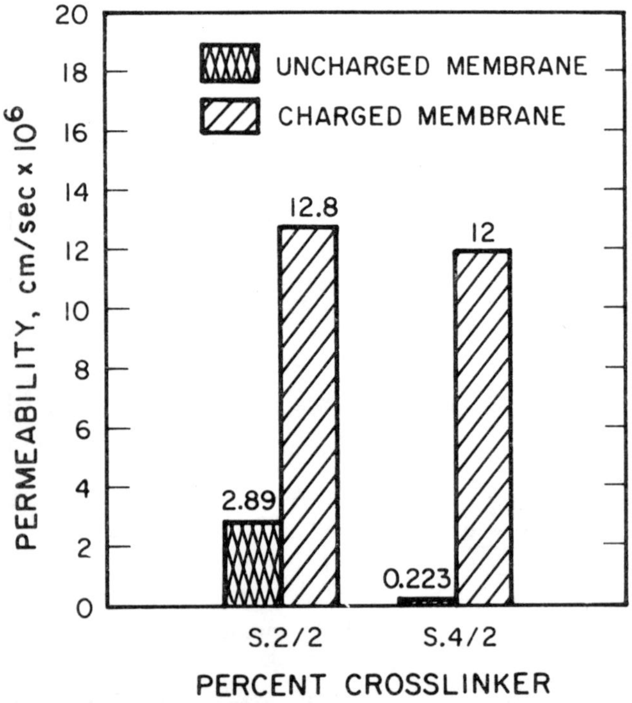

Figure 10. Variation of permeability with percent cross-linking agent (EGDMA) added. Solute: LIS-DEX (MW 6000, neutral). Membranes: PMAA S.2/2 and S.4/2; electrolyte baths: 0.02 M KCl.

base, or by means of an electrode reaction. These pH changes can alter the ionization state of the membrane macromolecules (Figure 3) and, hence, the electrostatic repulsion forces between molecules. On a macroscopic scale, these electrostatic forces mediate membrane swelling and hydration (Figure 4 and Table 2). On a microscopic scale, they help to determine the effective pore size of the membrane.

Evidence for Large, Size Specific Permeability Changes

The transport experiments of Figures 5 to 9 were specifically designed to isolate the effect of membrane hydration and fixed charge on transmembrane solute flux. Changes in membrane hydration can alter the solute flux and the apparent solute diffusivities in the membrane for a variety of reasons. First, hydration will affect the average path length which a diffusing particle follows when traveling through the membrane. Mackie and Meares (13) used a statistical argument applicable to homogeneous swollen gels, and showed that the average path length in the membrane exceeds that in an equivalent thickness of pure solvent by the tortuosity factor $\Theta = (1+V_p)/(1-V_p)$. Since the concentration gradient along a "pore" is reduced in proportion to Θ, the apparent intramembrane solute diffusivity $\bar{D}$ is less than that in infinitely dilute aqueous solution D_∞ by the amount $(\bar{D}/D_\infty = 1/\Theta^2)$ (13). This expression has been found to predict the diffusion of nonelectrolyte and co-ion solutes in lightly and moderately crosslinked gels, where the "pore size" is large compared to the solute diameter (14).

When solute diameter is not much less than "pore" diameter, solute transport can be significantly restricted by steric or "sieving" effects. In addition, the proximity of the membrane macromolecules to the solute can lead to hydrodynamic (viscous) drag on the solute that can further restrict solute flux. When steric and hydrodynamic hindrance are important, small changes in membrane hydration may produce large changes in the intramembrane solute diffusivity. Such restricted diffusion due to both steric and hydrodynamic hindrance has been described by Bean (15), Anderson and Quinn (16), and Deen and Smith (17). For the case of charged solutes, Donnan partitioning of the solutes between the charged membrane and the adjacent baths at each interface can have an additional, major effect on solute flux (18,19).

In order to determine which of the above hydration-dependent mechanism(s) were most important to the membrane responses of Figures 5 to 9, we had to distinguish between tortuosity and restricted diffusion effects. Therefore, an expression was derived (for neutral solutes) to compute that part of the rate of change of downstream dye concentration due specifically to hydration dependent changes in tortuosity, surface area, and thickness for isotropically swelling membranes (see Appendix B):

$$\frac{\Delta c_d}{\Delta t} = \frac{KD^* c_u}{V_d}\left[\frac{V_s}{K}\right]^{\frac{1}{3}}\left[\frac{H}{2+H}\right]^2(1+H)^{\frac{1}{3}} \qquad (4)$$

where c_u and c_d are the solute concentrations on the upstream and downstream sides, respectively, D^* is the intramembrane solute diffusivity in the absence of tortuosity effects, V_d is the downstream bath volume, V_s is the solid volume of the membrane, H is the membrane hydration and K is a constant which depends on the initial membrane dimensions (see Appendix B).

Equation (4) can be used to estimate the magnitude of the flux changes that would be expected from a given change in membrane hydration, independent of steric or hydrodynamic hindrance effects. A plot of $\Delta c_d/\Delta t$ versus membrane hydration, based on Equation (4), is shown in Figure 11. For the upper and lower bounds of membrane hydration corresponding to the bath pH conditions of Figures 5 to 9, flux changes of no more than a factor of 4 to 5 would be predicted by Equation (4). The small permeability changes observed with NSPA (Figures 6 and 9) are within the upper bound predicted by Equation (4). This leads us to conclude that a change in steric or hydrodynamic hindrance is not the dominant mechanism for the change in flux seen with our smaller solutes. This is consistent with the findings of other investigators who have studied the transport of small, non-ionic species and co-ion species across natural and synthetic membranes in the same hydration range as that used here (8,13,20). On the other hand, we attribute the large changes in solute flux (factors of 10 to 54) observed with the large solutes (LIS-MH, LIS-DEX and SRB) to changes in restricted diffusion. The experiments of Figures 5, 7 and 8 show large changes in permeability which have been demonstrated to occur rapidly and reversibly in real time.

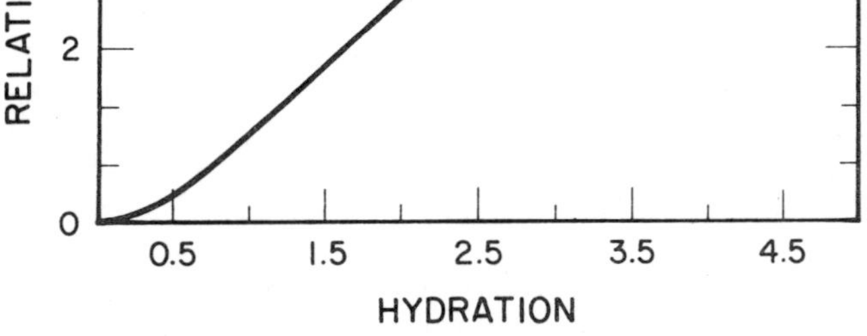

Figure 11. Theoretcial plot of solute transport ($\Delta c_d/\Delta t$) versus hydration H, based on equation 4. The transport is normalized to its value at H = 1.

Smaller pH mediated changes in membrane permeability have been observed by Kost, et al (2) and Kirstein et al (3). Kost et al observed a 40% change in the permeability of modified poly(HEMA) hydrogels to ^{14}C-ethylene glycol, as the swelling of the hydrogel varied over the pH range 4 to 10. Kirstein et al measured the permeabilities of a series of radiolabeled solutes in HEMA-MAA copolymer hydrogels at pH 5 and pH 8, and found that diffusivity of tritiated water changed by a factor of five for some membrane formulations. Kirstein et al also found that the diffusivities of larger solutes such as sucrose and glucose were very low at pH 5, and increased at pH 8. They attributed the change in observed diffusivity of sucrose and glucose to changes in the pore diameter of water filled channels within the polymer. It is not clear from their data that the change in membrane permeability to sucrose or glucose is larger than the change to tritiated water, since the data was not presented on a sufficiently sensitive scale. As seen from Equation (4) (Figure 11), changes in the membrane tortuosity and membrane surface area can themselves account for a 3 to 5 fold change in solute flux.

The swelling of polyelectrolyte gels like PMAA depends not only on the concentration of ionizable groups within the matrix, but also on the degree of crosslinking, the details of the network structure, the nature and concentration of the mobile ions in the solution phase, and the temperature (14). Kirstein et al. (3) varied the composition of their membranes by altering the ratio of crosslinker to monomer, and by adding the hydrophobic monomer BUMA to some of the membranes. They found that membrane transport properties changed as membrane composition was altered. The permeability of our PMAA membranes to a given solute decreased when the amount of crosslinker (EGDMA) added per unit monomer (MAA) was increased. An example of this is shown in Figure 10 for LIS-DEX. We believe that the increase in membrane crosslink density that should accompany the increased EGDMA content accounts for the observed decrease in permeability. The permeabilities of the charged forms of the membranes shown in Figure 10 are similar in magnitude. This may be related to the fact that a large component of the change in permeability to LIS-DEX between the charged and uncharged form of a given membrane is caused by changes in restricted diffusion. Swelling of an initially uncharged membrane will decrease steric and hydrodynamic hindrance. A further increase in hydration will cause the permeability to continue to increase, but by a relatively smaller amount. Therefore, even though the equilibrium hydration of the charged S.2/2 membrane of Figure 10 is higher than that of the more heavily crosslinked S.4/2, the hydration in both cases is probably high enough that restricted diffusion is of smaller consequence to solute transport. By comparison, the hydration of the uncharged forms of these membranes is low enough that small differences in hydration can lead to large restriction effects. Further experiments are needed to confirm this hypothesis. The effects of varying membrane composition (for example crosslink density and solvent content at the time of polymerization) will be reported in a separate publication.

Time Response

The response times observed in our transport experiments suggest strategies which may reduce the time required to effect changes in solute flux. In the following discussion, we focus on the effect of membrane thickness and fixed charge density on response time.

Since most chemical transport time constants are proportional to the square of the membrane thickness, a faster response to the stimuli of Figures 5 to 9 would have occurred if a thinner membrane had been used. More interesting is the possibility that for highly hindered solutes, a thin rate limiting surface layer of membrane may limit permeability almost as effectively as a thick membrane. In general, modulation of the properties of such a thin layer would require shorter times. Evidence for this possibility in our tests can be found in the response time for reducing the flux upon lowering the bath pH. In Figure 5, for example, addition of HCl to both baths at time T_b decreased solute flux within a few minutes. By comparison, the calculated diffusion-reaction time for the transient change in intramembrane pH (and the concomitant change in hydration) produced by the given change in bath pH should be on the order of 20 minutes to hours for a 100-150 µm thick membrane and an initial bath pH of 5 (21). The much more rapid decrease in flux which was observed leads us to believe that a decrease in hydration occurred within a thin surface layer of the upstream side of the membrane, which was responsible for the speed of the response. In contrast, addition of KOH to the baths at time T_a (Figure 5) led to a much slower change (increase) in solute flux. This latter change in flux was associated with an increase in the ionization state of the membrane, which typically required over 20-40 minutes. This lag time is consistent with a diffusion reaction time based on total membrane thickness. (The entire thickness of the membrane must ionize and hydrate to allow increased solute flux.) Since the main limitation to reduction of membrane thickness is mechanical strength, future studies will be performed with asymmetric membrane structures containing a thin rate limiting layer supported by a porous, mechanically rigid layer.

A reduction in the fixed charge density required to change membrane hydration will also decrease the response time, because the hydrogen diffusion-reaction time τ_{dr} is a function of the density of the fixed ionizable groups in the membrane (21):

$$\tau_{dr} = \frac{\delta^2}{\pi^2 D_H}\left[1 + \frac{c_m K}{(K + c_H)^2}\right] \qquad (5)$$

where δ is the membrane thickness, D_H is the hydrogen ion diffusivity, c_m is the concentration of fixed ionizable carboxyl groups in the membrane (moles/liter membrane water), K is the equilibrium dissociation constant for the fixed carboxyl groups, and c_H is the initial intramembrane hydrogen ion concentration. (Equation (5) is strictly applicable for a small change in c_H compared to its initial value.) For our PMAA membranes, at low pH (pH << pK), τ_{dr} is on the order of a minute. With intramembrane pH ~ pK, τ_{dr} can be on the order of hours, consistent with a value of c_m ~ 1 to 3 M. Therefore, as more fixed charge sites need to be ionized, the time required to do so will increase rapidly.

One may create a membrane with a lower density of ionizable sites c_m by copolymerizing a polar, uncharged monomer with methacrylic acid. An example of such a membrane is the HEMA-MAA copolymer membrane (3,22). Lowering c_m could lead to short response times even when operating the membrane at bath pH ~ pK of the fixed ionizable groups (Equation (5)). Varying membrane fixed charge density or the pK of the fixed ionizable sites provides a means of adjusting the pH at which the membrane most efficiently functions. This may be required, for example, in separations of pH sensitive proteins.

It is instructive to compare two cases in which dye flux was increased by raising bath pH, but which differ by the degree to which the membrane had to be ionized. In Figure 7, the final bath pH following the addition of KOH at time T_a was 4.3-4.5, which implies about 6 to 10% ionization of the membrane fixed charge groups, based on the titration curve of Figure 3. The flux change to SRB (MW 558) occurred in a few minutes, as seen in Figure 7. The rapid flux increase is evidence that very little fixed charge was needed in order to effect a large increase in solute flux. In Figure 8, over 40 minutes elapsed between the time that the upstream bath pH was raised at time T_b and the time that an appreciable increase in the flux of LIS-MH occurred. Furthermore, the flux was at least 25% away from its steady state value after an hour. In this case, the membrane had to be more heavily charged, in order to allow the MW 1711 LIS-MH to pass. The comparison of these two cases suggests that if membrane operation can be restricted to the region where the membrane fixed charged sites are less than 10% ionized, then fast response (on the order of minutes) should be obtainable.

Electric Field Control

Finally, we note that applied electric fields may be used to modulate membrane permeability using several mechanisms in addition to the electrode reaction-induced changes in bath pH described here. It has been shown that transmembrane electric fields can change the pH and salt concentration within a membrane that supports a pH or salt gradient, even when bath concentrations are held constant (20,23,24). This mechanism, referred to as electrodiffusion, has been found to selectively alter collagen membrane permeability to ^{14}C-sucrose in dual label transport studies (8). Successful adaptation of this technique to the membrane/solute combinations described here requires further optimization of membrane parameters to enable reasonable response times for practical values of applied current density. In addition, the simultaneous electrophoretic force on charged solutes may further contribute to large changes in solute flux controlled by an applied electric field.

CONCLUSIONS

A system has been developed which allows sensitive, real time characterization of membrane transport properties. Using this system, data have been obtained which show that real time, large and reversible membrane permeability changes can be obtained through chemical or electrical control. It is also clear that these permeability changes can be made very solute specific by matching the membrane to the solutes of interest. The effects of solute size and membrane composition on transmembrane flux are consistent with Eisenberg and Grodzinsky (8), Kirstein et al (3), Peppas and Lustig (25) and others. The present work extends previous work by demonstrating the contribution of size dependent processes which alter solute flux in real time. These can be used to make the induced permeability changes solute specific.

The real time nature of our flux measurements provided insight into the relationship between membrane parameters (formulation and thickness), solute size, and permeability changes. Certain hypotheses discussed in this paper (such as the evidence that a thin layer of the membrane can control the permeability) which are suggested by the data, require more experiments in order to quantify such characteristics as response time and membrane reproducibility.

Work is now in progress to quantify the effect of changes in membrane parameters such as thickness, crosslink density, and charge density on membrane performance (response time and magnitude of permeability changes). In addition to the empirical transport characterizations, a continuum theoretical model which relates membrane hydration to bath pH, ionic strength, and applied electric field has been developed (9).

APPENDIX A

Relationship of Permeability to Solute Flux

The one dimensional diffusion equation (Equation (26)) describing the intramembrane concentration profile $\bar{c}(x,t)$ of a solute within a membrane of thickness δ

$$\frac{\partial \bar{c}(x,t)}{\partial t} = \bar{D}\,\frac{\partial^2 \bar{c}(x,t)}{\partial x^2} \qquad \text{(A1)}$$

has, as its steady state ($\partial \bar{c}/\partial t = 0$) solution

$$\bar{c}(x) = \bar{c}(0^+) + \frac{(\bar{c}(\delta^-) - \bar{c}(0^+))}{\delta}\,x \qquad \text{(A2)}$$

where $\bar{D}$ is the intramembrane solute diffusivity, and $\bar{c}(0^+)$ and $\bar{c}(\delta^-)$ are the concentrations just inside the upstream and downstream surfaces, respectively. The intramembrane concentrations at the boundaries are related to the bath concentrations by a partition ratio:

$$\bar{c}(0^+) = \frac{c_u}{r_u} \qquad \text{(A3)}$$

$$\bar{c}(\delta^-) = \frac{c_d}{r_d} \qquad \text{(A4)}$$

where c_u and c_d are the upstream and downstream bath concentrations, respectively, and $r_{u,d}$ includes electrostatic (Donnan) as well as chemical and steric partitioning. Thus, r_u and r_d may not be equal if the fixed charge densities at $x=0^+$ and $x=\delta^-$ are unequal. The linear profile of Equation (A2) can be inserted in the flux equation

$$\Gamma = -\bar{D}\,\frac{\partial \bar{c}(x)}{\partial x} \qquad \text{(A5)}$$

where Γ is flux per unit area, to yield:

$$\Gamma = \frac{\bar{D}}{\delta r_d}\left(c_u\frac{r_d}{r_u} - c_d\right) \qquad \text{(A6)}$$

It is assumed that the solute concentration in the baths changes slowly enough compared to the time for diffusion of solute across the membrane to justify the use of steady state solutions (A2) and (A6).

The measured quantity in the transport experiments of Figures 5-9 is the change in downstream dye concentration with time, $\Delta c_d/\Delta t$. From continuity of mass,

$$\frac{\Delta c_d(t)}{\Delta t} = \Gamma\,\frac{A}{V_d} = \frac{A\bar{D}}{V_d \delta r_d}\left[c_u\frac{r_d}{r_u} - c_d\right] \qquad \text{(A7)}$$

where A is the area of the membrane, V_d is the volume of fluid on the downstream side. Since c_u is many orders of magnitude greater than c_d during the experiment, Equation (A7) can be approximated for the solute and bath conditions of Figures 5-9 and rewritten as

$$\frac{\Delta c_d(t)}{\Delta t} = \frac{A\bar{D}c_u}{V_d \delta r_u} \qquad \text{(A8)}$$

The permeability P is related to the diffusivity $\bar{D}$ by:

$$P = \frac{\bar{D}}{\delta r_u} = \left[\frac{\Delta c_d(t)}{\Delta t}\right]\frac{V_d}{A c_u(t)} \qquad \text{(A9)}$$

Equation (A9) is only valid during periods of steady transmembrane flux (after intramembrane diffusion transients have ended). Furthermore, $c_u(t)$ is approximately constant during the duration of any experiment compared to $c_d(t)$. Therefore, Equation (A9) is used to calculate permeabilities only in the regions of Figures 5 to 9 where the downstream fluorescence intensity is approximately linear with time.

APPENDIX B

Dependence of Solute Flux on Hydration

To relate $\Delta c_d/\Delta t$ to membrane hydration, surface area, and thickness, we start with the tortuosity factor Θ derived by Mackie and Meares (13) which gives the effective increase in diffusion path length across the membrane,

$$\Theta = \frac{1 + V_p}{1 - V_p} \qquad \text{(B1)}$$

where V_p is the volume fraction of the polymer. The intramembrane diffusivity $\bar{D}$ is then given by

$$\bar{D} = \frac{D^*}{\Theta^2} \qquad \text{(B2)}$$

where D^* is the intramembrane diffusivity that the solute would have in the absence of tortuosity effects (but including other solute membrane interactions) (13).

The relationship between θ and membrane hydration will now be developed by first relating V_p to the hydration H, and then substituting that result into Equation (B1).

$$V_p = \frac{V_s}{V} = \frac{1}{1+H} \qquad (B3)$$

where the second equality incorporates Eqs. (2) and (3). From (B1) and (B3),

$$\theta = \frac{2+H}{H} \qquad (B4)$$

$$\frac{1}{\theta^2} = \left[\frac{H}{2+H}\right]^2 \qquad (B5)$$

Substituting Equation (B5) into (B2) and using Equation (A8) for neutral solutes (r_u taken to be 1),

$$\frac{\Delta c_d}{\Delta t} = \frac{AD^* c_u}{V_d \delta \theta^2} = \left[\frac{AD^* c_u}{V_d \delta}\right]\left[\frac{H}{2+H}\right]^2 \qquad (B6)$$

where A is the membrane area and V_d is the volume of the downstream side.

For a membrane that responds to changes in bath pH and ionic strength by swelling isotropically, the three linear dimensions change by a constant percentage:

$$A = K\delta^2 \qquad (B7)$$

and $$V = K\delta^3 = (1+H)V_s \qquad (B8)$$

where K is a constant. Therefore,

$$\delta = \left[\frac{(1+H)V_s}{K}\right]^{\frac{1}{3}} \qquad (B9)$$

Substitution of Equations (B9) and (B7) into (B6) yields:

$$\frac{\Delta c_d}{\Delta t} = \frac{KD^* c_u}{V_d}\left[\frac{V_s}{K}\right]^{\frac{1}{3}}\left[\frac{H}{2+H}\right]^2 (1+H)^{\frac{1}{3}} \qquad (B10)$$

Equation (B10) is identical to Equation (4).

ACKNOWLEDGEMENT

This research was supported in part by NSF Grant CBT-8512814 and by the NSF Engineering Research Center Initiative (Grant CDR-8500003) for the Biotechnology Process Engineering Center at MIT. AMW was supported in part by a Fannie and John Hertz Foundation fellowship. The authors also gratefully acknowledge the use of the facilities of the MIT Regional Laser Research Center, and the important contributions made by our colleagues Jeremy H. Nussbaum and Paul E. Grimshaw.

LITERATURE CITED

1. Grodzinsky, A.J. and A.M. Weiss, "Electric Field Control of Membrane Transport and Separations", Separation and Purification Methods 14, 1 (1985).

2. Kost, J., T.A. Horbett, B.D. Ratner and M. Singh, "Glucose-sensitive Membranes containing Glucose Oxidase: Activity, Swelling and Permeability Studies", J. Biomed. Mater. Res., (in press).

3. Kirstein, D., H. Braselmann, J. Vacik and J. Kopecek, "Influence of Medium and Matrix Composition on Diffusivities in Charged Membranes", Biotech. Bioeng. 27, 1382 (1985).

4. Katchalsky, A. and Zwick, M., "Mechanochemistry and Ion Exchange," J. Polymer Sci., 16, 221, (1955).

5. Osada, Y. and Takeuchi, Y, "Water and Protein Permeation Through Polymeric Membrane Having Mechanochemically Expanding and Contracting Pores. Function of Chemical Valve I., J. Polymer Sci., Poly. Let. Ed., 19, 303, (1981).

6. Pasechnik, V.A., A.N. Cherkasov, V.P. Zhemkov, V.V. Chechina and R.S. Alimardanov, "Effect of Direct Current on the Permeability of Ultrafiltration Membranes", Kolloidnyi Zhurnal 41, 586 (1979).

7. Burgmeyer, P. and Murray, R.W., "An Ion Gate Membrane: Electrochemical Control of Ion Permeability Through a Membrane With an Embedded Electrode," J. Am. Chem. Soc., 104, 6139 (1982).

8. Eisenberg, S.R. and A.J. Grodzinsky, "Electrically Modulated Membrane Permeability", J. Mem. Sci. 19, 173 (1984).

9. Nussbaum, J.H., "Electric Field Control of Mechanical and Chemical Properties of Polyelectrolyte Gel Membranes", PhD thesis, Department of Electrical Engineering and Computer Science, M.I.T. (1986).

10. Weiss, A.M., "Real Time Control of the Permeability of Crosslinked Polyelectrolyte Membranes to Fluorescent Solutes", Ph.D. Thesis, Department of Electrical Engineering and Computer Science, MIT (1986).

11. Sefton, M.V. and E. Nishimura, "Insulin Permeability of Hydrophilic Polyacrylate Membranes", J. Pharmaceutical Sciences 69, 208 (1980).

12. Brummer, S.B. and M.J. Turner, "Electrochemical Considerations for Safe Electrical Stimulation of the nervous System with Platinum Electrodes", IEEE Trans. Biomed. Eng. 234, 436 (1977).

13. Mackie, J.S. and P. Meares, "The Diffusion of Electrolytes in a Cation-Exchange Resin Membrane. I. Theoretical", Proc. Royal Soc. A 232, 498 (1955).

14. Meares, P., "Transport in Ion Exchange Polymers", chapter 10 in Diffusion in Polymers , J. Crank and G.S. Park editors, Academic Press, London (1968).

15. Bean, C.P., "The Physics of Porous Membranes - Neutral Pores", chapter 1 in Membranes volume 1, G. Eisenman editor, Marcel Dekker, Inc., New York (1972).

16. Anderson, J.L. and J.A. Quinn, "Restricted Transport in Small Pores", Biophys. J. 14, 130 (1974).

17. Deen, W.M. and F.G. Smith III, "Hindered Diffusion of Synthetic Polyelectrolytes in Charged Membranes", J. Mem. Sci. 12, 217 (1982).

18. Teorell, T., "Transport Processes and Electrical Phenomena in Ionic Membranes", Prog. Biophys and Biophys. Chem. 3, 305 (1953).

19. Meares, P., "The Physical Chemistry of Transport and Separation by Membranes", chapter 1 in Membrane Separation Processes, P. Meares editor, Elsevier Scientific Publishing Company, Amsterdam (1976).

20. Helfferich, F., Ion Exchange, McGraw-Hill, New York (1962).

21. Nussbaum, J.H. and A.J. Grodzinsky, "Proton Diffusion-reaction in a Protein Membrane and the Kinetics of Electromechanical Forces", J. Membrane Sci. 8, 193 (1981).

22. Vacik, J. and J. Kopecek, "Specific Resistances of Hydrophilic Membranes Containing Ionogenic Groups", J. Appl. Poly. Sci. 19, 3029 (1975).

23. Schoenfeld, N.A. and A.J. Grodzinsky, "Contribution of Electrodiffusion to the Dynamics of Electrically Stimulated Changes in mechanical Properties of Collagen Membranes", Biopolymers 19, 241 (1980).

24. Valleton, J.M., J.C. Vincent and E. Selegny, "Monitoring of Artificial Enzyme Membrane Systems by Electric Currents I. Analytical Treatment with Direct Current and Buffered Conductivity", Biophys. Chem. 15, 235 (1982).

25. Peppas, N.A. and S.R. Lustig, "The Role of Crosslinks, Entanglements, and Relaxations of the Macromolecular Carrier in the Diffusional Release of Biologically Active Materials. Conceptual and Scaling Relationships", Ann. NY Acad. Sci. 446, 26 (1985).

26. Crank, J., The Mathematics of Diffusion second edition, Oxford University Press, Bristol (1975).

APPLICATION OF FUEL CELLS TO CHLORINE-CAUSTIC TECHNOLOGY

A. Gelb and E. J. Taylor ■ Physical Sciences Inc., Dascomb Research Park
P.O. Box 3100, Andover, MA 01810

The results of a systems analysis and preliminary experimental investigation of the application of fuel cell technology to the chlorine-caustic electrolytic industry are presented. A coupled electrolytic/alkaline fuel cell system designed to concentrate caustic and generate electricity shows large potential energy savings (30-50 percent reductions) and rapid return on investment.

INTRODUCTION

Electrochemical industries consume about seven percent of all the electric power produced in the United States. The production of chlorine and caustic soda accounts for about 20 percent of this consumption (1). The daily production of chlorine in the United States is about 37,000 tons (2) with about 70 domestic plants accounting for this production. In this paper, we present an analysis of a new chlorine-caustic production system involving a coupled membrane electrolysis cell and a membrane alkaline fuel cell reactor. The predicted energy consumption of this system is compared with existing technologies and an analysis of the economics of this system is presented.

Three types of electrolysis cells currently dominate the electrolytic chlorine-caustic industry. mercury, diaphragm and membrane cells (3). Diaphragm cell technology is well developed and accounts for 75 percent of the U.S. electrolytic chlorine-caustic production. Membrane cell technology is an emerging technology which is being used extensively in Japan (4). Mercury cells are finding increasingly limited application because of pollution problems and will be phased out in the near future.

In the diaphragm process the cell is divided into anode and cathode compartments by an asbestos diaphragm. Feed brine, i.e. saturated NaCl solution, enters the anode compartment where chlorine gas is liberated. The brine convects through the diaphragm into the cathode compartment where H_2 is produced. The anode and cathode reactions are:

$$2Cl^- \rightarrow Cl_2 + 2e^- \quad \text{(anode)}$$

$$2H_2O + 2e^- \rightarrow 2OH^- + H_2 \quad \text{(cathode)}$$

Sodium ions flow through the asbestos diaphragm by convection and under the influence of the imposed electric field. The catholyte exits the cell with 10 to 15 wt percent NaOH and diluted brine. The caustic is concentrated in a series of evaporation procedures to approximately 50 wt percent NaOH. The H_2 gas produced at the cell cathode may be used as an energy source or as a commercial product. A brief schematic of the diaphragm process is shown in Figure 1. Discussion of each process element is given in Refs. 1 and 5.

In the membrane process the electrolysis cell is divided into anode and cathode compartments by a cation-transporting membrane. The membrane allows only the transport of cations. Brine is fed into the anode compartment where it is electrolyzed to produce Cl_2 gas. Dilute caustic enters the cathode compartment where H_2 gas is produced at the cathode and NaOH is concentrated in the catholyte. The NaOH concentration must be maintained between 15 and 30 wt percent for proper membrane function. The electrode reactions are the same as for the diaphragm cell. The catholyte effluent is typically 25 to 30 wt percent NaOH with approximately 50 ppm NaCl. Again, H_2 is produced as a by-product. A brief schematic is given in Figure 2.

The membrane cell process somewhat reduces the total process energy requirements. However, several penalties must be paid. The presence of the membrane in the electrolysis may require higher cell voltages. Back migration of hydroxyl ion through the membrane at high NaOH concentration reduces faradaic efficiency. Also, high cell current densities required for single step caustic purification diminish membrane life. Abam Engineers, Inc.[6] has recently compared membrane and diaphragm cell technologies. They compared capital costs and energy consumption for current membrane and diaphragm cells.

The steps in chlorine/caustic production requiring the greatest energy usage are: 1) electrolysis of the brine solution; and 2) concentration of the catholyte effluent to 50 wt percent caustic soda. The development of membrane cell technology was largely motivated by its promise of producing higher concentration caustic free of chloride, thus reducing the energy required for caustic production. Energy saving has been a major focus of the chlorine/caustic industry.

Efficient usage of the by-product hydrogen from the electrolysis of the brine solution has received considerable attention as a strategy for reducing process energy requirements. Currently, most by-product hydrogen is burned to provide steam for caustic concentration. An alternative use of the by-product hydrogen is as fuel in a hydrogen/air fuel cell.

Fuel cells are direct conversion devices, i.e. they directly convert chemical energy to electrical energy by electrochemical reactions. Thus they do not suffer from Carnot cycle efficiency limitations. In addition to efficient electricity generation, high temperature fuel cells are capable of producing steam by cogeneration.

Several investigators (3,7,8) have proposed using fuel cells in the chlorine-caustic process to utilize the by-product hydrogen from brine electrolysis to generate electricity. The use of fuel cells has obvious merit since the power generated for the process by conventional technologies is produced at about 33 percent efficiency, whereas a phosphoric acid fuel cell operating on pure hydrogen would have a generating efficiency of about 47 percent, depending, of course, on the operating cell voltage. The waste heat from a phosphoric fuel cell, which operates at about 400°F, could also provide steam for the thermal operations of the process. The savings which can be realized by such an approach are quite large. A similar analysis may be made for an alkaline fuel cell.

Recently several investigators (8) have developed a coupled diaphragm/ alkaline fuel cell system capable of significantly reducing the process energy consumption. In their approach, an alkaline fuel cell, operating on sodium hydroxide, is coupled to the diaphragm cell process to both provide process power and to electrochemically separate the caustic from the contaminant brine. This process could potentially eliminate most of the boiler energy used to concentrate caustic, and further reduce the electrical power requirement.

Coupled Membrane Electrolysis/Alkaline Fuel Cell System

PSI has developed a new system based on a membrane electrolysis cell coupled to a membrane alkaline fuel cell. The alkaline fuel cell is coupled with the electrolysis cell to produce power and further concentrate the caustic soda. This system is schematically shown in Figures 3 and 4. There are several significant advantages to this configuration over the usual membrane cell systems and coupled diaphragm/alkaline fuel cell systems.

Diluted caustic solution leaving the anode side of the fuel cell may be recycled directly to the cathode compartment of the membrane electrolysis cell. Hence no product is lost as in the coupled diaphragm/alkaline fuel cell system. Additionally, in the brine preparation step the purification is already carried out to eliminate cations harmful to the membrane in the electrolysis cell. Thus no significant degradation of the alkaline fuel cell membrane will occur.

The significant advantages of the PSI system are that caustic concentration is carried out in the fuel cell without any need of external energy input and the by-product hydrogen is used efficiently to generate electricity.

In this report we discuss the results of a system study and preliminary experimental results regarding the application of fuel cells to chlorine-caustic technology. Several systems have been examined and compared. These include: 1) conventional diaphragm cell technology; 2) conventional membrane cell technology; 3) diaphragm and membrane cell technologies with phosphoric acid fuel cell for electrical power generation and steam generation; and 4) the PSI coupled membrane/-alkaline fuel cell system.

System Studies

Models were constructed for the various components of the five aforementioned systems. These include models for diaphragm and membrane electrolysis cells, caustic evaporators, phosphoric acid and alkaline fuel cells. Performance characteristics for the alkaline and phosphoric acid fuel cells were taken from the literature (9).

For convenience, we label the systems under consideration as the following:

Case 1 - diaphragm cell

Case 2 - diaphragm cell and phosphoric acid fuel cell

Case 3 - membrane cell

Case 4 - membrane cell and phosphoric acid fuel cell

Case 5 - membrane cell and alkaline fuel cell.

Baseline calculations were performed to compare each with regard to the following quantities: 1) net energy consumption, i.e., energy consumed in the plant; 2) gross energy expenditure; and 3) gross margin, defined as the difference between the (current) market value of products minus the plant operating costs.

A series of parametric studies of the PSI system (Case 5) were performed to investigate system characteristics with regard to the alkaline fuel cell and membrane performance. The varied parameters included: 1) membrane electrolysis cell exit caustic concentration; 2) water transport number of the alkaline fuel cell membrane, and 3) alkaline fuel cell performance, i.e., power output.

Baseline Calculations

A summary of the baseline case conditions is listed in Table 1 for the brine electrolysis and fuel cells. Membrane cell characteristics were those of a cell using a Nafion®* 901 membrane with a water electro-osmotic flux equal to 3.6. The water electro-osmotic flux is the ratio of water molecules to sodium ions transported across the membrane. The water electro-osmotic flux in the alkaline fuel cell was equal to 1.5 for the baseline case calculations. The concentration of caustic in the diaphragm electrolysis cell was 10 wt percent. Caustic concentration from the membrane electrolysis cells was 30 wt percent. In all cases, final product caustic was 50 wt percent. For Cases I and II, triple effect evaporators were used for caustic concentration. Single effect evaporators were used for Cases III and IV using membrane electrolysis cells.

Net energy consumption for each of the five cases is listed in Table 2. The values are computed on a per ton of 50 wt percent caustic produced. Electrical power is consumed in the electrolysis cell to produce chlorine, hydrogen and dilute caustic. The thermal energy used is defined as the additional fuel required for caustic concentration and other processes. In Cases 2, 4 and 5 by-product hydrogen is used to generate electricity in an acid (Cases 2 and 4) or alkaline (Case 5) fuel cell, and the electrical power shown is the net value consumed after subtracting the power generated in the fuel cells. In Cases 2 and 4 the acid fuel cell is also used to generate steam for caustic concentration and this contribution reduces the net thermal energy requirement. However, it should be noted that by using the by-product hydrogen in a phosphoric acid fuel cell there is an increase in the required thermal energy. In the PSI system (Case 5) the hydrogen is used in the alkaline fuel cell both to generate electricity and concentrate caustic.

As is readily seen, the largest energy cost in all cases is the electrical energy required for the electrolysis cell. Caustic evaporation requires the next largest energy expenditure (except for Case 5). All other energy expenditures, such as those required for chlorine product treatment, brine preparation (not listed) and hydrogen drying, are relatively minor.

The most energy intensive system is the conventional diaphragm technology. It

*Nafion is a registered trademark of the E.I. DuPont deNemours and Company.

requires a net energy expenditure of 4148 KWH/ton 50 percent NaOH. By adding a phosphoric acid fuel cell to the system (Case 2) we improve the efficiency of the system, somewhat reducing the net energy consumption to 4037 KWH/ton 50 percent NaOH, or a 2.7 percent decrease. Note again that using the hydrogen in an acid fuel cell results in a reduction of net electrical energy consumption but an increase in thermal energy consumption.

Conventional membrane electrolysis technology (Case 3) is substantially less energy intensive than either diaphragm cell case (1 or 2). The decrease in energy consumption is due to an 11 percent decrease in electrolysis cell electrical consumption and a 54 percent decrease in the energy required to concentrate caustic. Conventional membrane technology requires 26 percent less energy than conventional diaphragm technology. The addition of a phosphoric acid fuel cell to membrane electrolysis cell system (Case 4) results in a 5 percent decrease in net energy consumption. This saving is slightly larger than that for adding an acid fuel cell to a diaphragm electrolysis cell system. This is due to the incorporation of multistage evaporators into the caustic concentration process for the diaphragm technology. As mentioned earlier, we consider single stage evaporators for membrane technology caustic concentration.

The PSI system (Case 5) shows the lowest net energy consumption of all the systems considered (Figure 5 and Table 2). As is readily seen, this results from the elimination of any energy requirements for caustic concentration. This is made possible by the use of the cation permeable membrane in the alkaline fuel cell to separate the water from the caustic. Only about 55 percent of the hydrogen by-product is used for caustic concentration. Approximately 35 percent of the hydrogen is used to generate steam in chlorine processing and 10 percent is available as an additional product. It may also be possible to use an acid fuel cell to consume the hydrogen not used in the alkaline fuel cell. This would produce electricity and the necessary steam for chlorine processing (by cogeneration).

We now compare each system on the basis of gross energy expenditure. Gross energy expenditure is defined as the total fuel energy expended to produce products. In effect the gross energy is the sum of the fuel energy consumed in the plant and the fuel energy used at the electrical utility to generate the necessary electricity for the brine electrolysis. This quantity has a direct bearing on the national energy consumption picture. The difference between net and gross energy consumption is due to the efficiency of generating electricity at the utility. To compute gross energy consumption we assume a 35 percent electrical power generation efficiency. This assumed efficiency is equivalent to assuming 1 kWh of electricity requires 9749 Btu of fuel energy. This is somewhat optimistic. The results of our analysis for the baseline cases are shown in Figure 6.

Conventional diaphragm technology (Case 1) is seen to be most energy intensive. The addition of a phosphoric acid fuel cell (Case 2) results in an 8.5 percent decrease in gross energy consumption. This value is larger than the computed decrease for net energy consumption because of the weighting of electrical energy consumption by its generating efficiency. Case 3, conventional membrane electrolysis cell technology, has a lower gross energy consumption than either of the diaphragm cell cases. Conventional membrane cell technology (Case 3) consumes 13.5 percent less energy than Case 1. By adding a phosphoric acid fuel cell to membrane cell technology a gross energy consumption decrease of 23 percent is achieved compared to conventional diaphragm cell technology.

The lowest gross energy consumption system is Case 5, the PSI system. Its largest single energy savings arises from the elimination of fuel for caustic concentration. In fact even if the alkaline fuel cell produced no electrical energy it would still have substantially lower gross (or net) energy consumption than any of the other systems. The baseline calculation shows a decrease in gross energy consumption of 36 percent compared to diaphragm cell technology and 26 percent compared to membrane cell technology.

We can estimate the decrease in the total national energy consumption which would result from replacing all existing diaphragm cell systems with the PSI membrane electrolysis cell/alkaline fuel cell system. Our estimate is probably low since we use state of the art diaphragm cell performance in our calculations for that technology. Undoubtedly, older plants use energy less efficiently. We assume a daily production in the U.S. of 37,000 tons of chlorine. By installing the PSI system in every existing chlorine-caustic plant in the U.S. there could be a decrease of annual national energy consumption of approximately

4.8 x 10^{10} KWH or 1.6 x 10^8 MBtu or 0.16 quad. This corresponds to approximately 0.2 percent of the total national energy consumption per year.

As a prelude to evaluating the economic viability of these systems we compute the gross margin defined as the value of saleable products, determined by market price, minus the energy costs of production. This quantity does not take into account capital costs, plant maintenance, taxes, depreciation, interest payments, etc. However, it is useful in a differential sense, i.e., to rank the profitability and economic viability of the five systems considered here. For these calculations we use:

$5.00/MBtu fuel gas (methane),

$0.049/kWh electricity,

$0.05/1000 gal cooling water,

$0.20/1000 lb condensate,

$195/ton chlorine,

$250/ton 50 wt percent NaOH, and

$2.00/1000 SCF hydrogen.

The computed values of gross margin are shown in Table 3 for each baseline case. Several general observations may be made. Membrane cell technology reduces the cost of chlorine and caustic production compared to diaphragm technology. The addition of an acid fuel cell to either diaphragm or membrane cell technologies reduces production costs; this is due to the efficient generation of electrical power in the acid fuel cell. The PSI system has the lowest energy costs; this is due primarily to the reduction of fuel consumption for caustic concentration.

Parametric Studies

Parametric studies were conducted to investigate the sensitivity of the PSI system (Case 5) to variations in operating conditions and membrane parameters. The parameters which were varied are: 1) effluent caustic concentration from the membrane electrolysis cell, 2) electroosmotic water flux across the alkaline fuel cell membrane, and 3) alkaline fuel cell performance.

Figure 7 shows the variation of net energy consumption with changes in the concentration of caustic produced in a membrane electrolysis cell. For Cases 3-5, net energy consumption decreases as the caustic concentration increases. The dominant factor in this variation for Cases 3 and 4 is the decrease in thermal energy needed to concentrate the caustic produced by the electrolysis cell to 50 wt percent NaOH.

The PSI system (Case 5) shows substantially less variation with caustic concentration. This is because this system eliminates the thermal energy required to concentrate caustic. The dominant factors for this case are the variation of electrolysis faradaic efficiency and anode overpotential with electrolysis cell caustic concentration. These variations are quite small. Clearly the advantage of the PSI system is somewhat reduced as higher concentrations of caustic are produced in the membrane electrolysis cell. However, even at 40 wt percent NaOH effluent the PSI system reduces net energy consumption by about 20 percent compared to conventional membrane cell technology.

The water electroosmotic flux is defined as the ratio of water molecules transported across the membrane to that of sodium ions. In order to use an alkaline fuel cell with a membrane to concentrate caustic, it is necessary that the water electro-osmotic flux not be too large.

Presently, no data are available for water flux under the conditions required for the alkaline fuel cell operation. Existing data indicate a value of 1.5 may be expected. This value was used in our baseline calculations. We computed exit caustic concentrations from the alkaline fuel cell as a function of water electroosmotic flux. The results in Figure 8 show clearly that values of water electroosmotic flux below about 2.8 lead to acceptable values of product caustic concentration. The reason for the relative insensitivity of caustic concentration to water flux is the ability of the fuel cell to remove water by convection out of the cell through the gas channels. This factor substantially increases the flexibility of the system because it broadens the allowable membrane performance requirements.

We now consider variations in the net energy consumption for Case 5 due to variations in the alkaline fuel cell performance. Available commercial membranes have not been tested under operating conditions similar to those of the alkaline fuel cell in this system. In particular, the electrical resistance of suitable membranes has not been determined.

In Table 4 the variation of net energy consumption as a function of the alkaline fuel cell voltage at constant current density for the baseline calculation considered earlier. This calculation shows the net energy consumption variation when high values of membrane resistance are assumed.

It is clear from the values presented in Table 4 that the net energy consumption is insensitive to the membrane resistance. This is because the largest energy savings comes from the elimination of the required caustic concentration energy. In the worst possible case when no electrical energy is extracted from the alkaline fuel cell the PSI system still consumes 28 percent less energy than conventional membrane cell technology.

The results of these parametric variations point out the relative insensitivity of the PSI system net energy consumption to variations in alkaline fuel cell performance. The critical factor for the viability of the PSI system is the ability of the membrane to concentrate caustic as fuel and oxidant gases are consumed.

Economic Analysis

A differential economic analysis was performed to compare Case 4 (membrane cell/acid fuel cell) with Case 3, (membrane cell) and Case 5 (PSI system) with Case 3. We suppose a manufacturer is currently operating a chlorine-caustic production plant with a membrane electrolysis cell and natural gas-fired boiler to generate process steam for caustic concentration. The object of this analysis is to assess the effect on net cash flow from converting from Case 3 to Case 4 and from Case 3 to Case 5.

The PSI Financial Analysis Code was exercised to calculate key financial indicators for the proposed plant investments. It was assumed that each retrofit would be 100 percent equity financed by the investing party. The discounted return rates are therefore equivalent to a return on the manufacturer's investment. The key assumptions are given in Table 5. A more detailed discussion of this analysis has been given earlier (9).

Capital and operating costs of the PSI membrane cell/alkaline fuel cell are difficult to estimate at present. However, we will show that within the estimated range of uncertainty of certain key unknown variables, the retrofit of an alkaline fuel cell to an existing membrane electrolysis cell plant appears to be very attractive. The principal unknowns at present which affect the economic analysis include: 1) electrode precious metal loadings, 2) fuel cell operating current density, and 3) alkaline fuel cell capital costs and lifetime.

To estimate the capital costs of the alkaline fuel cell, the following cost components were considered: 1) raw materials for cell, 2) cathode platinum catalyst, and 3) installation and delivery. The largest single materials cost for the alkaline fuel cell is the anion impermeable membrane. Its cost was estimated from the retail price of the DuPont Nafion 901 membrane ($50/ft^2). Other component costs were estimated from typical raw material costs. Total materials costs are about $70/ft^2 of fuel cell electrode area.

Estimating the installation and delivery price is rather more difficult in that alkaline cells are not yet commercially available for industry. The price at high market supply is about $30/ft^2. The ultimate price of delivered alkaline cells will depend on market penetration. For a conservative estimate of alkaline fuel cell construction and installation cost in the future, we considered installation and delivery prices of between $40 and $100 per ft^2.

It remains to calculate the total area of the alkaline fuel cells required to achieve the desired chlorine and caustic outputs. The number of cells required to form the stack is inversely proportional to the achievable current density. The cell current density is expected to be in the range of 200 to 300 A/ft^2.

With these assumptions and estimates, discounted rates of return were computed for the retrofitting of a membrane chlorine-caustic plant by a membrane/alkaline fuel cell system. For the most favorable case: $40/ft^2 construction and 300 ASF the discounted rate of return exceeded 65 percent. For the least favorable case ($100/ft^2 construction and 200 ASF) the value was computed to be in excess of 30 percent. It is clear that this system is quite favorable from an economic point of view. Calculations were also performed to test the sensitivity of the analysis to the lifetime of the alkaline fuel cell. The analysis showed a relative insensitivity to the fuel cell lifetime. This is due primarily to depreciation allowances. The results indicated a slight inverse dependence of profitability on fuel cell lifetime.

A similar analysis was made for the addition of phosphoric acid fuel cell to membrane chlorine-caustic plant, i.e. retrofitting Case 3 to Case 4. The results indicate large potential return on investment similar to those for retrofitting Case 3 to Case 5. However the absolute profitability (and investment) are substantially smaller than in the earlier case and much more sensitive to construction and installation costs.

Preliminary Experimental Results

The membrane properties which need to be determined in the PSI system environment are: 1) conductivity, 2) Na^+ transport, and 3) electroosmotic H_2O flux. Additionally, the oxygen reduction behavior of the gas diffusion cathode in the PSI system environment must be determined. Membrane conductivity measurements are just now beginning. The work to date and the future research direction are reported below.

A schematic diagram of the membrane conductivity measurement cell is shown in Figure 9. This cell was modeled after a design reported in the literature (10). The cell is constructed entirely of Teflon®* and the membrane is positioned between the two electrolyte compartments. Provision is made for electrolyte stirring and cell heating and temperature control via Teflon-coated immersion heaters and RTD temperature controllers. Water cooled condensors are positioned above the gas vents to prevent water loss by evaporation. A constant current is applied across the membrane via the platinum screen electrodes on either side using a PAR Model 371 Potentiostat/ Galvanostat. The potential drop as a function of current density across the membrane is measured via the two Luggin capillary probes on both sides of the membrane providing solution contact to the dynamic hydrogen reference electrodes (DHE) (11). The potential drop in absence of the membrane is subtracted from that in the presence of the membrane to correct for solution resistance. The specific conductivity, $K(\Omega^{-1}cm^{-1})$, is then calculated from:

$$K = \frac{i \cdot d}{(E_{m+e} - E_e)}$$

where i is current density (A cm^{-2}), d is membrane thickness (cm), E_{m+e} is the voltage drop of the membrane plus electrolyte (V), and E_e is the voltage drop of the electrolyte. It is noteworthy that A-C techniques are not appropriate for measuring conductivity under actual electrolysis conditions.

To date Nafion® 324 and 901 as well as two proprietary membranes obtained under a secrecy agreement have been obtained for evaluation under conditions relevant to the PSI system. The Nafion membranes were treated prior to evaluation by expanding them in boiling NaOH (2 wt percent) for 45 min. After assembling the membrane in the conductivity cell, the electrolytes were heated to the desired temperature at a rate of 15°C/hr and held at the measurement temperature for 1 hr prior to measurements. This procedure allowed the membranes to equilibrate.

Conductivity data for N-901 in 3M NaCl anolyte and 9.5M NaOH catholyte at 90°C are shown in Figure 10. These conditions were selected for the first evaluation because voltage drop data for this membrane under these conditions were previously reported. At 200 mA/cm^2, a voltage drop of approximately 300 mV was observed compared to that of 400 mV reported in the literature (12). Since the literature report did not mention if an expansion procedure was employed, we decided to ascertain the effect of expansion on the membrane conductivity. The data for a non-expanded membrane (N-901(N)) and an expanded membrane (N-901(Ex)) in 20 percent NaOH are shown in Figures 11 and 12. A dramatic reduction in the voltage drop is observed for the expanded membrane and may explain some of the discrepancies between our data and those reported in the literature for Nafion 901 in the NaCl:NaOH environment.

In order to select the most promising membrane(s) for detailed conductivity evaluation, screening in 30:30 and 40:25 percent NaOH (catholyte: anolyte), was initiated. The 40:25 percent NaOH solution environment corresponds to that which would be encountered in the PSI system at 30 percent NaOH inlet concentration and at the mid-point of caustic concentration to 50 percent. In the 30:30 percent NaOH environment, N-324 was

*Teflon is a registered trademark of the E.I. Dupont deNemours and Company.

observed to be unstable and no further work on this membrane is anticipated. The data for proprietary membrane designated A1 are shown in Figures 13 and 14. Included in these figures is the performance of a single cell alkaline fuel cell operating on pure hydrogen and air (13). This system is being developed by the Institute for Hydrogen Systems (IHS) in Canada. The electrolyte is 12 N KOH at 65°C and the cathode electrocatalyst is Pt supported on carbon at a loading of 0.5 mg Pt per cm^2 of electrode (geometric area). In the previously mentioned systems study, Pt cathode catalyst loadings of 5, 10, and 20 mg per cm^2 were assumed. The Pt loading of the IHS alkaline fuel cell is an order of magnitude lower than the lowest value assumed. The IHS fuel cell data are qualitatively useful to predict the power output/requirements of the membrane alkaline fuel cell. However, since the IHS data are in 12 N KOH at 65°C, the performance of the alkaline fuel cell cathode in the environment of interest to the PSI system will be determined.

For the membranes studied here, the voltage drop does not increase linearly with increasing current density, i.e., these membranes do not exhibit ohmic behavior. This behavior is advantageous to the PSI fuel cell reactor since at higher current densities the measured voltage drop is lower than what would be expected from extrapolation of the low current density data. Similar observations have been reported previously (14) for a perfluorosulfonate membrane in 10M NaOH (~30 wt percent) at elevated temperatures (100° to 180°C) and pressures. The specific conductivity of this membrane was observed to increase or reach a plateau as the current density was increased to 500 mA/cm^2. For the Nafion 901 membrane, the voltage drop at a given current density decreased as the temperature was increased from 60° to 90°C. However, in the case of the proprietary membranes a more complicated temperature dependence is observed. A similar observation has been reported in the literature when the NaOH concentration was increased from 5 to 10M (14).

Recently, conductivity data for Nafion-901 in NaOH (7 to 40 percent) at 20° to 90°C have been reported (15). These measurements were conducted with an A-C bridge at 1 kHz in 31 percent NaOH and areal conductivities of 0.2 and 0.4 ohm^{-1} cm^{-2} at 60° and 90°C, respectively, were reported. From the data in Figure 14, areal conductivities in the range of 0.14 to 0.24 ohm^{-1} cm^{-2} at 60°C and 0.22 to 0.30 ohm^{-1} cm^{-2} at 90°C may be calculated. Although the agreement between the two types of measurements is fairly good, the direct current method employed in our work should yield a more realistic value for the membrane voltage drop in the membrane alkaline fuel cell.

The intersection of the membrane voltage drop data with the IHS alkaline fuel cell performance data represents the operating current density which would result in zero net power. At higher current densities, external power is required by the membrane alkaline fuel cell. At lower current densities, power is generated by the membrane alkaline fuel cell. The optimal operating current density of the membrane alkaline fuel cell is defined by a tradeoff of energy power production and capital cost considerations. At low current densities, the electric power output which may be used in the electrolysis cell is increased but the membrane alkaline fuel cell area (and hence capital cost) required to concentrate a given amount of caustic soda is also increased. Conversely, at high current densities the capital cost is decreased but the electric power output is also decreased. These issues will be addressed in a later systems study. In Tables 6 and 7, a summary of the membrane voltage drop data and the projected membrane alkaline fuel cell at 100 and 200 mA/cm^2, respectively, are presented. Based on these data, a membrane alkaline fuel cell using proprietary membrane A1 operating at 60°C yields the highest power output (427 to 553 mV at 100 mA/cm^2 and 214 to 296 mV at 200 mA/cm^2). These results are quite promising for the PSI system.

The specific conductivity of an ion exchange membrane is determined by the concentration and mobility of charge carriers within the membrane (16). In dilute solutions the specific conductivity of an ion exchange membrane increases with increasing external electrolyte concentration due to increasing electrolyte sorption (14). However, in more concentrated solutions, membrane dehydration leads to decreasing specific conductivity. Electrolyte and water sorption studies for a perfluorosulfonate membrane in 5 to 12.5M NaOH solutions have been reported at room temperature (17) and in the temperature range of 70° to 90°C (12,18,19). At 80°C the water to sodium ion ratio decreased by ~30 percent as the equilibrating solution concentration increased from 5 to 10M NaOH. The sodium ion self-diffusion decreased by an order of magnitude over the same concentration range.

(Membrane conductance and ionic diffusion are related phenomena).

The performance of Nafion 901 is further complicated by the fact that it is an "alloy" membrane consisting of a perfluorosulfonate layer and a perfluorocarboxylate layer. It has been postulated that carboxylate membranes undergo a structural change at external NaOH (15) solution concentrations greater than 9M NaOH whereas sulfonate membranes do not. This postulate is supported by sodium ion transport and self-diffusion data. Sodium ion transport decreases monotonically with increasing caustic concentration (5 to 13M) for the sulfonate membrane but exhibits a complicated concentration dependence for the carboxylate membrane. The activation energy for sodium ion self-diffusion is lower than 20 kJ/mol for the sulfonate membranes in the concentration range studied and for the carboxylate membranes at concentrations below 9M. Above 9M, the activation energy for the carboxylate membrane increases from 85 to 130 kJ/mol (17). The existence of two linear regions in the Arrhenius conductivity data for Nafion 901 has been attributed to the differences in the properties of the carboxylate and sulfonate layers. Electrolyte sorption studies for the three membranes studied here in higher sodium hydroxide solution environments (up to 50 percent) should further the understanding of the membrane performance data.

Conclusions

The study presented here shows that the application of fuel cell technology to current chlorine-caustic production can lead to substantial reductions in energy consumption. In particular, the PSI system, which couples an alkaline fuel cell to a membrane electrolysis cell shows net energy consumption reductions of between 30 and 50 percent compared to current membrane and diaphragm cell technologies. The feature of this system which leads to these large energy reductions is its ability to concentrate electrolysis effluent caustic in the alkaline fuel cell to commercial grade and to recycle diluted caustic directly into the electrolysis cell.

Studies of the sensitivity of the PSI system to variations in alkaline fuel cell and membrane performance indicate satisfactory behavior over wide ranges of system parameters. Experimental determinations of membrane resistance, sodium ion transport, and water electroomostic flux in the fuel cell environment are currently being conducted. Additionally, cathode electrocatalyst evaluation in the membrane alkaline fuel cell environment will be determined.

The economic analysis shows that the installation of the coupled membrane electrolysis alkaline fuel cell is an extremely attractive investment. The PSI system holds great potential for economic reward due to its large energy savings.

LITERATURE CITED

1. Kirk, J. and Othmer, D., Encyclopedia of Chemical Technology, Volume 1, 3rd edition, John Wiley and Sons, New York (1978).

2. North American Chlor-Alkali Industry: Plants and Production Data Book, Pamphlet No. 10, The Chlorine Institute, Inc., New York (1976).

3. "Proceedings of the Workshop on Energy Conservation in Industrial Electrochemical Processes," Aug. 10-12, 1976, ANL/OEPM-77-1, Argonne National Laboratory, Contract W-31-109-Eng-38, Ed. N.P. Yao et al.

4. Greek, B. F. and Fallwell, W. F., Chemical and Engineering News, March 1 (1982), p. 10.

5. Caldwell, D. L., in Comprehensive Treatise of Electrochemistry, Vol. 2 edited by J. O'M. Bockris, B. E. Conway, E. Yeager, and R. E. White, Plenum, New York (1981).

6. "Final Report on Process Engineering and Economic Evaluation of Diaphragm and Membrane Chloride Cell Technologies," Dec. 1980, ANL/OEPM-80-9, Abam Engineers for Argonne National Laboratory, Contract 31-109-38-5747.

7. Beck, T., "Final Report on Improvements in Energy Efficiency of Industrial Electrochemical Processes," ANL/OEPM-72-2, Electrochemical Technology Corp., for Argonne National Laboratory, Contract W-31-109-Eng-38 (January 1977).

8. Emergy, A. T., et al., "Hybrid Cell - A New Process for Improving the Energy Efficiency and Reducing the Cost of Chlor-Alkali Production," Abstract No. 47, The Electrochemical Society, Extended Abstracts Vol. 81-1, Spring Meeting, Minneapolis, Minn., May 10-15 (1981).

9. Gelb, A., Garman, A., Bloomfield, D., and Piperopoulous, H., Application of Fuel Cells to Chlorine-Caustic Technology, Topical Report, prepared for DOE (February 1985).

10. Yeager, H.L., O'Dell, B., and Twardowski, Z., "Transport Properties of Nafion® Membranes in Concentrated Solution Environments," J. Electrochem. Soc., 129(1), 85 (1982).

11. Giner, J., "A Practical Reference Electrode," J. Electrochem. Soc., 111(3), 376-7 (1964).

12. Yeager, H.L. and Malinsky, J.D., "Power Consumption of a Chlor-Alkali Membrane as a Function of Cell Parameters," Extended Abstract No. 237, 165th Meeting of the Electrochemical Society, Cincinnati, Ohio (1984), p. 345.

13. Tomantschger, K., Srinivansan, S., Findlay, R., Joanes, J. McClusky, F., Reid, A., Kordesch, K., "Development of a 500 Watt Bipolar Alkaline Fuel Cell," in Proceedings of the 20th Intersociety Energy Conversion Engineering Conference, Society of Automotive Engineers, Warrendale, PA (1985).

14. Steck, A. and Yeager, H.L., "Conductance Measurements in an Electrolytic Membrane Cell at Elevated Temperatures and Pressures," J. Electrochem. Soc., 130(6), 1297 (1983).

15. Chandran, R.R., Yeo, R.S., Chin, D-T., "Sorption and Transport Behavior of Perfluorinated Ionomes Membranes in Concentrated NaOH Solution," Presentation 35th ISE, Berkley, CA, submitted for publication in Electrochmica Acta, (1984).

16. Helfferich, F., Ion Exchange, Chapter 7 McGraw-Hill Book Co., New York, (1962).

17. Mauritz, K.A., Branchick, K.J., Gray, C.I., and Lowry, S.R., Polym. Prepr. Am. Chem. Soc. Div. Polym. Chem., 21, 122 (1980).

18. Yeager, H.L., Kipling, B., and Dotson, R.L., "Sodium Ion Diffusion in Nafion® Ion Exchange Membranes," J. Electrochem. Soc., 127(2), 303 (1980).

19. Twardowski, Z., Yeager, H.L., and O'Dell, B., A Comparison of Perfluorinated Carboxylate and Sulfonate Ion Exchange Polymers II. Sorption and Transport Properties in Concentrated Solution Environments," J. Electrochem. Soc., 129(2), 328(1982).

Table 1. Basline Cases Operating Conditions.

	Current Density (amps/ft^2)	Cell Voltage (volts)
Diaphragm Cell	200	3.85
Membrane Cell	200	3.4
Phosphoric Acid Fuel Cell	300	0.56
Alkaline Fuel Cell	200	0.70

Table 2. Net Energy Expenditures for Baseline Cases.

Case	1	2	3	4	5
ENERGY DISTRIBUTION SUMMARY		KWH/TON NaOH			
Electrolysis Cell	2438.	2438.	2169.	2169.	2170.
Caustic Evaporation	1521.	2171.	706.	1279.	0
Auxiliary Energy Consumption					
Cl_2 compression	38.	38.	45.	45.	45.
Cl_2 drying	150.	150.	151.	151.	151.
Energy Production					
Fuel Cell Energy to Process	0	339.	0	340.	349.
Steam to Process	0	422.	0	426.	
H_2 Value	0	0	0	0	93.
Electric Energy	2476.	2137.	2214.	1874.	1965.
Thermal Energy	1672.	1900.	857.	1004.	0.
Total Net Energy	4148.	4037.	3071.	2878.	1965.

CASE 1: Diaphragm cell

CASE 2: Diaphragm cell + acid fuel cell

CASE 3: Membrane cell

CASE 4: Membrane cell + acid fuel cell

CASE 5: Membrane Cell + alkaline fuel cell (PSI system)

Table 3. Energy Costs and Product Value for Baseline Case per ton Cl_2.

Case	1	2	3	4	5
Product Values ($)/ton$Cl_2$					
Cl_2	195.03	195.03	195.03	195.03	195.03
H_2	0	0	0	0	2.68
50% wt NaOH	265.71	265.71	265.71	265.71	265.71
Product/ ton Cl_2	460.74	460.74	460.74	460.74	463.42
Energy Cost ($)/ton Cl_2					
Electricity @ $0.049/kWh	144.46	124.67	123.98	104.97	110.08
Fuel @ $5/MBtu	32.60	37.05	16.72	19.55	0
Total Energy Costs	177.06	161.72	140.70	124.52	110.08
"Gross Profit"/ ton Cl_2	283.68	299.02	320.04	336.22	353.34

CASE 1: Diaphragm cell

CASE 2: Diaphragm cell + acid fuel cell

CASE 3: Membrane cell

CASE 4: Membrane cell + acid fuel cell

CASE 5: Membrane cell + alkaline fuel cell (PSI system)

Table 4. Net Energy Consuption for Varying Alkaline Fuel Cell Voltage.

Volts	Net energy consumption (kWh/ton 50 wt percent NaOH)
0.3	2114.4
0.4	2080.6
0.5	2046.8
0.6	2013.0
0.7	1979.2

Table 5. Chlorine-Caustic System Financial Analysis Assumptions.

- 100% equity financing in retrofit
- 1 ton/h plant operating continuously
- Negligible differences in operating and maintenance costs between cases
- Negligible working capital requirements
- No salvage value for boiler or displaced cuastic concentration equipment
- No additional property taxes, insurance fees, or legal fees for retrofit.
- Less than one year construction period
- By-product hydrogen gas is sold at $2.00/1000 standard cubic foot.
- Purchased electricity costs are $0.049 kWh in 1985 with 6% annual escalation rate
- Fuel cell installation costs escalate at 6% annually
- Manufacturer's operating income taxed at 50% rate

Table 6. Summary of Membrane Voltage Drop Data at 100 mA/cm² and Membrane Alkaline Fuel Cell Voltage.

	30:30 NaOH			40:25 NaOH		
Temp (°C)	N-901 (Ex)	A1*	B1*	N-901 (Ex)	A1	B1
60	552 (+298)	297 (+553)	564 (+286)	634 (+216)	423 (+427)	596 (+254)
75	477 (+373)	330 (+520)	476 (+374)	535 (+315)	463 (+387)	514 (+336)
90	378 (+472)	298 (+552)	465 (+385)	458 (+392)	568 (+282)	471 (+379)

*Proprietary membranes

**In parenthesis

Table 7. Summary of Membrane Voltage Drop Data at 200 mA/cm² and Membrane Alkaline Fuel Cell Voltage.

	30:30 NaOH			40:25 NaOH		
Temp (°C)	N-901 (Ex)	A1*	B1*	N-901 (Ex)	A1	B1
60	962 (-162)	504 (+296)	884 (-84)	1044 (-244)	586 (+214)	736 (+64)
75	734 (+66)	528 (+272)	702 (+98)	832 (-32)	616 (+184)	600 (+200)
90	676 (+124)	594 (+206)	727 (+73)	752 (+48)	976 (-176)	632 (+168)

*Proprietary membranes

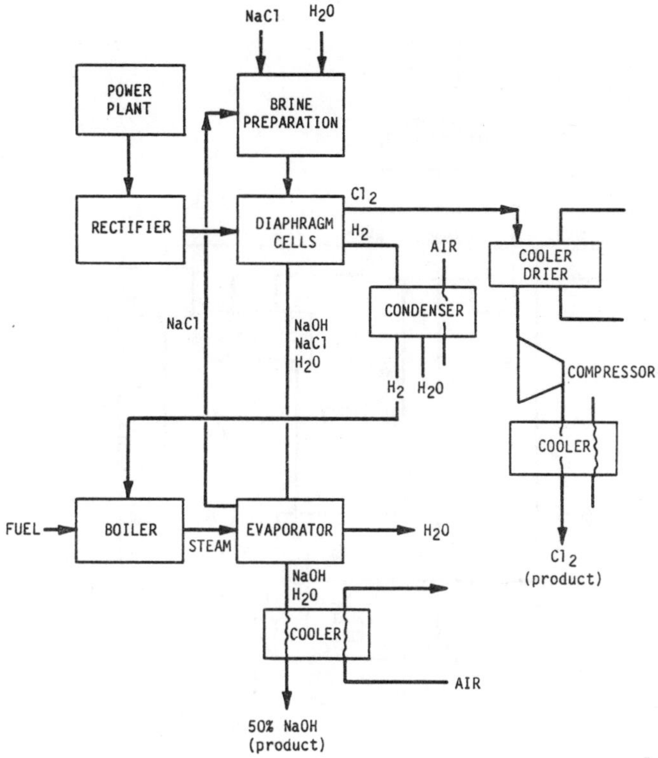

Figure 1. Schematic for Diaphragm Chlorine/Caustic System.

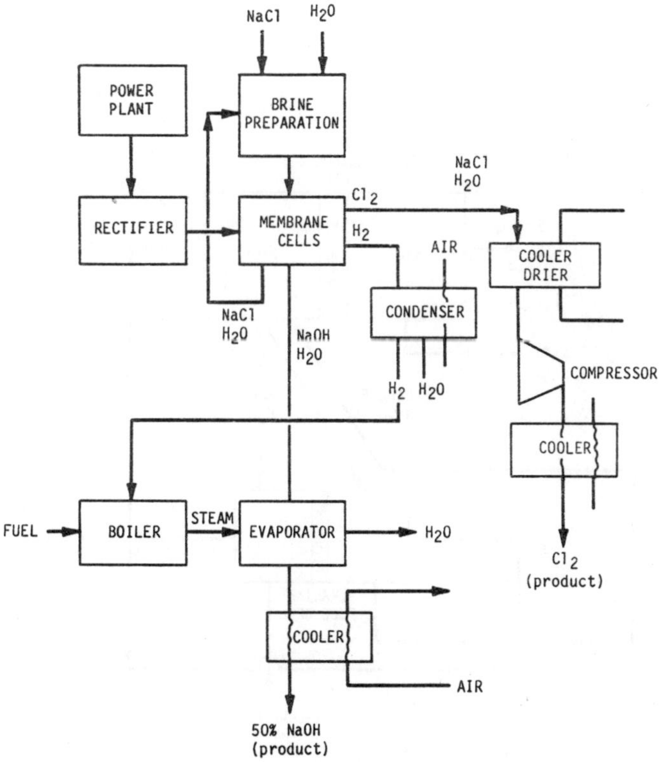

Figure 2. Schematic for Membrane Chlorine/Caustic System.

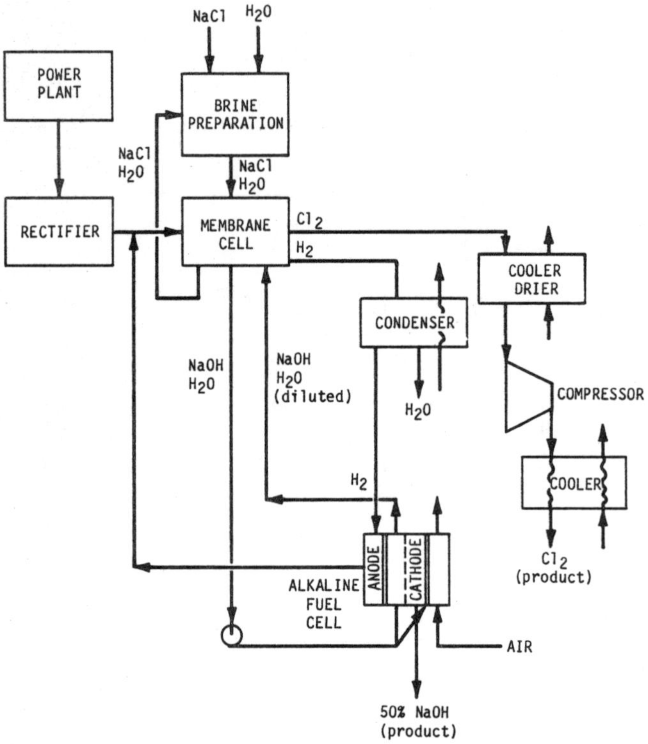

Figure 3. Schematic for PSI Chlorine/Caustic System.

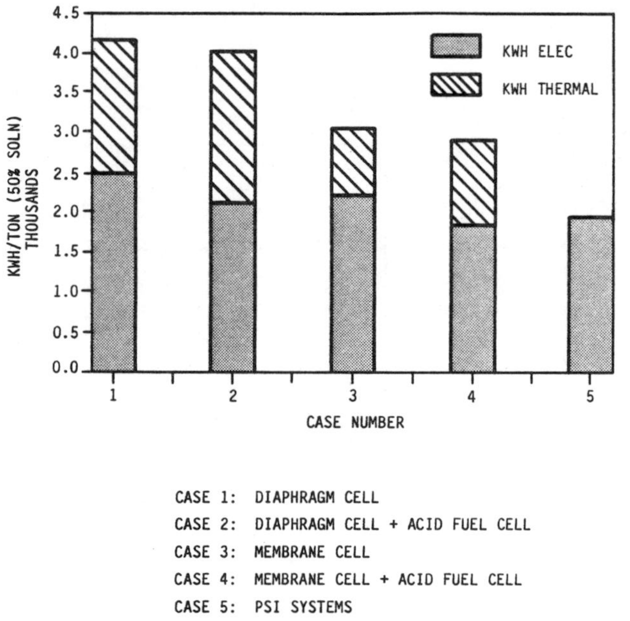

Figure 5. Net Energy Consumption (per ton NaOH).

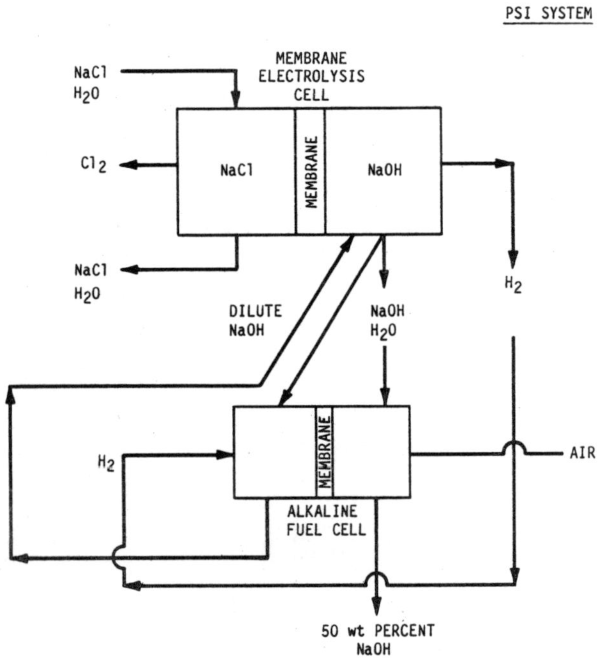

Figure 4. Schematic for Alkaline Fuel Cell Coupled to Membrane-Type Chlorine-Caustic Cell.

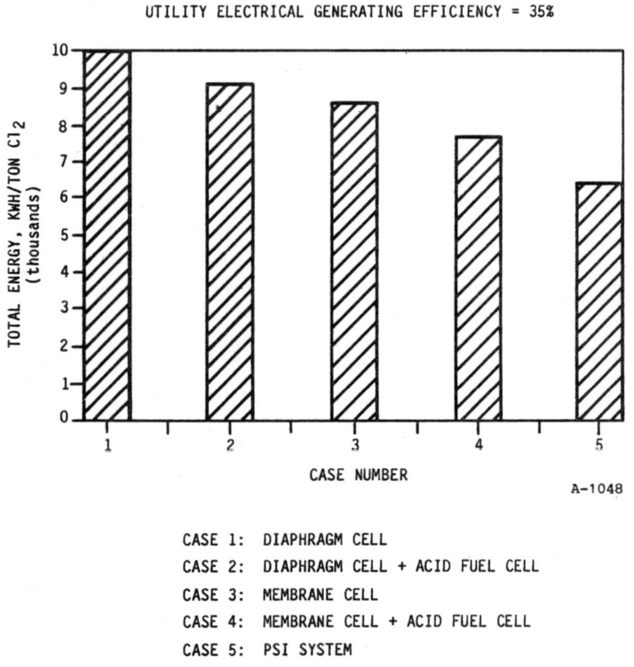

Figure 6. Gross Energy Consumption (per ton Cl_2).

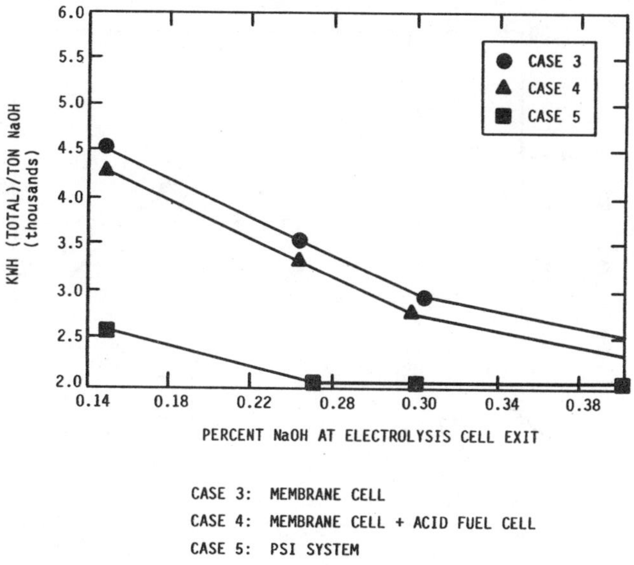

Figure 7. Net Energy Consumption as Function of Caustic Concentration Produced by Membrane Electrolysis Cell.

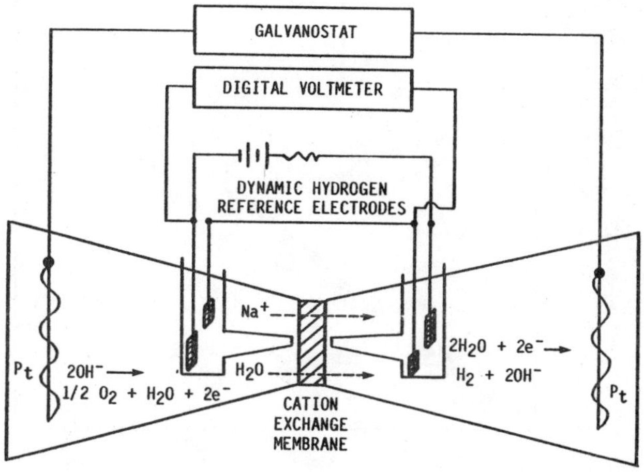

Figure 9. Schematic of Membrane Conductivity Measurement Cell.

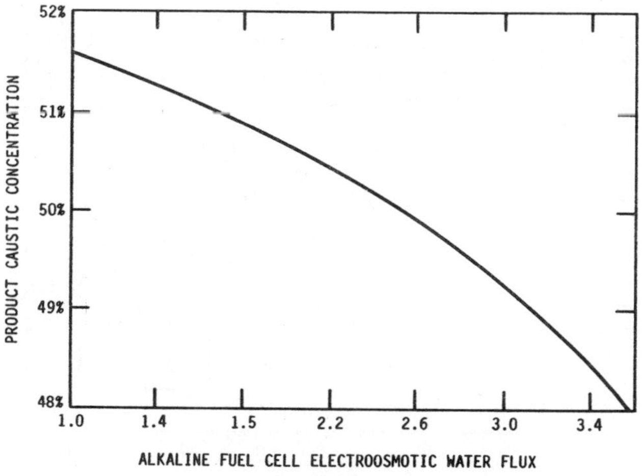

Figure 8. Product Caustic Concentration as a Function of Alkaline Fuel Cell Membrane Electroosmotic Water Flux for PSI System.

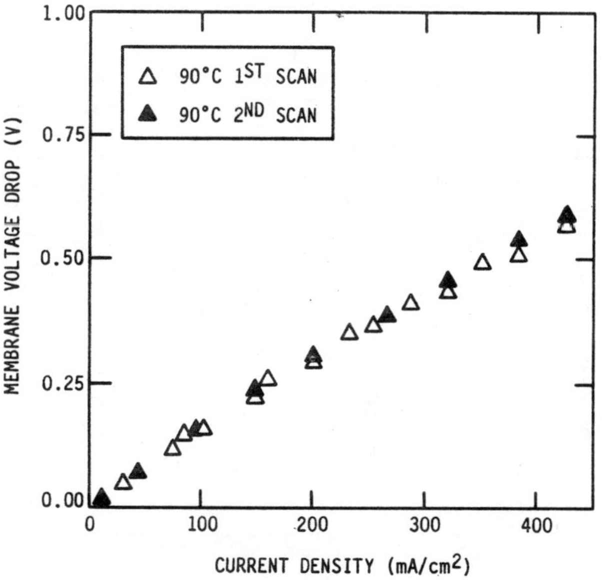

Figure 10. Conductivity Data for N-901(Ex) in NACℓ Anolyte: NaOH Catholyte.

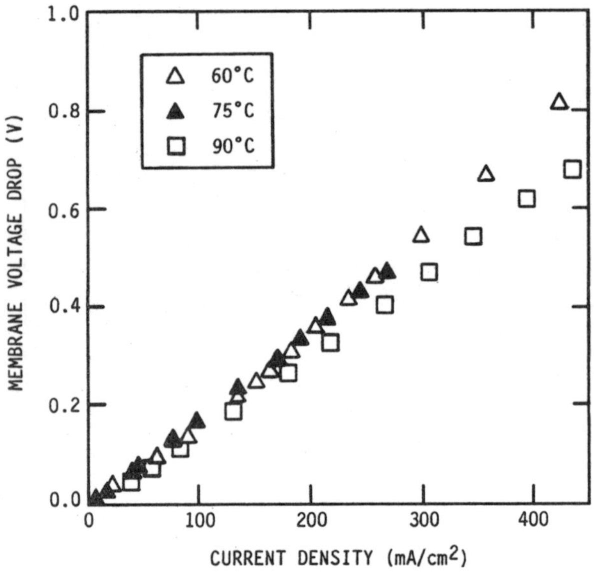

Figure 11. Conductivity Data for N-901(N) in 20 wt Percent NaOH.

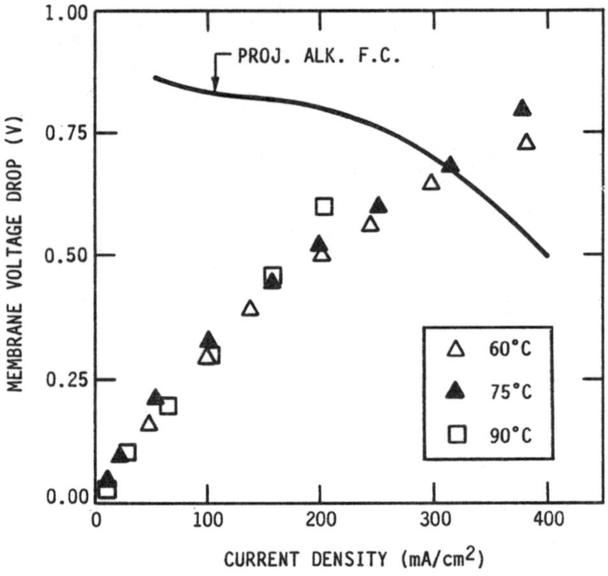

Figure 13. Conductivity Data for Membrane AI in 30 wt Percent NaOH.

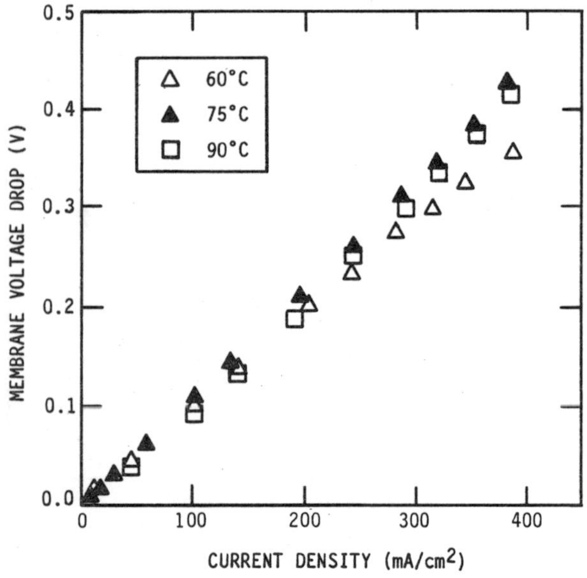

Figure 12. Conductivity Data for N-901(Ex) in 20 wt Percent NaOH.

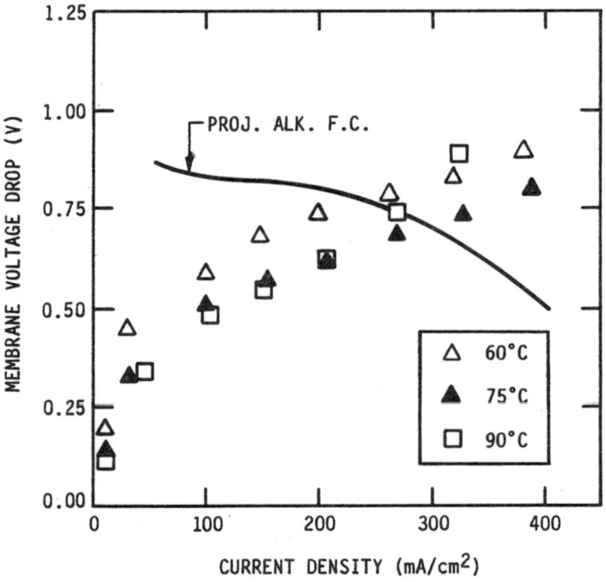

Figure 14. Conductivity Data for Membrane AI in 40:25 wt Percent NaOH (Catholyte:Anolyte).

QUANTITATIVE MODELS OF AFFINITY CHROMATOGRAPHY

Ben J. McCoy ■ Department of Chemical Engineering, University of California, Davis, CA 95616

Among separations, the biospecific adsorption of affinity techniques sets a standard for selectivity. In this paper, mass transfer and affinity adsorption are included in a quantitative description of affinity chromatography. Limiting special cases are examined: mass-transfer controlled processes, linear and nonlinear adsorption equilibrium, and irreversible equilibrium. While the most comprehensive model requires numerical solution, the useful special cases are readily solved.

Large numbers of chemical engineers may be involved directly with biotechnology in the not too distant future. Innovations and progress in genetic engineering methods ensure opportunities for a variety of new products and processes. Vital to bioprocessing, the separation and purification of high-value bioproducts will be a substantial part of the investment in biomanufacturing.

A number of recent publications review chromatographic separations of biochemical components. Regnier (1) discussed liquid chromatography of biopolymers in general. Parikh and Cuatrecasas (2) reported on the several uses of affinity chromatography, particularly for separating proteins. Chaiken et al. (3), in their collection of papers on affinity chromatography, showed that the basic principle of this separation technique is biological recognition. A discussion of progress in affinity separations utilizing immobilized monoclonal antibodies was given by Chase (4).

The essence of affinity adsorption is the strong and selective attachment of a macromolecule to a ligand that is covalently bonded to a solid support. The molecule-ligand pairs can be, for example, enzyme-inhibitor, antigen-antibody, or hormone-receptor pairs, The cost of the liquid-particle materials may be a significant economic factor (5). An adsorbed molecule can be eluted with a solution that weakens the specific macromolecule-ligand interaction. Typically the desorption is complete and irreversible: when the elution fluid, of different chemical property from the feed solution, is passed over the adsorbate, it completely desorbs. The solute can be eluted as a concentrated pulse, avoiding the dilution that accompanies some other chromatographic processes.

For large-scale chromatographic units it will be important that the solid support be a rigid material that will resist compression and deterioration due to the high pressure gradient along the column (5). Radial compression methods (6) that have been developed for high performance liquid chromatography will likely improve the separation by making the packing more homogeneous, thus reducing channeling.

Success in scale-up and optimization of affinity chromatography will depend on the accurate mathematical modelling of the process. Graves and Wu (7), Wankat (8), and Dunn and Chaiken (9), recognizing that affinity adsorption isotherms are usually nonlinear, developed theories based on mass action equilibrium relations. Scouten (10),

Department of Chemical Engineering, University of California, Davis.

in his monograph on affinity chromatography, discussed these approaches to mathematical modelling. Yang and Tsao (11) surveyed the application to affinity chromatography of methods for describing adsorption chromatography, including constant pattern development. Arnold et al. (12) assumed that mass transfer controlled their experiments for affinity chromatography, and applied traditional constant pattern methods (13) to model their data. McCoy (14) considered an irreversible equilibrium relation in the absence of mass transfer to examine the periodic adsorption and desorption in semicontinuous affinity chromatography.

The purpose of this paper is to provide an overview of the quantitative analysis of affinity chromatography. We demonstrate the mathematical complexity of the governing differential equations for affinity chromatography and their relationship to more conventional adsorption models. The special limiting cases that are discussed may facilitate the first-estimate design of these separations.

GENERAL MODEL

Consider a column packed with spherical porous particles with affinity ligands attached at the pore surfaces. Adsorbate macromolecules flow through the interparticle voids, transfer to the outer surface of particles, diffuse into the particles, and adsorb to the ligands. If we ignore the effect of rate of adsorption, i.e., assume adsorption is relatively very fast, the governing partial differential equations for the interparticle species concentration, $c(z,t)$, for the intraparticle concentration, $c_i(r,t)$, and for the adsorbed species concentration, $c_a(z,t)$, are

$$\alpha\partial c/\partial t + v_0\, \partial c/\partial z = D_z\partial^2 c/\partial z^2 - k_p a\left[c-c_i(R)\right] \quad (1)$$

$$\beta\partial c_i/\partial t = D_i\frac{1}{r^2}\frac{\partial}{\partial r}\left(r^2\frac{\partial c_i}{\partial r}\right) - \beta\partial c_a/\partial t \quad (2)$$

To account for the possibility that all sites can be filled with the adsorbent, we consider the equilibrium relation to be of the Langmuir type,

$$c_a/q_s = Kc_i/(1 + Kc_i) \quad (3)$$

The initial conditions are

$$c(t=0) = 0 \quad (4)$$

$$c_i(t=0) = 0 \quad (5)$$

$$c_a(t=0) = 0 \quad (6)$$

The boundary conditions are

$$c(z=0) = c_o(t) \quad (7)$$

$$c(z \to \infty) = \text{finite} \quad (8)$$

$$c_i(r=0) = \text{finite} \quad (9)$$

$$D_i(\partial c_i/\partial r)_R = k_p\left[c-c_i(R)\right] \quad (10)$$

From the equilibrium relation, Equation (3), one sees that if $Kc_i >> 1$ we have $c_a \simeq q_s$, the rectangular or irreversible isotherm; if $Kc_i << 1$ we have the linear isotherm, $c_a \simeq q_s Kc_i$, which is readily analysed by the moment technique (15,16). Because the equilibrium relation is almost always nonlinear for affinity adsorption, the solution of the governing equations (1)-(10) will necessarily have to be numerical. Usually, the breakthrough curve, i.e. the response to a step input, is required. Chromatography is inherently a transient process, thus the problem requires a dynamic response analysis. Ramachandran and Dudukovic (17) recommend a moving element collocation method for solving steep-gradient systems of this kind. Yoshida et al. (18) included mass transfer in a numerical solution for adsorption with a rectangular isotherm. Ruthven (19) aptly surveyed numerical, as well as analytical, solutions of column adsorption problems.

SPECIAL CASES

Mass transfer controlling. If movement of the large macromolecules in the pores is restricted significantly, intraparticle diffusion limitations can dominate the process. Intraparticle diffusion can be reduced as a resistance if particles are small and/or pores are large. According to boundary condition (10) if $Bi = Rk_p/D_i >> 1$, intraparticle diffusion resistance dominates over fluid-to-particle mass transfer.

Dispersion and fluid-to-particle mass transfer may also affect the separation. The main effect of axial dispersion is only to round off sharp corners (i.e., to smooth shock fronts). Allowing $Pe = vL/D_z$ to be much smaller than unity gives the well-mixed limit, and $Pe >> 1$ is the plug flow limit.

Some affinity separations will apply to biochemical species that are extremely dilute in the feed to the column; the isotherm for this situation may be in the linear range of the isotherm. Razavi et al. (16), using the moment approach, treated the case when the isotherm is linear. They found good agreement with approximate estimates of dispersion, fluid-to-particle mass transfer, and intraparticle diffusion. By superposition methods, the response to an input other than the delta function is an integral over the impulse response for linear systems.

Ruthven (19) discussed how for favorable isotherms the concentration profile under mass transfer control will reach an asymptotic form that is constant in shape (constant-pattern behavior). The solutions tabulated by Ruthven (19) should be useful in design of many kinds of adsorption systems, including affinity chromatography.

Arnold et al. (12) assumed that mass transfer will control the affinity adsorption of bioproducts from dilute solutions. For breakthrough data from a monoclonal antibody-antigen affinity column, they found agreement with approximate models (13) based on diffusion control.

Local equilibrium. Optimal utilization of adsorption columns requires that mass transfer effects be minimized. When all mass transfer resistances are negligible, the particle-volume average of Equation (2) combined with Equation (1) by means of the boundary condition (10) yields

$$\varepsilon\, \partial c/\partial t + v_0 \partial c/\partial z = -\beta(1-\alpha)\partial c_a/\partial t \qquad (11)$$

where $\varepsilon = \alpha + \beta(1-\alpha)$. The method of characteristics serves to provide solutions for the dynamic response (20). For favorable isotherms, e.g. the Langmuir isotherm, the the slope for characteristic lines is

$$v_c = v_0/[\varepsilon + (1-\alpha)df/dc] \qquad (12)$$

in terms of the derivative of the isotherm $c_a = f(c)$. The same approach yields a relationship for the line (i.e. the shock front) on which characteristic lines terminate:

$$dz_s/dt = v_0/[\varepsilon + (1-\alpha)f(c)/c] \qquad (13)$$

Rectangular isotherm. The key attributes of affinity adsorption are specific and strong binding. McCoy (14) focused on these in presenting an idealized model for which the adsorbate was bound irreversibly, and other species were totally unadsorbed. This picture ignores hydrophobic or ionic interactions, which interfere with affinity adsorption. The usual local equilibrium approximation, for which mass transfer is assumed very efficient, allowed a solution by the method of characteristics. The equilibrium isotherm was considered irreversible, i.e.,

$$q = \begin{cases} q_s; & c > 0 \\ 0; & c = 0 \end{cases} \qquad (14)$$

According to this representation, all adsorbate sites are filled to capacity, q_s, as adsorbent molecules come into contact with the sites. The desorption was likewise assumed complete and irreversible when elution fluid passed over the filled sites. Site-limited isotherms, such as the Langmuir isotherm, are asymptotic to the saturation concentration, q_s, ensuring that the rectangular isotherm, Equation (14), will be a satisfactory approximation for relatively large concentrations of strongly adsorbed species.

This model may be considered a special case of the local equilibrium approach to nonlinear adsorption. From Equation (13) we obtain for the irreversible isotherm, Equation (14), the expression for the slope, which, for $(1-\alpha)q_S/\varepsilon c \gg 1$, i.e. when adsorption capacity is much larger than volume capacity, becomes

$$dz_s/dt = v_o\, c/q_s(1-\alpha) \qquad (15)$$

This Equation (15) was the principal result upon which McCoy (14) based an analysis of semicontinuous operation of affinity chromatography. An attractive feature of the model is that it involves only one parameter, q_S, the adsorption capacity of the column. The elution likewise is imagined to be irreversible, and thus eluted-solute band spreading, affected only by mass transfer, can be accounted for by a temporal moment description. This simplified and approximate model is useful (14): (a) for timing valves for solution and feed steps in semicontinuous operation of multiple columns and, (b) for adjusting column flow rates to avoid interference between adsorption fronts and elution peaks in a simulated moving bed.

CONCLUSION

Affinity chromatography, though similar to standard column adsorption processes, is based on specific biological recognition and selective binding described by a nonlinear isotherm. In general, mass transfer effects, i.e. convective dispersion, fluid-to-particle mass transfer, and intraparticle diffusion, will have a significant influence on the dynamics of the process. Under certain conditions, simple special cases apply that have been solved. The objective of this article has been to examine these cases, and their relationship to a comprehensive model.

NOMENCLATURE

- a — $=3(1-\alpha)/R$, mass transfer area (per column volume) of spherical particles, 1/m
- c — concentration in column voids, mol/m^3
- c_a — adsorbed concentration (per pore volume), mol/m^3
- c_i — intraparticle concentration (per pore volume), mol/m^3
- $c_o(t)$ — inlet concentration, mol/m^3
- D_z — effective axial dispersion coefficient, m^2/s
- D_i — effective intraparticle diffusion coefficient, m^2/s
- K — equilibrium parameter in Langmuir isotherm, $m^3/mole$
- k_p — fluid-to-particle mass transfer coefficient, m/s
- q_s — saturation adsorbate concentration (concentration of adsorption sites per pore volume), mol/m^3
- R — radius of particles, m
- r — radial coordinate in particles, m
- t — time, s
- v_o — superficial velocity (volume flow rate/column cross-sectional area), m/s
- z — column length coordinate, m
- z_s — position of adsorption front, m
- α — column void fraction (1 - volume of particles/volume of column)
- β — particle porosity (volume of pores/volume of particle)
- ε — $=\alpha + \beta(1-\alpha)$, total void volume/column volume

LITERATURE CITED

1. Regnier, F.E., Science, 222, 245 (1983).
2. Parikh, I. and P. Cuatrecases, Chem. Engr. News, 17, August 26, 1985.
3. Chaiken, I.M., M. Wilchek, I. Parikh, Eds., Affinity Chromatography and Biological Recognition, (Academic Press, 1983).

4. Chase, H.A., Chem. Eng. Sci., 39, 1099 (1984).

5. Hamman, J.P., G.J. Calton, in Purification of Fermantation Products: Applications to Large-Scale Processes, ACE Symposium Series 271, 105, (1985).

6. Heckendorf, A.H., E. Ashare, C. Rausch, in Purification of Fermentation Products: Applications to Large-Scale Processes, ACS Symposium Series 271, 91, 1985.

7. Graves, D.J. and Y.-T. Wu, in Methods Enzymol., 34, 140 (1974).

8. Wankat, P.C., Anal. Chem. 46, 1400 (1974).

9. Dunn, B.M. and I. M. Chaiken, Biochemistry, 14, 2343 (1975).

10. Scouten, W.H., Affinity Chromatograph -- Bioselective Adsorption on Inert Matrices (Wiley, New York, 1981), Chapter 4.

11. Yang, C.-M., G.T. Tsao, in Chromatography (A.Fiechter, ed.) Advs. in Biochem. Engr., vol. 25, page 1, Springer-Verlag, Berlin (1982).

12. Arnold, F.H., J. J. Chalmers, M. S. Saunders, M. S. Croughan, H. W. Blanch, and C. R. Wilke, in Purification of Fermentation Products: Applications to Large-Scale Processes, ACS Symposium Series 271, 113 (1985).

13. Vermeulen, T., G. Klein, N.K. Hiester, in Chemical Engineer's Handbook, McGraw-Hill, 5th Ed., McGraw-Hill, 1973. Chapter 16.

14. McCoy, B.J., Biotech. Bioengr. 27, 1477 (1985).

15. Carbonell, R.G., B.J. McCoy, Chem. Engr. J. 9, 115 (1975).

16. Razavi, M.-S., B.J. McCoy, R.G.Carbonell, Chem. Eng. J., 16, 211 (1978).

17. Ramachandran, P.A., M.P. Dudukovic, Chem. Eng. Sci. 39, 1321 (1984).

18. Yoshida, H., T. Kataoka, D.M. Ruthven, Chem. Eng. Sci. 39, 1489 (1984).

19. Ruthven, D.M., Principles of Adsorption and Adsorption Processes, Wiley, 1984. Chapter 8.

20. Sherwood, T.K., R.L. Pigford, C.R. Wilke, Mass Transfer, McGraw-Hill, 1975, p. 556.

HIGH PERFORMANCE LIQUID CHROMATOGRAPHY AS A PROCESS TOOL

James L. Dwyer ■ Systems Division, Millipore Corporation, 80 Ashby Road, Bedford, MA 01730

The analytical technique of High Performance Liquid Chromatography is being adapted to chemical isolation process needs. The advantages of process HPLC include high resolution of similar species, a broad range of applicability, and short process times. Techniques of scale-up from a laboratory separation to process quantities are illustrated by the purification of a beta-lactam antibiotic. Based on laboratory experiments, it is not only possible to predict larger scale process performance, but also to model the economics of this unit operation.

INTRODUCTION

High Performance Liquid Chromatography (HPLC) has become one of the primary analytical chemical methodologies. The speed, sensitivity and resolution of separation, coupled with its broad range of applicability provide the basis for widespread analytical use of HPLC.

HPLC has proved equally useful as a preparative tool for isolating gram quantities of a synthetic or biologic compound. This paper deals with the use of HPLC as a process unit operation for the isolation of various chemical species. The example data presented are taken from the purification of various fermentation and synthetic antibiotics, but the technique of HPLC purification is being used for a broad range of biological and therapeutic products.

PRINCIPALS

HPLC, as other chromatographic procedures, provides a separation mechanism based on the interaction of a solute mixture to be separated with the packed bed of the chromatographic column (stationary phase) and the carrier solution (mobile phase). The use of very small diameter rigid packing materials in HPLC provides a high frequency of solute/stationary/mobile phase interactions, thus permitting a rapid, high plate count separation, albeit at high operating pressures.

Various packing chemistries provide a variety of separation modes that can take advantage of differential characteristics of product molecule polarity, charge, hydrophile/lipophile balance, or molecular size. As such, the method has a broad scope in separation/isolation of natural and synthetic compounds. Increasingly, HPLC has become a primary tool in the biological area, both for analysis and isolation.

The use of HPLC for preparative/process separations is governed by different objectives than the analytical applications. In the latter case, desirable method attributes are a maximum resolution of species, speed, linearity and maximum sensitivity of detection. By contrast, the process chromatographer aims to achieve bulk purification of a compound to some specified purity level and at a minimum cost/kilogram. In a chromatographic separation, load (hence) favorable operating economics) is inevitably achieved at the expense of decreased resolution. Despite the perceived loss of separation resolution, one can judiciously cut fractions of acceptable product purity at loads orders of magnitude higher than used in analytical HPLC. As seen below, the development of a process separation

Millipore Corporation, Bedford, Massachusetts, James L. Dwyer consists of:

1. Development of a satisfactory analytical separation on conventional, laboratory HPLC equipment.

2. Replacement of the analytical packing with the material to be used in the process-scale separation.

3. Maximizing product loading on this column.

4. Direct scale-up to process size columns.

EXPERIMENTAL: CEPHALOSPORIN-C PURIFICATION

The following data illustrated the approach to scale-up. The structure of Cephalosporin-C, a metabolic product from the fermentation Cephalosporin-accremonium, is shown in Figure 1. It is the precursor to the family of cephalosporin antibiotics. In our laboratory, we have by fermentation produced Cephalosporin-C in quantity. We have then isolated the product utilizing only membrane and chromatographic processes.

The effectiveness of microfiltration for separating cells from soluble compounds in fermentation broth is demonstrated by the two vials shown in Figure 2. The left vial contains the whole fermentation broth as it is discharged from the fermenter. The ferment mass was processed directly through a tangential flow microporous membrane unit to yield the clear broth (right vial). Although all suspended solids are removed, the broth is still rich in soluble nutrients and metabolic products of fermentation. An analytical chromatogram of this material is shown in Figure 3. The separation is done on a C_{18} analytical microBondaPak^R reverse phase column operated in an isocratic mode. The mobile phase in the separation is simply buffered water. Control of pH is required to prevent degradation of the labile Cephalosporin-C.

In this separation, the product is eluting in approximately 12 column volumes. Preferably, the elution of product should occur in three to six column volumes. As load is increased retention volumes tend to decrease, and one wishes to preserve a maximum separation from other species. Large elution volumes are to be avoided because of longer separation times and consumption of larger mobile phase volumes per cycle.

Figure 4 shows the separation performed in analytical sized columns, but packed with somewhat larger particle size reverse phase packings that are used in process. The larger particle size and, more particularly, the higher column loading, results in some loss of resolution, but the Cephalosporin-C fraction is still adequately resolved. In Figure 5, the separation was performed at a much higher loading (120 mg broth solids/gm of packing). Chromatography is markedly distorted and there is only partial resolution of the product molecule.

The same separation was performed on a Waters Prep 500^R chromatograph (Figure 6). The sample, containing 46 grams of solids in 1100 ml of clear broth, was loaded onto a column with bed dimensions of 5.7 cm diameter by 30 cm length. The fractions indicated in Figure 6 were collected and assayed. Results are summarized in Table 1.

Separation was then scaled up to a Kiloprep^{tm} 1000 process HPLC unit. The HPLC columns of this unit have bed dimensions of 20 cm diameter by 60 cm length and contain approximately 10 kilograms of C_{18} reverse phase material. The injection load contained 1150 grams of crude solids in 32 liters of broth. A KILOPREP 1000 is shown in Figure 7. in Figure 7.

In Figures 8, 9, and 10, two chromatograms are shown. The right hand cureves are the detector output of the Kiloprep system. Indicated fraction cuts are shown in dark shading. The left side curves are analytical chromatograms of the fraction cut, also performed on a reverse phase column.

Note first the similarity of the Kiloprep, Prep 500, and high-load small column chromatographs. HPLC is a process amenable to direct scale-up. One can predict large-scale performance on the basis of methods developed on laboratory HPLC instruments.

The analytical chromatogram in Figure 8 demonstrates that fraction #3 of the process run, although only partially resolved from other components, is primarily the Cephalosporin-C molecule. In fraction #4 (Figure 9) contaminating species are at trace level, and fraction #5 (Figure 10) is homogeneous as indicated by the analytical HPLC assay. The mass quantities and recoveries of the run are summarized in Table 2. There was no mass loss of product. Independent Kirby-Bauer assays of the collected fractions showed complete biological activity of the Cephalosporin-C.

In developing process separations of this type, particularly on novel compounds, it is important for the process engineer to be able to confirm the mass balance and activity recovery of the process. This can be a challenge. Parameters such as the chemical structure of the product, or its molecular weight, may be uncertain. Typically, the product is found in a complex mixture, and pure reference standards do not exist.

In this situation, analytical chromatography techniques can be used to develop a simple but effective approach to mass balance on incompletely characterized materials. As a first step, a milligram quantity of the product is purified by appropriate methods for use as a reference standard. This "standard" should satisfy the criterion of the purest material available by presently known technology.

Using this material, an analytical HPLC separation is developed and a quantitative calibration curve of peak area vs mass of standard is established.

To perform a mass balance on the process HPLC separation, samples of crude feed materials plus all fraction cuts are gravimetricly assayed. (If the HPLC mobile phase contains non-volatiles, these components must be factored into the gravimetric assays.) The total solids from the gravimetric assays are re-solubilized for HPLC assay. Known aliquots are injected and peak areas determined. The purity of each fraction can be determined by the following formula:

$$\%\ \text{Purity} = \frac{(\text{AREA}_{\text{sample}}/\text{WT}_{\text{sample}})}{(\text{AREA}_{\text{standard}}/\text{WT}_{\text{standard}})}$$

Using this technique, it is practical to establish a simple and rapid mass balance calibrated against the standard material. During the development phase of a product, when compound identity is only partially known, such an assay method is extremely helpful.

DISCUSSION

The Cephalosporin data cited above is an isolation of product from a very complex mixture. From the perspective of the chromatographer, only moderate resolution is required to achieve satisfactory purity. What are the consequences of process HPLC applied at higher levels of chromatographic resolution? The chromatographer defines column resolution in terms of the height of an equivalent theoretical plate (HEPT). While this is a useful concept in analytical chromatography, it has limited utility for process HPLC.

In Figure 11, the measured number of plates of a normal phase HPLC column as a function of the mass loading of the sample is shown. Both axes have logarithmic scales. Two features should be noted:

1. Plate count drops markedly as a function of load.

2. Plate count is a strong function of particle size at low loads typical of the analytical chromatography regime.

Process chromatography separations are most often loaded in the 1 to 100 milligram/gram packing range. Note that the small particle resolution decays most rapidly as a function of load; the net result being little performance difference in the region of process interest. Similar data on reverse phase packing are shown in Figure 12.

Two conclusions can be drawn. First, there is little resolution benefit obtained by using analytical particle sizes for process load chromatography. The much higher costs of small diameter packings and their greater hydraulic resistance to mobile phase flow argue against their use.

Secondly, when very high resolution <u>is</u> required in a process separation, it can be obtained by using smaller particle size packings; but only by severely reducing column loadings. For most process separations, 20 to 100 micron packings are most suitable. At larger particle sizes, inter-pore diffusion predominates and satisfactory resolution can only be obtained at very low flow rates.

What about the product dilution aspects of a chromatographic process? Chromatographic techniques are commonly perceived as highly dilutive. This is not necessarily the case. In the Cephalosporin-C experiment, the product molecule was loaded on-column at a concentration of approximately 4 milligrams/milliliter. The data presented in Table 2 show that the pooled fractions were recovered at a concentration of approximately 8 milligrams/milliliter. In a second example, Figure 13 shows analytical HPLC curves of a preparative HPLC concentration of the anitbiotic Erythromycin. This compound was loaded on-column and then eluted with a step gradient, thus concentrating the material. This is an important technique used for the production purification of therapeutic peptides. A purified peptide fraction is loaded onto a reverse phase column in a weak solvent mobile phase. The peptide is adsorbed strongly to the stationary phase, and several column volumes may be loaded before breakthrough. Salts can be washed off during this step. Switching to an ethanol-rich mobile phase causes spontaneous elution of the peptide, resulting in both a desalting and liquid concentrating of the product.

In these examples, product concentration was performed on-column. There are situations where product dilution can occur, especially when the product has limited solubility in the mobile phase. In this situation, the solute partitions strongly to the packing, typically eluting only after many columns of mobile phase passage and with considerable peak tailing. The preparative separation of Cephalexin (Figure 14) demonstrates this situation.

Many of these compounds are heat labile, and evaporative concentration results in significant activity loss. Recently, it has proved practical to concentrate various low molecular weight solutes, such as peptides and antibiotics, by using reverse osmosis membrane systems.

EQUIPMENT DESIGN

The process chromatography equipment used in these experiments are configured basically similar to an analytical chromatograph. Differences of note include injection techniques. Because of the relatively large volumes of crude loaded, injector loops are seldom required. Crude can be loaded directly through the mobile phase pump. At the column sized used, equipment must satisfy pressure vessel codes. Typically, the electrical system (including detectors) must satisfy explosion-proof codes. Both requirements result in equipment which, while compact, is quite heavy.

Mobile phase flow rates are often several liter/minute. In a production operation, adequate tankage and blending equipment nust be provided, both for collected fractions and for mobile phase feed. Mobile phase cost frequently will dictate solvent recovery operations, particularly with normal phase separations.

The next section describes process economics. As would be expected, the HPLC columns are a significant contributor to cost. Satisfactory economics dictates that columns must last through a hundred or more injections before replacement. It is critical to prevent any bed voiding in a large column. The system described utilizes a technique called radial compression to maintain bed integrity and prevent voiding or channeling of the packed bed. The chromatographic bed consists of a pre-packaged cartridge which is inserted into a high pressure housing. A cylinderical elastomeric bladder surrounds the cartridge. A hydraulic working fluid, pumped to the outside of this bladder applies a uniform pressure to the cartridge cylinder wall, effectively compressing the bed and eliminating any voiding of the bed caused by particle shifts or fracture. In the production of pharmaceuticals, it is not acceptable practice to use a bed for multiple products, because of concerns about cross contamination. Cartridges permit multiple use of process equipment in this situation.

ECONOMICS

The data demonstrate that it is practical to predict large-scale operating parameters based on bench-scale experiments. We have developed a spreadsheet-based economic model that provides estimated operating economics. The first input section of the model accepts the experimental data (crude load, flow rates, separation times) plus plant costs (labor rates, operating hours, amoritization, etc.). The second input section of the program is concerned with solvent costs. It accepts multiple component cost/composition data for up to four separate elution steps. The algorithms used also compute integrated costs of solvent gradients.

An example of the model output is presented in Table 13. Many of the input variables are restated to the process scale. The direct operating cost components are calculated and displayed. This example displays a moderate load on a single chromatographic cartridge. Under these conditions, the system should have a capacity of several thousand kilograms/year and an operating cost of $170/kg.

The "what-if" capabilities of such a model are quite useful. Initial laboratory work may have yielded a satisfactory separation, but at unacceptable projected costs. One can rapidly explore various options (solvent recovery, load, mobile phase flow rate) to suggest those separation modifications that may achieve the goal. This can apparently shorten the bench work required.

CONCLUSION

HPLC provides a novel unit operation of great versatility that can be used for the isolation and purification of many products. The examples cited in this paper have centered around reverse phase chromatography, but the use of other separation modes, such as ion exchange, normal phase, and size exclusion, are equally practical. Since economic product purification is the driving criterion, loads/mass of packing are much higher than are employed in analytical chromatography.

HPLC shows linear scaling behavior. Consequently, it is practical to use analytical size columns for preliminary experiments. With such coloumns, it is possible to predict process performance and economics of a separation.

Process scale HPLC is currently being used for two functions: one is the production purification of various high value therapeutic products including peptides, proteins, antibiotics, steroids, and various optical isomers. The other application is as a general pilot plant tool. The broad utility of HPLC separations can eliminate or at least reduce the need to develop complex multi-step fractionation sequences typical of the traditional product isolation methods.

Table 1
CEPHALOSPORIN-C PREP RUN:
SUMMARY OF RESULTS

Injection Volume: 1100 mls
Flow Rate: 140 ml per minute

Fraction #	Vol (ml)	Ceph-C Conc. (mg/ml)	Ceph-C Recovered (grams)	Purity
1	110	2.5	0.275	46
2	65	7.8	0.507	87
3	95	11.0	1.045	92
4	115	9.5	1.093	96
5	540	2.8	1.512	94
Total	925		4.432	

% Purity (Fractions 2 to 5): 93%
% Recovery: 100%

Table 2
CEPHALOSPORIN-C KILOPREP RUN:
SUMMARY OF RESULTS

Injection Volume: 33 liters
Flow Rate: 1.8 liters per minute
Amount Loaded: 150 grams Ceph-C in 1200 grams total mass

Fraction #	Vol (ml)	Ceph-C Conc. (mg/ml)	Ceph-C Recovered (grams)
1	940	2.98	2.80
2	1660	8.94	14.84
3	1870	14.23	26.61
4	2500	16.27	40.67
5	12079	5.70	68.80
Total	19040		153.72

% Purity (Fractions 2 to 5): 93%
% Recovery: 100%

Table 3
KILOPREP OPERATING CONDITIONS

Column/Load Cycle	450	grams/cycle
No. Kiloprep Cartridges	1	
Feed Purity	55	%
Recovery	98.5	%
Mobile Phase Flow Rate	3	liters/minute
Cycle Time	35	minutes/cycle
Cartridge Life	600	cycles
Operation Time	3720	hr/yr
Mobile Phase Cost	0.12	$/liter
Labor Rate	18	$/hr
System Useful Life	5	years
Value Crude	120	$/kg
Value Pure Product	550	$/kg
Rejuvenation Solvent Cost	0.25	$/liter
Rejuvenation Volume	75	liters
Cycles/Rejuvenation	10	

SYSTEM OUTPUT

System Feed Rate	0.771	kg/hr crude
System Product Rate	0.418	kg/hr
System Product Rate	1555	kg/hr

SYSTEM ECONOMICS

Value, Purified Product	$550.00	/kg product
Value of Production:		
Labor	$43.07	/kg product
Kiloprep Cartridge	$51.27	/kg product
Solvent Cost	$59.38	/kg product
Depreciation (Line)	$16.08	/kg product
Operating Costs	$169.80	
Value of Crude	$221.50	/kg product
Total Value of Prod.	$391.30	/kg product
Contribution	$158.70	/kg product
Contribution (Annual)	$246,718.10	

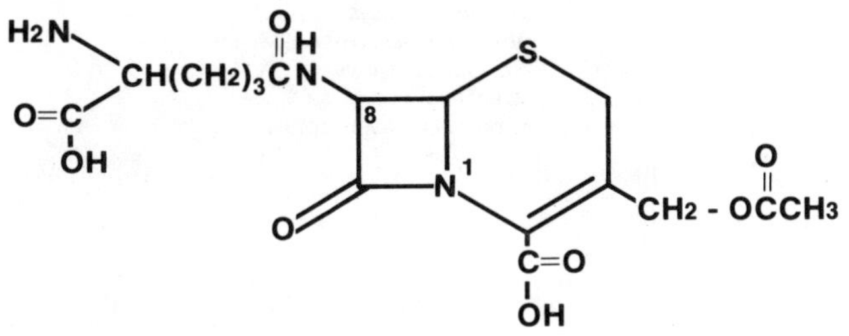

Figure 1. Cephalosporin-C

Figure 2.

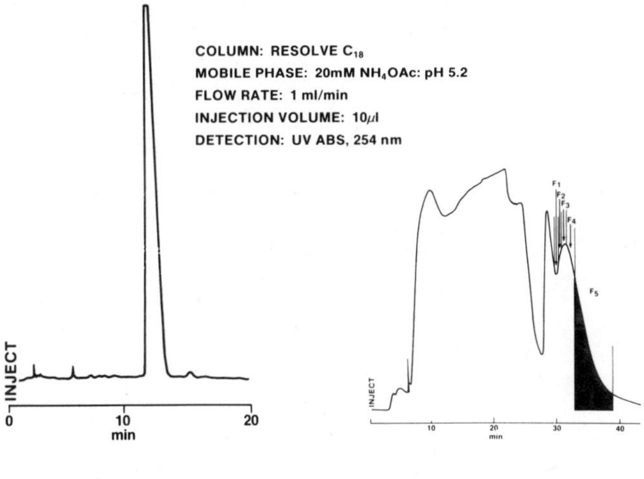

Figure 3. Fraction #5 from kiloprep run of cephalosporin-C.

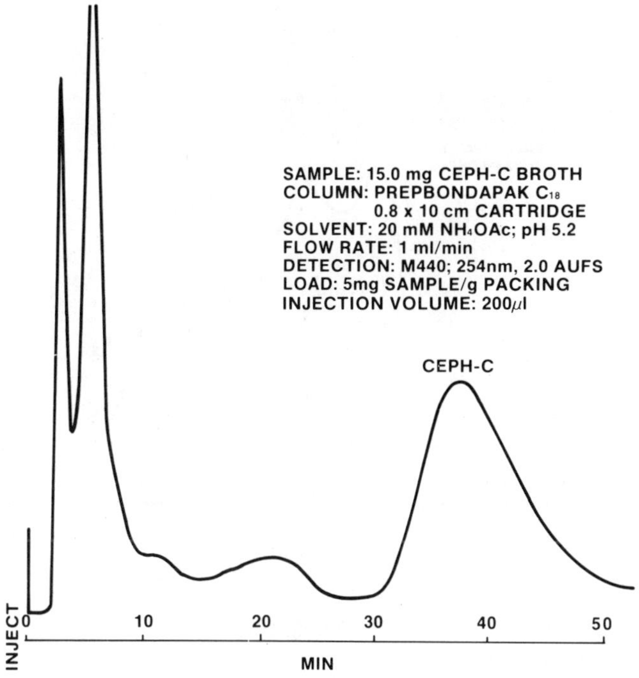

Figure 4. Prep simulation of cephalosporin-C.

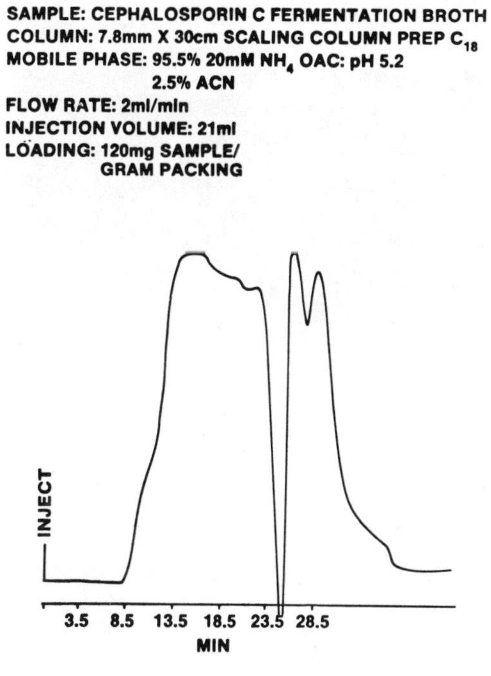

Figure 5. Loading study of cephalosporin-C.

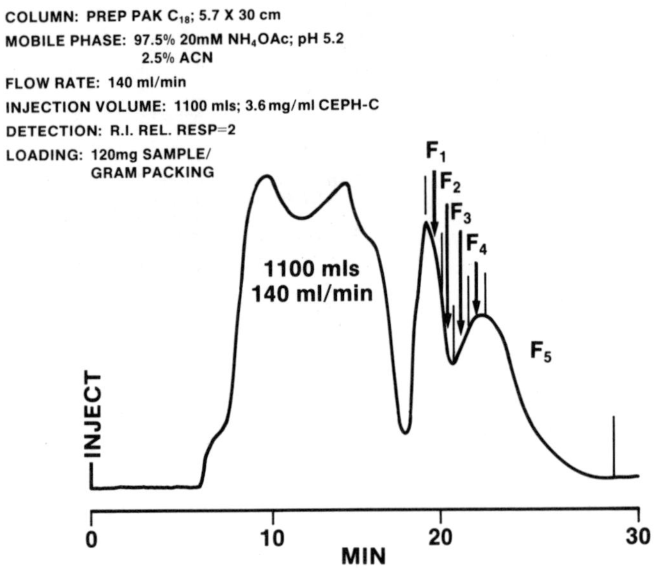

Figure 6. Prep run of cephalosporin-C.

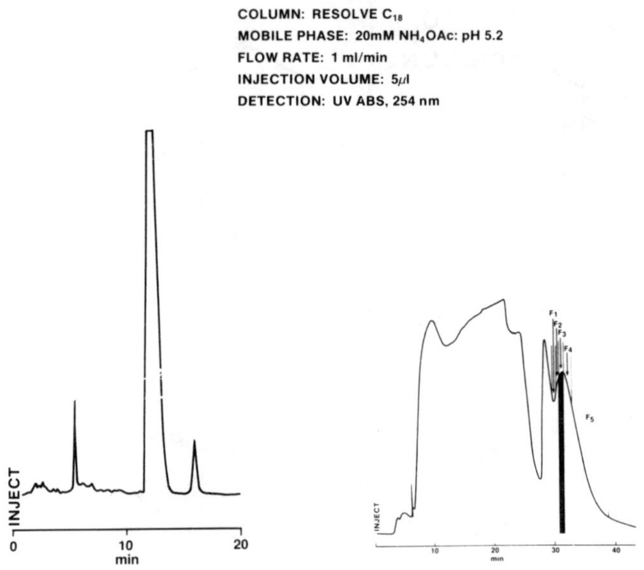

Figure 8. Fraction #3 from kiloprep run of cephalosporin-C.

Figure 7.

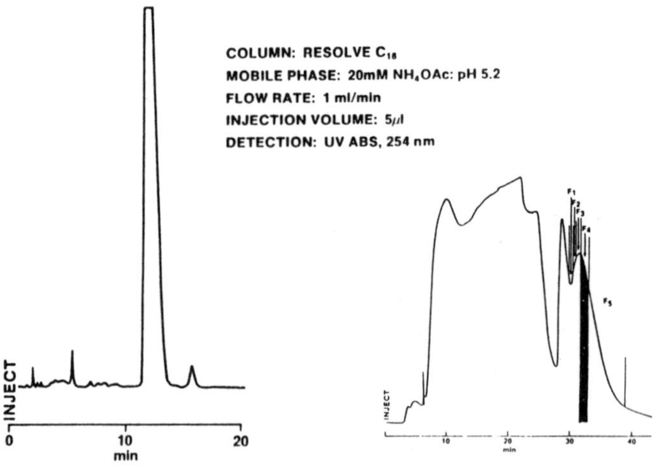

Figure 9. Fraction #4 from kiloprep run of cephalosporin-C.

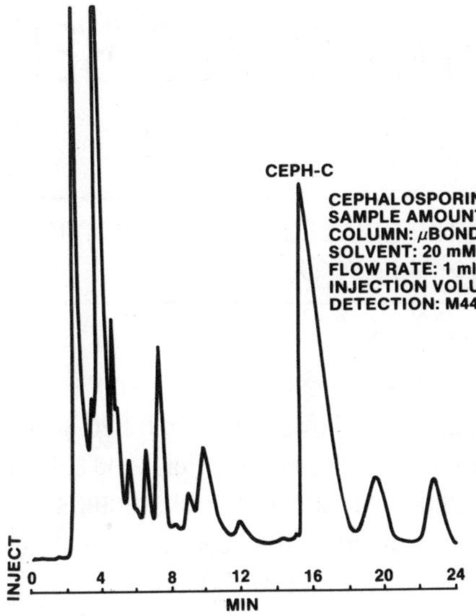

Figure 10. Ceph-C.

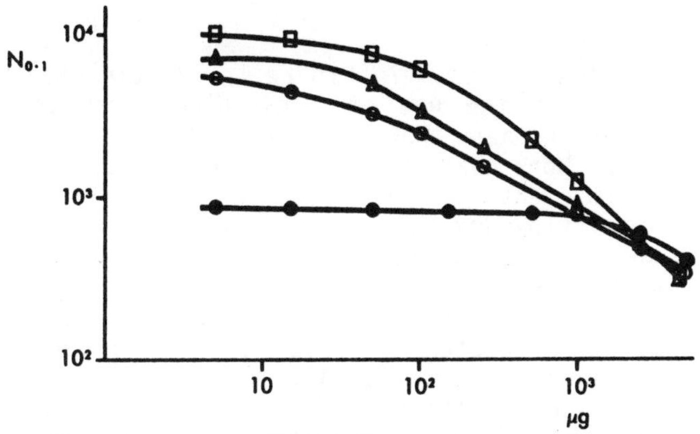

Figure 11. Column efficiency as a function of load. N calculated with σ, measurement at 0.1 of peak height (No. 1) ■, Δ, ○, 5-8 μm silica; ●, 20-25 μm silica.

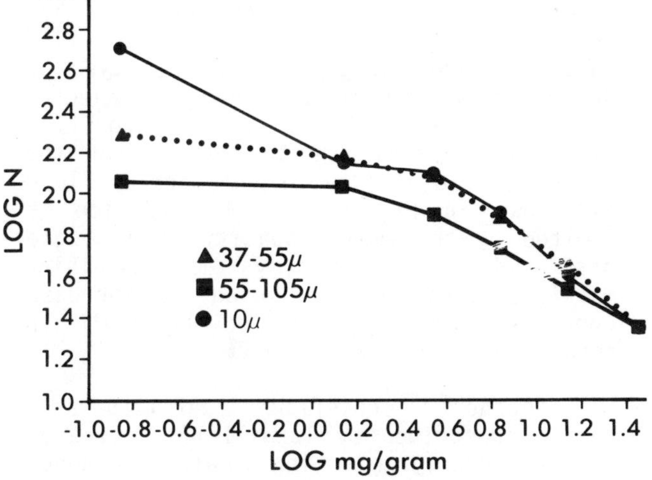

Figure 12. Load vs. column efficiency.

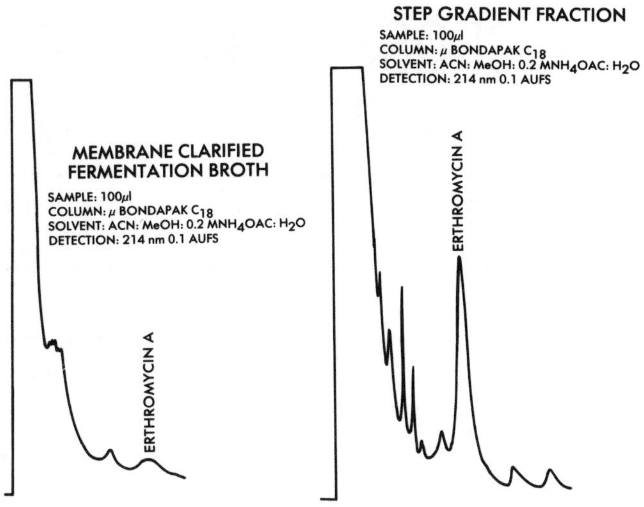

Figure 13. Concentration of erythromycin A.

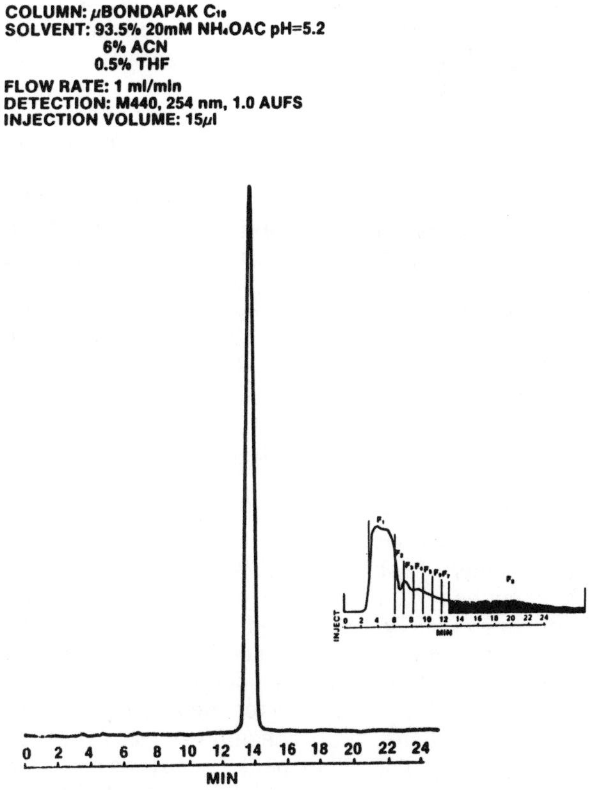

Figure 14. Purification of cephalexin fraction #8.

CHROMATOGRAPHIC SEPARATIONS IN A CROSSFLOW MAGNETICALLY STABILIZED BED

Jeffrey H. Siegell, Gerald D. Dupre and J. Carl Pirkle, Jr. ■ Exxon Research and Engineering Company
Florham Park, NJ 07932

Several devices for the continuous chromatographic separation of multicomponent feeds are reviewed and a new technique is described. This new technique consists of a magnetically stabilized fluidized bed (MSB) in which bed solids move crossflow to the fluidizing fluid. Adsorbable components of the feed move in two directions; with the fluidizing carrier fluid to the bed surface, and with the solids across the bed to different positions depending on their sorption characteristics. The feature that makes this MSB chromatograph more attractive than the previously developed devices is its ability to continuously separate feed mixtures into multiple products in a single vessel with no moving mechanical parts.

Chromatographic separation makes use of the difference in sorption equilibria of components in a feed when contacted with a suitable stationary phase. This stationary phase is usually a porous or granular solid which in some cases may be coated with a liquid substrate. The conventional chromatographic process consists of introducing a pulse feed into a packed column of adsorbent solid with a continuous flow of carrier fluid. The components of the feed are stratified in the column by selective adsorption and desorption on the solid, causing them to exit from the column with the carrier fluid at different times. The components that were the least adsorbed will exit first and the components that were most strongly adsorbed will exit last. This process, however, is only suitable for batch operation.

CONTINUOUS CHROMATOGRAPHY DEVICES

There have been numerous attempts to operate chromatographic separators in a continuous fashion(1-3). In particular, the concept of crossflow chromatography has been previously attempted in at least three different operational modes. The initial disclosure of crossflow operations was made by Martin(4). He described an annular space between two concentric cylinders, packed with adsorbent solids, for continuous separations. Carrier fluid was fed axially while the bed was slowly rotated as shown in Figure 1. Martin also suggested the use of a circular array of tubes rather than the continuous packed bed. Work has continued in recent years on improvements to this mechanically rotated axial flow device(5-12). A disadvantage of this type of device is that it is mechanically complicated due to the difficulty of having large systems of tubes or beds rotated in order to obtain high capacities. In addition, high separation efficiencies cannot be obtained because, as in other packed chromatographic systems, particle size cannot be substantially reduced due to pressure drop limitations.

Using two rotating discs placed closely together and coated with a thin layer of solvent, Sussman and co-workers(13-16) operated a radial flow continuous chromatograph. In this type of device, carrier fluid and feed are injected at the center of the discs and crossflow separation occurs due to rotation of the discs and radial flow of the carrier fluid as shown in Figure 2. This device is limited by the small capacity of a single disc system. Larger systems containing multiple discs can be used to increase capacity, but they may be difficult to rotate mechanically.

Another crossflow chromatography device, described by Tuthill(17), operates by continuously injecting a multicomponent feed into the corner of a rectangular slab which is packed with adsorbent, and then changing the direction of carrier flow at right angles. This results in the products moving to different positions at the end of the slab as shown in Figure 3. The inability to use very fine particles, due to excessive pressure drop, limits the separation efficiency of this type of device.

MAGNETICALLY STABILIZED BEDS

The quiescent, fluid like, magnetically stabilized bed (MSB) is free of bubbles or pulsations which are normally present in fluidized beds, and results when a magnetic field is applied to a fluidized bed of magnetizable solids(18-23). A comparison of conventional fluidized beds and MSBs is shown in Figure 4. The magnetic stabilization produces a non-bubbling fluid state with operating velocities ranging between the normal minimum fluidization velocity in the absence of the applied magnetic field, and an upper limit given by the superficial fluid velocity required to cause bubbling to occur in the bed. The stably fluidized solids resemble a liquid and are easily transported.

The fluid/solid/magnetic system exists in one of three regimes as shown in Figure 5. Below the minimum fluidization velocity the pressure drop across the bed is less than the bed weight per unit area and the bed is packed. In the stabilized regime the bed is free of mixing motions. In the regime denoted bubbling magnetized the bed bubbles even though magnetized. The transition between these regimes is sharp and reproducible. The MSB regime existing at velocities between the minimum fluidization velocity and the transition velocity to bubbling is useful in performing chromatographic separations. Applications of countercurrent MSBs for the affinity chromatography separation of binary mixtures has recently been described(24-26).

When increasing the gas velocity in an MSB, as shown in Figure 6, the breakpoint of the pressure drop curve corresponds to a minimum fluidization velocity. Thereafter, as in unmagnetized bubbling fluidized beds, the pressure drop through the MSB equals the weight of bed solids per unit bed area

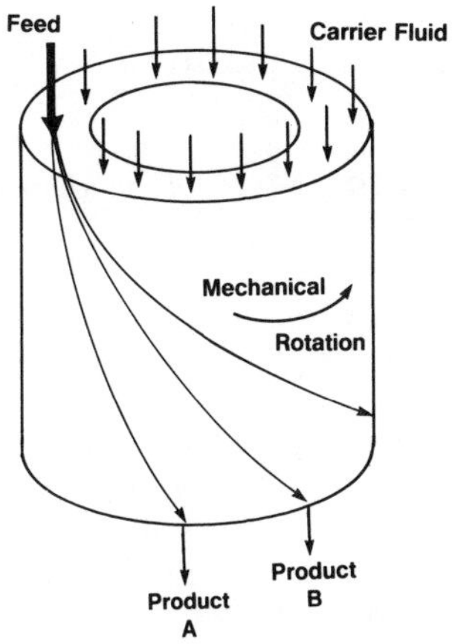

Figure 1. Mechanically rotated axial flow separation devices.

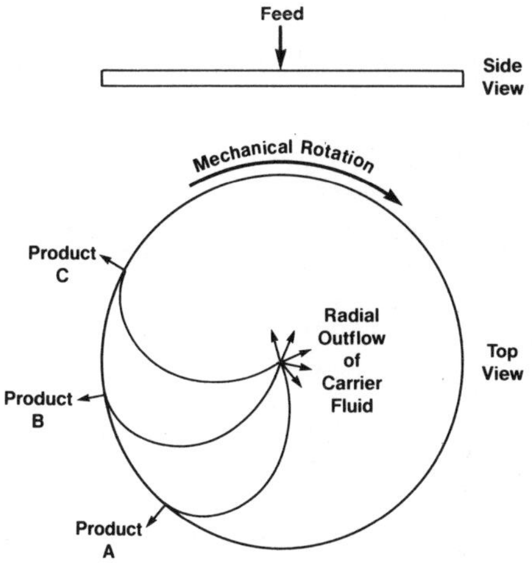

Figure 2. Mechanically rotated radial flow separation devices.

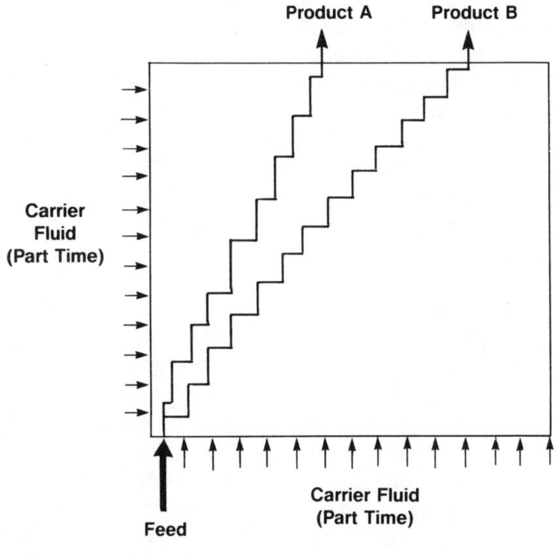

Figure 3. Alternating crossflow separation devices.

independent of particle size or gas velocity.

CROSSFLOW MAGNETICALLY STABILIZED BED CHROMATOGRAPHY

The Crossflow Magnetically Stabilized Bed Chromatograph is a device for continuous separation of multicomponent feeds in which the components are separated by the MSB solids and the fluidizing carrier fluid moving in different directions(27). The device is attractive for process separation applications due to its ability to continuously separate feed mixtures into multiple products in a single vessel with no moving mechanical parts.

As shown in Figure 7, the crossflow MSB chromatograph consists of a magnetically stabilized bed with solids flowing horizontally. These solids can be made to flow in plug fashion without a significant vertical or horizontal gradient. Carrier fluid is introduced to the stabilized bed as fluidization medium and ascends in essentially plug flow through the bed. A multicomponent feed is injected continuously in a zone at the bottom of the bed. The components in this feed have varying adsorption/desorption characteristics with the bed solids. In the crossflow MSB, the feed components have velocities in two directions. They move in the direction of the carrier fluid toward the top surface of the bed and also, due to adsorption on the solid, in the direction of the movement of solids. Weakly adsorbed components come to the surface of the bed close to the feed point, and strongly adsorbed components are transported along the horizontal axis with the solids and reach the top of the bed at different positions. Collection of the different components may be achieved by removing them at the bed surface at these different positions.

Properties of a crossflow MSB which make it advantageous for use as a separation tool include the plug flow of both gas and solids, without which the device would not work efficiently. The pressure drop is limited to the weight of solids per unit bed area, independent of the particle size and fluid throughput, allowing the use of smaller size adsorbent particles. The beds are mechanically simple since they are comprised of relatively free flowing fluidized solids.

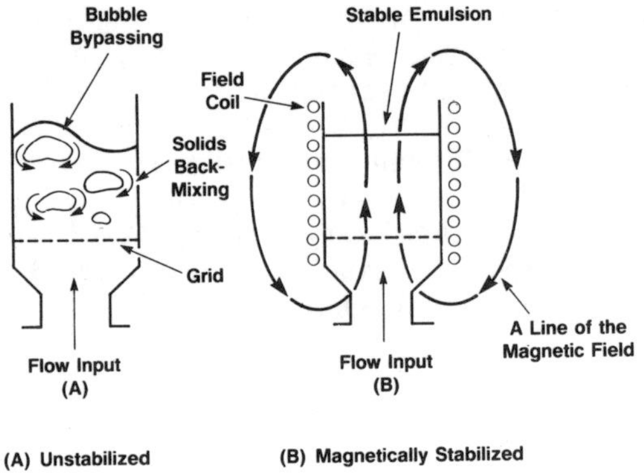

Figure 4. Comparison of bubbling and stabilized fluidized states.

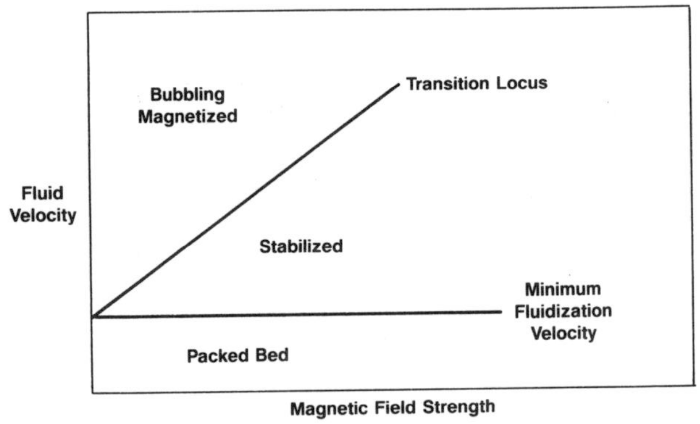

Figure 5. Phase diagram of fluid/solids/magnetic phenomena.

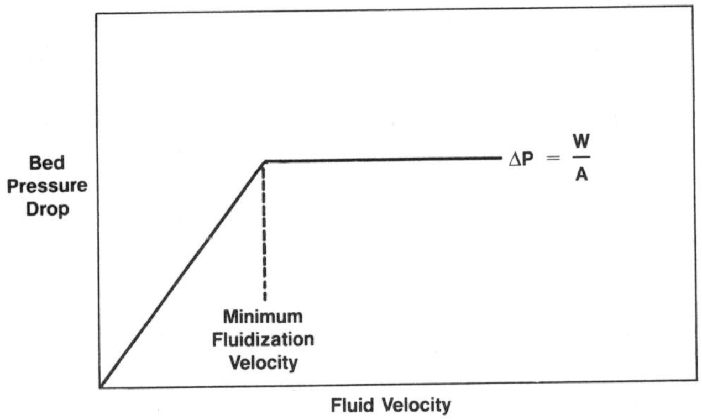

Figure 6. Pressure drop behavior of magnetically stabilized fluidized beds.

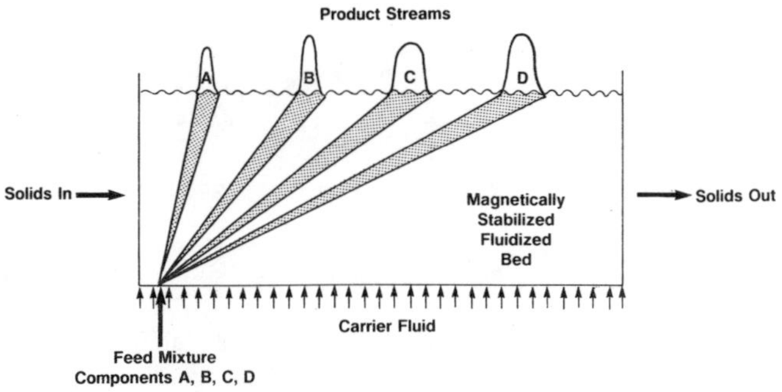

Figure 7. Chromatographic separations in a crossflow magnetically stabilized bed.

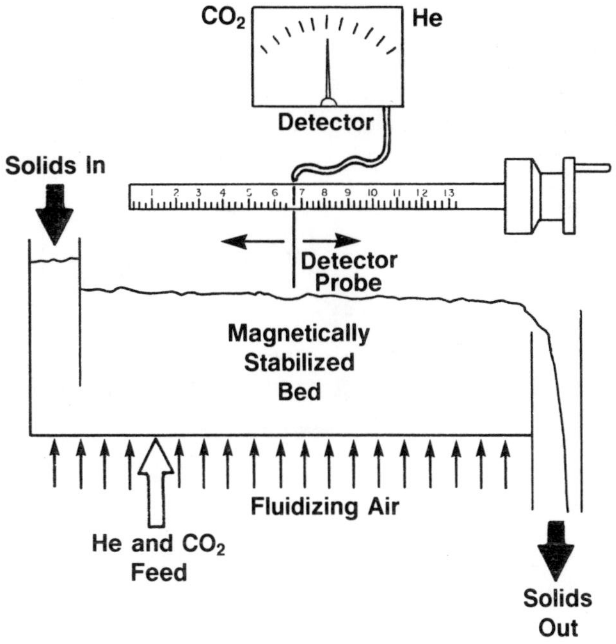

Figure 8. Experimental crossflow unit.

Also, the attrition normally associated with fluidized beds is virtually eliminated.

EXPERIMENTAL TECHNIQUES

Experimental data for the crossflow MSB chromatograph were obtained with the equipment shown in Figure 8. The crossflow MSB was approximately 69 cm long and 7.6 cm wide, with means to convey solids pneumatically from an overflow weir back to the other end of the bed. It was filled with -20 + 60 U.S. Sieve composite material of 30 wt % alumina and 70 wt % stainless steel to a depth of about 14 cm. Air was used as the fluidizing carrier fluid and also as a means of conveying the solids. The density of the solids was 1.7 g/cc; the superficial air velocity was 30 cm/sec; and the crossflow velocity of solids was 0.56 cm/sec. The applied magnetic field was approximately 192 Oersted and was supplied by two solenoidal electromagnets surrounding the MSB. Additional details on the apparatus for crossflow MSB experiments can be found in Reference 22.

RESULTS OF SEPARATION EXPERIMENTS

To compare the separation efficiency of MSBs to packed beds, initial tests were conducted using a feed mixture of argon and carbon dioxide in a 2.5 cm diameter vessel containing composite solids of alumina and stainless steel. The results of these experiments, using three different particle sizes, are shown in Figure 9. The height equivalent to a theoretical plate (HETP), a measure of separation efficiency, was found to be the same for MSBs as for packed beds. The advantage of smaller particle size resulting in a smaller HETP and thus higher separation efficiency is shown in the Figure.

To check that a separation could be achieved in a crossflow MSB, a mixture of helium and carbon dioxide was used as a tracer gas with a thermal conductivity detector (GOW MAC, leak detector, Model 21-150) used as a monitor. The detector probe could be moved along the top surface of the bed to determine the elution points of the tracers. With continuous injection of the helium and carbon dioxide at the bottom of the bed, Figure 10 shows the profile at the top of the bed as a function of the position in the direction of solids flow. Since helium causes a negative deflection of the detector indicator due to its high thermal conductivity relative to air, and carbon dioxide cause a positive deflection due to its low value of thermal conductivity, it is clear that separation is taking place. By repeating this experiment with each of the pure tracer gases individually, the amount of overlap of the peaks was determined.

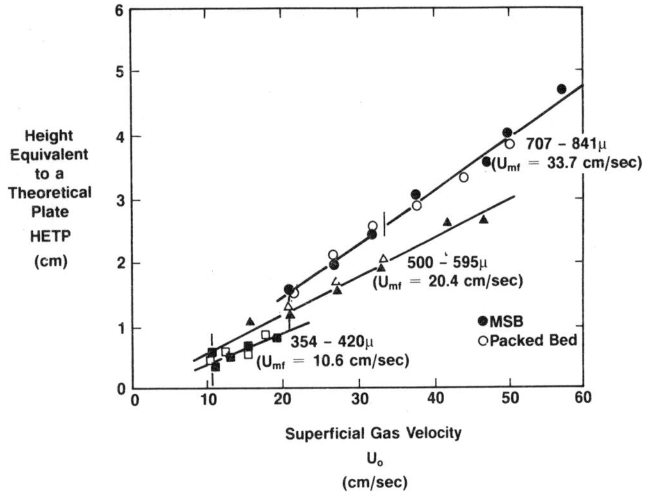

Figure 9. Comparison of HETP in magnetically stabilized and packed beds.

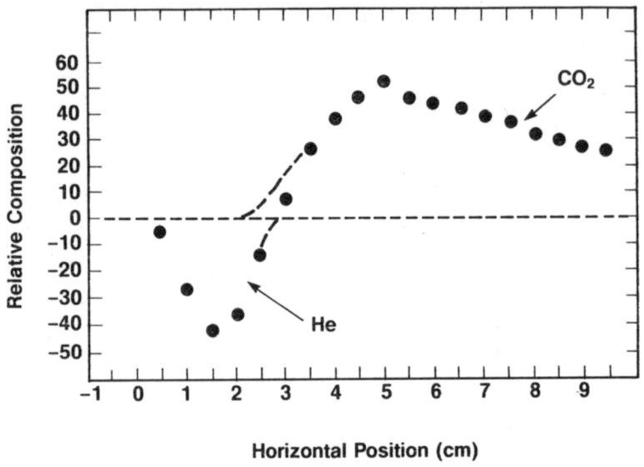

Figure 10. Continuous chromatographic separation of helium and carbon dioxide in a crossflow magnetically stabilized bed.

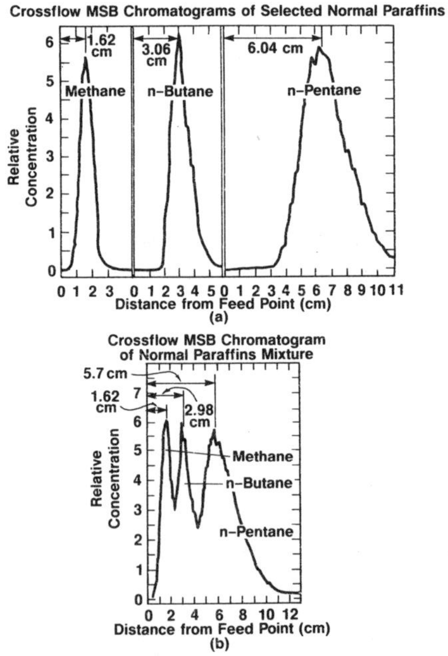

Figure 11. Comparison of peak locations for single components and mixtures.

Confirmation that for light loading the peak locations for the components in the mixture would be the same as when they were injected alone was done using a mixture of methane, n-butane, and n-pentane. The results in Figure 11 show that the location for each of the components is nearly the same for individual injection (a) as for the mixture separation (b). It is expected that this may not be the case for higher concentrations. In the experiments where hydrocarbons were separated, the detector was a flame ionization organic vapor analyzer (Century Systems, Model OVA-128).

It was expected that as the height of the bed was increased, the separation of individual components would improve. Figure 12 shows what happens to the separation of methane, n-butane, and n-pentane when the bed height is increased from 14 to 22 cm. Although the components that are strongly adsorbed tend to elute at a further distance from the feed point, the separation achieved has been improved and is what would be termed "baseline". Higher bed heights would certainly improve the separation even further. Thus, high purity components can be recovered without contamination due to peak overlap.

Figure 9 showed that the separation efficiency decreased in batch MSBs with increasing gas velocity as indicated by

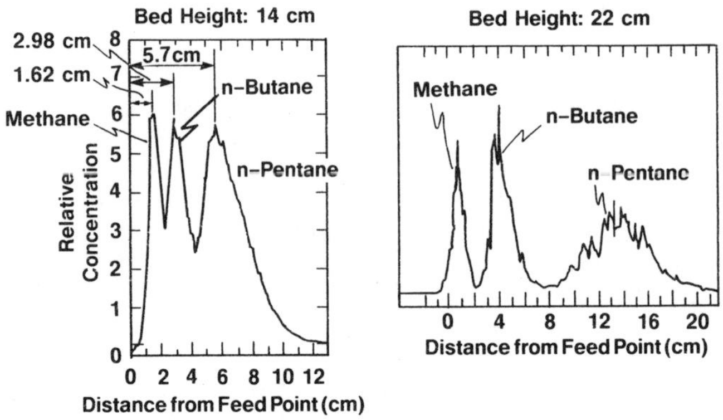

Figure 12. Effect of bed depth on separation.

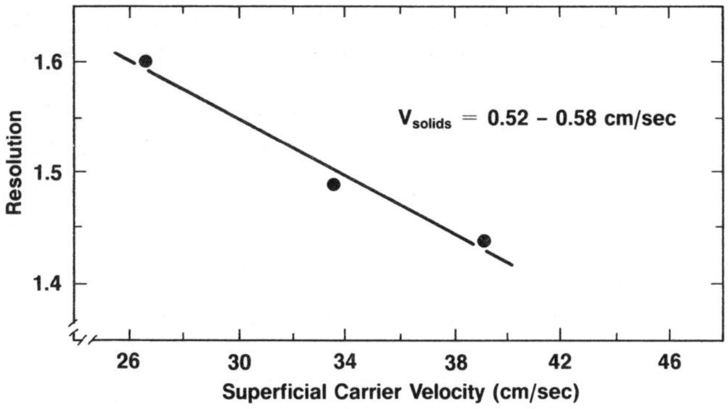

Figure 13. Effect of superficial carrier velocity on resolution of peaks

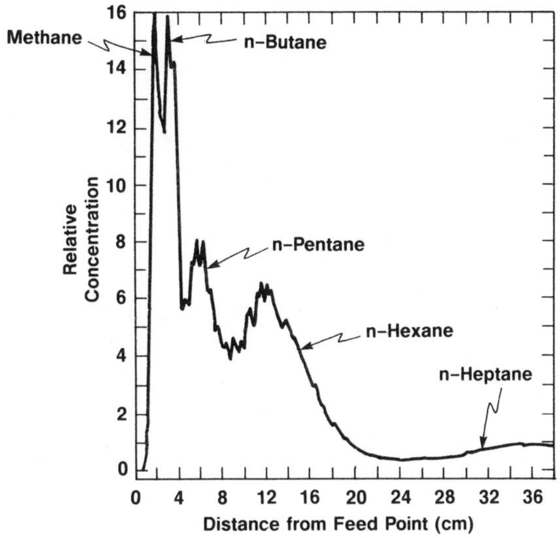

Figure 14. Separation of five normal paraffins in a crossflow magnetically stabilized bed.

higher values of the HETP. The same happens for crossflow beds as shown in Figure 13. The resolution between peaks of argon and carbon dioxide can be seen to decrease as the carrier fluid velocity is increased. The resolution of peaks is a direct measure of the separation efficiency.

A more rigorous test of the crossflow MSB chromatograph was the continuous separation of a multicomponent stream of five normal hydrocarbons. The results of this test are shown in Figure 14. The elution positions for the components in the mixture are nearly the same as for the injections of the individual components. This experiment was performed in the 14 cm high crossflow MSB used for the separation of helium and carbon dioxide, at comparable operating conditions.

DISCUSSION

The separations of helium and carbon dioxide and of mixtures of hydrocarbons demonstrate the feasibility of MSB crossflow separations. Although other available separation techniques will provide efficient separation of these test mixtures, the present study has demonstrated that a mixture can be introduced continuously to a crossflow magnetically stabilized bed and can be effectively separated into individual components.

Other applications of the crossflow MSB chromatograph may include the recovery and separation of trace components in waste streams and the general area of separation of organic from inorganic compounds. It may also be useful as an analytical tool for laboratory preparative chromatography or for continuous process monitoring. A broad literature base already exists on the analytical chromatographic separation of most compounds. In many cases, these data are directly applicable from batch analytical to continuous crossflow.

LITERATURE CITED

1. Sussman, M. V. and R. N. S. Rathore, Chromatographia, 8 (2), 55 (1975).

2. Rendell, M., Process Engineering, 66 (April 1975).

3. Sussman, M. V., Chemtech, 6, 260 (1976).

4. Martin, A. J. P., Discuss. Faraday Soc., 7, 332 (1949).

5. Svensson, H., C. E. Agrell, S. O. Dehlen and L. Hagdahl, Sci. Tools, 2 (2), 17 (1955).

6. Dinelli, D., S. Polezzo and M. Taramasso, J. Chromatogr., 7, 477 (1962).

7. Polezzo, S., and M. Taramasso, J. Chromatogr., 11, 19 (1963).

8. Taramasso, M. and D. Dinelli, J. Gas Chromatogr., 150 (May 1964).

9. Scott, C. D., R. D. Spence and W. G. Sisson, J. Chromatogr., 126, 381 (1976).

10. Canon, R. M., and W. G. Sisson, J. Liq. Chromatogr., 1 (4), 427 (1978).

11. Canon, R. M., J. M. Begovich and W. G. Sisson, Sep. Sci. Technol., 15 (3), 655 (1980).

12. Begovich, J. M. and W. G. Sisson, AIChE J., 30 (5), 705 (1984).

13. Sussman, M. V., and C. C. Huang, Science, 156, 974 (1967).

14. Sussman, M. V., K. N. Astill, R. Rombach, S. S. Chen and A. A. Cerullo, AIChE Symp. Ser., 68 (120), 72 (1972).

15. Sussman, M. V., K. N. Astill, R. Rombach, A. A. Cerullo and S. S. Chen, Ind. Eng. Chem. Fundam., 11 (2), 181 (1972).

16. Sussman, M. V., K. N. Astill and R. N. S. Rathore, J. Chromatogr. Sci., 12, 91 (1974).

17. Tuthill, E. J., J. Chromatogr. Sci., 8, 285 (May 1970).

18. Rosensweig, R. E., Science, 204, 57 (1979).

19. Lucchesi, P. J., W. H. Hatch, F. X. Mayer and R. E. Rosensweig, "Proc. 10th World Petrol. Cong., Bucharest", Heyden and Sons, Philadelphia, PA (1979).

20. Rosensweig, R. E., J. H. Siegell, W. K. Lee and T. Mikus, AIChE Symp. Ser., 77 (205), 8 (1981).

21. Arnaldos, J., J. Casal and L. Puigjaner, Powder Technol., 36, 33 (1983).

22. Siegell, J. H., and C. A. Coulaloglou, Powder Technol., 39, 215 (1984).

23. Arnaldos, J., J. Casal, A. Lucas and L. Puigjaner, Powder Technol., 44, 57 (1985).

24. Burns, M. A., and D. J. Graves, "Continuous Affinity Chromatography Using a Magnetically Stabilized Fluidized Bed", presented at AIChE Annual Meeting, San Francisco (1984).

25. Burns, M. A., and D. J. Graves, Biotechnol. Prog., 1 (2), 95 (1985).

26. Burns, M. A., and D. J. Graves, "Application of Magnetically Stabilized Fluidized Beds to Bioseparations", presented at AIChE Annual Meeting, Chicago (1985).

27. Siegell, J. H., J. C. Pirkle and G. D. Dupre, Sep. Sci. Technol., 19 (13-15), 977 (1984-85).

MOLTOX™* SYSTEM INTEGRATION FOR THE COGENERATION OF OXYGEN AND STEAM

W. R. Brown, A. A. Cassano, B. R. Dunbobbin ■ Air Products and Chemicals, Inc.
P.O. Box 538, Allentown, PA 18105
R. G. Massey ■ U.S. Department of Energy, Office of Industrial Programs, Washington, DC 20585

ABSTRACT

Oxygen produced by integrating the MOLTOX™ system with a high-temperature, oxygen-using process can be cost-competitive with cryogenically produced oxygen. This novel air separation system uses an oxygen-accepting molten nitrate/nitrite salt mixture and heat from flue gases or low-Btu fuels to produce high-purity oxygen and steam.

INTRODUCTION

Oxygen ranks sixth in volume on the list of chemicals produced in the United States, amounting to 15.5 million tons in 1984. Air separation by cryogenic distillation was introduced 90 years ago by Carl von Linde of Germany, and continues to be the choice for tonnage oxygen production. Currently, the cryogenic process is used for 99 percent of industrial oxygen production.

In 1979, Donald C. Erickson of Energy Concepts Company received a U.S. patent for a chemical air separation process for tonnage oxygen production. This new method of producing oxygen uses a molten mixture of alkali nitrates and nitrites to chemically react with oxygen in compressed air. Heating or depressurization then releases 99.8% purity oxygen in a reversible reaction. The major portion of the energy used to compress or heat the air is recovered from the waste nitrogen exhaust.

The process requires up to 40% less energy consumption than today's commercial oxygen technology. This translates into a five to thirty percent oxygen cost advantage. This includes both capital and energy costs.

With support from the U.S. Department of Energy, Mr. Erickson proved his concept in a bench-scale unit that produced 6 liters/per minute of oxygen.(1) Three patents now cover the basic invention.(2,3,4) In 1982 Air Products and Chemicals, Inc., in a cost-sharing, cooperative agreement with the U.S. Department of Energy, undertook the continuing development of the process, now known as the MOLTOX oxygen system. This $5 million, four-year project is currently in its third year.(5)

This paper describes the process, the projected applications, the research and development, and the plans of Air Products leading to commercialization of the MOLTOX system.

PROCESS DESCRIPTION

The MOLTOX system is based on the reversible reaction of oxygen with sodium and potassium nitrite to form nitrate. This reaction can be used in one or both of two

*MOLTOX is a trademark of Air Products and Chemicals Inc.

MOLTOX system salt loop types, PSA (Pressure Swing Absorption) and TSA (Thermal Swing Absorption). The basic operating modes are:

1. Pure Pressure Swing - For integration with the pressurized gases of gas turbine power plants.
2. Pure Thermal Swing - For integration with the heat recovery and steam generation section of industrial and utility steam boilers.
3. Combined Pressure - and Thermal Swing For any application in which heat and pressure energy are available.

Simplified process diagrams of the two salt loop types are shown in Figure 1. These diagrams show the absorber and desorber salt flows, the gaseous air and product flows and the integration of the salt loop with external processes.

For either MOLTOX system salt loop type, dry, CO_2 free air enters the absorber at the absorber temperature of 783 to 922K (950 to 1200°F) and pressure 0.41 to 1.2 MPa (60 to 186 psia) and is contacted with the molten salt. The oxygen reacts chemically with the salt ($NO_2^- + 1/2O_2 \rightarrow NO_3^-$) and is removed with the salt from the bottom of the absorber. The molten salt from the absorber flows to the desorber where the chemical reaction is reversed ($NO_3^- \rightarrow NO_2^- + 1/2O_2$), and gaseous oxygen is released from the salt and removed as product. The reversal of the chemical reaction in the desorber requires either pressure reduction (PSA) or temperature increase (TSA). The compressed air entering the absorber contains about 21 wt % oxygen and 79 wt % chemically inert (mainly nitrogen) gases. These inert gases, along with some unreacted oxygen, are removed from the top of the absorber at essentially the same pressure and temperature as they entered.

For the <u>TSA salt loop</u>, the temperature of the molten salt from the absorber is increased across the salt heater. The salt is heated to 922K (1200°F) and then enters the desorber where the gaseous oxygen is released and removed from the salt. The 922K (1200°F) salt is withdrawn from the bottom of the desorber and is cooled to 783K (950°F) in the salt cooler before entering the absorber for recontact with the hot air. The salt is circulated around this TSA loop by a pump operating at the 783K (950°F) salt temperature.

For the <u>PSA salt loop</u>, the pressure of the molten salt from the absorber is reduced across the pressure letdown valve before the salt enters the desorber. The pressure of the desorber is controlled by the gaseous oxygen pressure through a vacuum oxygen compressor. The salt is removed from the desorber by a salt pump and recirculated back to the absorber at the necessary pressure for recontact with compressed air.

The pure TSA MOLTOX system requires 8.7 MM Btu of heat through the salt heater for each ton of oxygen product. Limited heat availability for certain TSA MOLTOX process applications led to the development of process designs for combined PSA/TSA and TSA/PSA MOLTOX systems. Figure 2 graphically summarizes the design flexibility of these MOLTOX systems by showing the energy required per ton of oxygen versus the heat throughput for the various designs. This figure also shows that the net energy required for MOLTOX system oxygen production is substantially below that for the cryogenic process. Existing plants were largely built in the 1960 to 1970 period and require in excess of 450 KWH per ton of oxygen, while new oxygen cryogenic plants can be designed for about 350 KWH per ton of oxygen. The thermal swing MOLTOX system utilizes the least net energy and cogenerates steam.

The thermal swing process, which will be described in more detail in this paper, is applicable at any site where fuel or high-quality heat is used to generate steam. The pressure swing process is applicable when fuel will be used in a gas turbine for mechanical or electrical power.

MOLTOX SYSTEM HEAT INTEGRATION WITH STEAM BOILERS

The MOLTOX system is projected to be competitive with cryogenic air separation plants in those applications in which significant process heat integration can be achieved.(5) Specifically, the salt heater integration can be used in the radiant and convective heat transfer sections of fired or unfired steam boilers. Figures 3 and 4 show respectively the heating and cooling curves for a typical industrial high-pressure steam boiler and a typical industrial medium-pressure steam boiler, with the proposed heating and cooling curves of an integrated TSA MOLTOX system superimposed. In the high-pressure steam boiler, metallurgical constraints limit the tube wall temperatures in the boiler to about 1088K (1500°F); thus, only saturated steam generation takes place in the radiant and convective section where the flue gas temperature ranges from about 1588K (2400°F) to 2033K (3200°F). Steam superheating and boiler feed water preheating occur in the convective section at flue gas temperatures of about 477K (400°F) to 1588K (2400°F). The most efficient utilization of the high-quality radiant heat, which would be obtained by heating the hottest heat accepting stream by the hottest heat supplying stream, is therefore not obtained economically by steam generation alone in this type of boiler. In the medium-pressure steam boiler, shown in Figure 4, the cooling curves are very open, giving the potential for recovery of significant amounts of lost work. Placing the MOLTOX salt heater in the radiant and convective section of either boiler type then allows more efficient use of this high-quality heat by transferring it to the MOLTOX salt at about 922K (1200°F). Oxygen is generated and the heat is then rejected at about 783K (950°F) in the salt cooler of the MOLTOX plant to generate steam as before.

Examples of fired steam boilers for which this integration may be done include coal-fired utility or industrial boilers, gas or oil-fired boilers, or special boilers using waste fuels. The salt heater can also be integrated with unfired heat recovery boilers. Heat recovery steam generators are used in partial oxidation reactors, in coal gasifiers, and in cogeneration applications, such as in combined-cycle plants recovering heat from the exhaust of gas turbines.

CASE STUDY - HEAT INTEGRATION WITH BLAST FURNACE OFFGAS BOILERS

A large potential MOLTOX system oxygen market is the replacement of cryogenic oxygen plants currently serving the steel and chemical industries. Good examples of heat integration opportunities in replacement oxygen markets are blast furnace offgas steam boilers at integrated steel mills. With 20% of the heat throughput at this source, MOLTOX systems could supply approximately half of the domestic steel industry's oxygen requirements while reducing the industry's energy consumption. The 1990 oxygen use for steelmaking in the U.S. is forecasted to be 22,000 tons/day. Therefore, the process design case selected for analysis was MOLTOX system integration using the combustion energy available from the blast furnace offgas at integrated steel mills. The oxygen produced would be used in the Basic Oxygen Furnace (BOF) for steelmaking.

DESCRIPTION OF MOLTOX SYSTEM INTEGRATION PROCESS MODIFICATIONS

The integration required to retrofit a steel mill with a TSA/PSA MOLTOX oxygen plant is described below. The blast furnace air/steam/offgas flow sheet in Figure 5 shows typical utilization of 65% of the offgas in fired boilers. Currently, the offgas from the blast furnace is sent to a scrubber which removes particulates and reduces the temperature to about 305K (90°F). This gas is then split into two streams: 35% is consumed in the stoves to preheat the blast air, and the remaining 65% is fired in boilers to raise 3-MPa (440 psia), 644K (700°F) steam. The steam is partially used to drive the blast air turboblowers, and the remainder is exported for general process needs. The operation of the blast furnace itself and the net steam export are unchanged by integration with the MOLTOX system. The integration is made only to the blast furnace auxiliaries, requiring the following process conditions:

1. Twenty percent of the offgas to the boilers must be diverted to a new steam/salt heater boiler in the MOLTOX oxygen plant. The remaining 80% is burned in the existing boilers, which unloads the boilers and results in more steam generation per unit of heat and a lower stack gas temperature (433K [320°F] vs. 477K [400°F]).

2. The new steam/salt heater boiler in the MOLTOX oxygen plant is also designed with a 433K (320°F) stack gas temperature. In this way, the diverted offgas burned in this boiler supplies the heat for the MOLTOX salt heater, as well as generates steam that is returned to the integrated steel mill supply header.

3. Air to the MOLTOX oxygen plant is provided from the surplus capacity in the blast air turbo-blowers. The extra steam, generated from items 1 and 2 above, is sufficient to provide the additional required energy for the turbo-blower driver.

NEW MOLTOX PLANT RETROFIT PROCESS DETAILS

The new MOLTOX oxygen plant as retrofit at the blast furnace site is described in Figure 6. The salt loop, comprising an absorber column, desorber, new salt heater/steam boiler, salt cooler/steam boiler, salt/salt economizing heat exchanger, and salt pump, is shown in the "separation" block. Blast furnace offgas diverted from the existing boiler is burned in the new salt heater/steam boiler, which incorporates the salt heater of the TSA/PSA salt loop as well as new steam generation. Steam is also generated in the salt cooler/steam boiler.

The air feed, which was compressed to 0.5 MPa (70 psia) in the blast furnace turbo-blowers, first passes through the molecular sieve air purifiers for water and CO_2 removal to the 1-ppm level. The air is then preheated to 730K (855°F) in the packed bed regenerators of the heat exchange section against returning vent gases. After being preheated, the air is passed through the absorber column where molten salt chemically absorbs oxygen from the air at about 783K (950°F) and 0.41 MPa (60 psia). The waste nitrogen exits the absorber and is cooled to 591K (605°F) by reheating the packed bed regenerators. The nitrogen is then expanded in an expander and used as regeneration gas in the pressure swing molecular sieve air purifiers before being vented to the atmosphere. Excess electric power (30 KWH/ton O_2/day) from the expander is returned to the steel mill power grid for general plant use.

The oxygen-rich molten salt leaves the absorber column and is reduced in pressure to 0.14 MPa (20 psia) and preheated to about 866K (1100°F) against the oxygen-lean molten salt exiting the desorber. The final stage of heating to 922K (1200°F) is supplied by a boiler which burns the imported blast furnace offgas. The heated, oxygen-rich salt is then passed to the desorber where high-purity oxygen is generated. The final cooling of the oxygen-lean molten salt is achieved in a second steam generator. The molten salt is then pumped back to the absorber. The product oxygen is cooled against the incoming air before being compressed to 4.2 MPa (615 psig) in the product oxygen compressor. The oxygen is then available for use in BOF steel-making.

SAVINGS REALIZED FROM MOLTOX SYSTEM RETROFIT

Energy savings through the replacement of existing oxygen plants by the MOLTOX system is due to several factors:

- MOLTOX system energy advantage over cryogenic plants.
- Better compressor efficiency through utilization of the blast furnace air compressor rather than the old cryogenic plant compressor.
- Boiler stack gas temperature decrease due to better utilization of boiler surface area.

With an existing cryogenic plant energy usage of 450 KWH/ton O_2/day, an energy saving of 480 KWH/ton O_2/day, or 106%, is realized by the integration described above. Forty-six percent of this energy saving is due to the MOLTOX system energy advantage and to the use of the blast furnace air compressor, and 54% is due to the increase in the boiler surface area which lowers the stack gas temperature.

MOLTOX AIR SEPARATION SYSTEM DEVELOPMENT PLAN

The MOLTOX system is being developed jointly by Air Products and Chemicals, Inc. and the U.S. Department of Energy. Process economic studies have shown a five to thirty percent oxygen cost advantage for the MOLTOX system over modern, efficient cryogenic distillation. This cost advantage is dependent upon the specific application. A 0.25-ton/day oxygen pilot plant is under construction with start-up scheduled for early 1986. A nominal 50-ton/day oxygen semi-works unit is planned following successful pilot plant operation. Commercial oxygen plants, in excess of 500 ton/day oxygen, should be available in the early 1990s.

The areas of technical uncertainty to be addressed by pilot plant operation include:

- TSA/PSA and TSA modes of operation
- Salt losses (vapor, corrosion, salt stability.)
- Absorption/desorption kinetics
- Salt loop equipment designs
- Adequacy of materials of construction
- Long-term operability, and
- Gas purity, impurities, and by-products.

Salt losses, kinetics, air purification, alternative salts, and better materials of construction will be addressed by continuing laboratory and bench-scale experimentation. Integration and process optimization will also be addressed by further engineering studies.

ACKNOWLEDGEMENTS

We would like to thank Air Products and Chemicals, Inc. and the U.S. Department of Energy for permission to publish this paper.

LITERATURE CITED

1. "Oxygen Production by the MOLTOX™ Process." Final report DOE/CS/40287-T2 (DE83012847); February 1983.

2. "Separation of Oxygen From Gaseous Mixtures With Molten Alkali Metal Salts." U.S. Patent 4,132,766; 2 January 1979.

3. "Nitrogen and Oxygen Via Chemical Air Separation." U.S. Patent 4,287,170; 1 September 1981.

4. "Oxygen Production By Molten Alkali Metal Salts." U.S. Patent 4,340,578; 20 July 1982.

5. "Pilot Plant Development of A Chemical Air Separation Process." Final Task I DOE Report DE-AC07-82CE40544; May 1985.

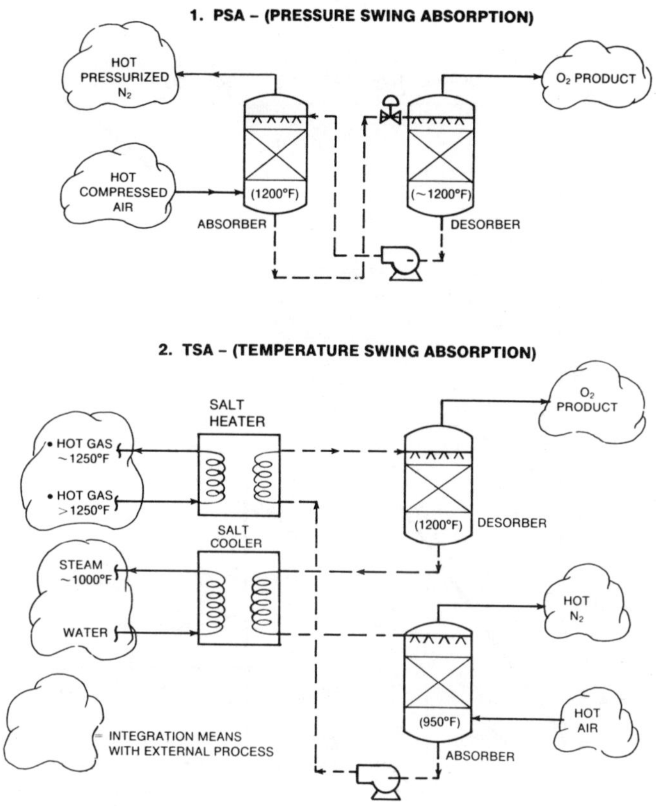

Figure 1. Integrated Moltox™ system salt loops.

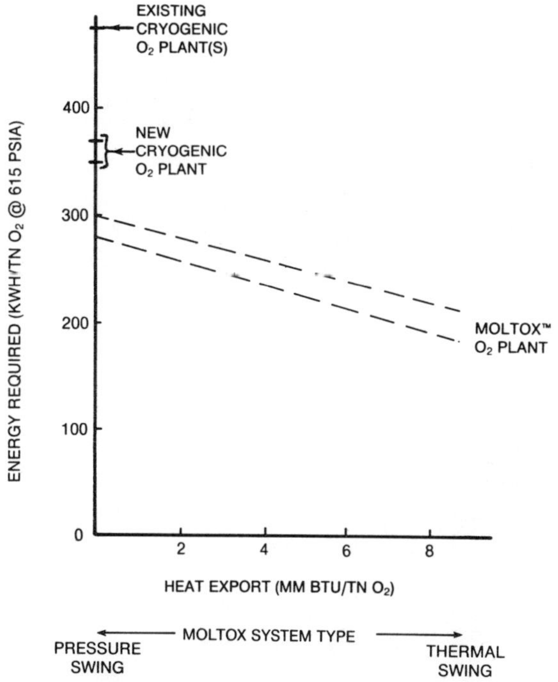

Figure 2. Energy advantage for integrated Moltox™ plant cases vs. cryogenic plants.

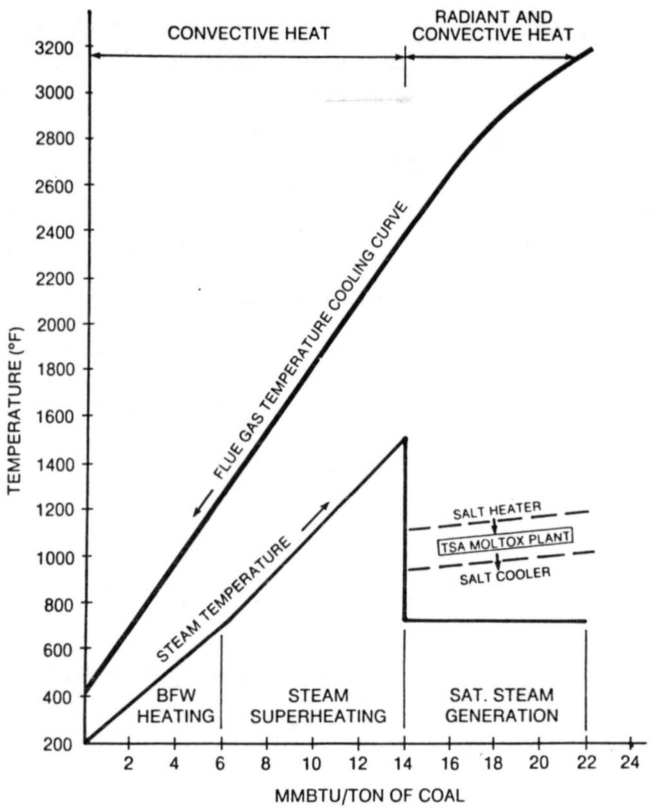

Figure 3. Thermal swing Moltox™ O_2 plant integration with a high pressure steam boiler. (2500 PSIA, 1500°F, coal fired)

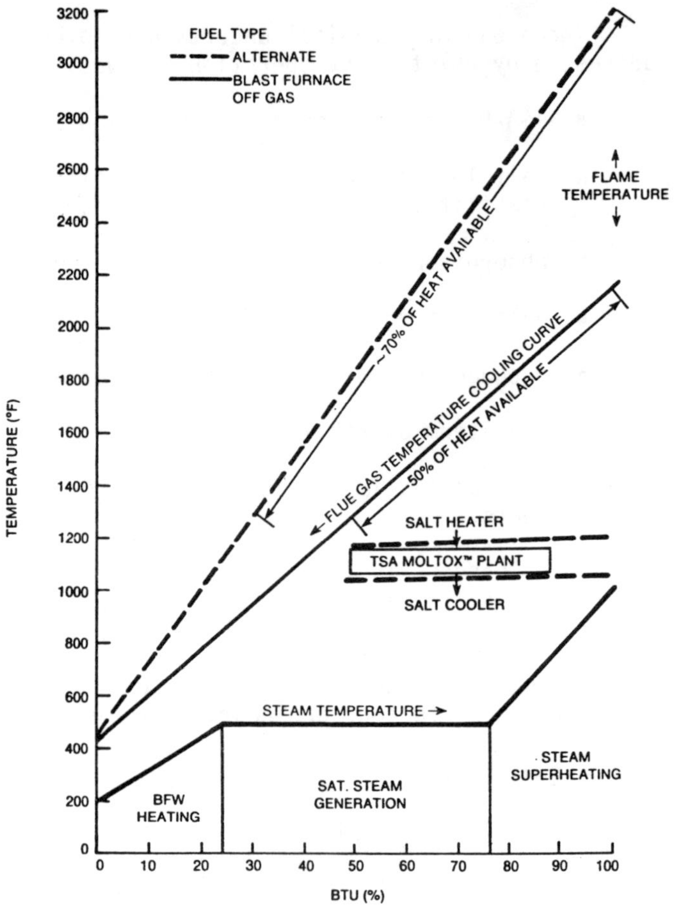

Figure 4. Thermal swing Moltox™ O_2 plant integration with a medium pressure steam boiler.

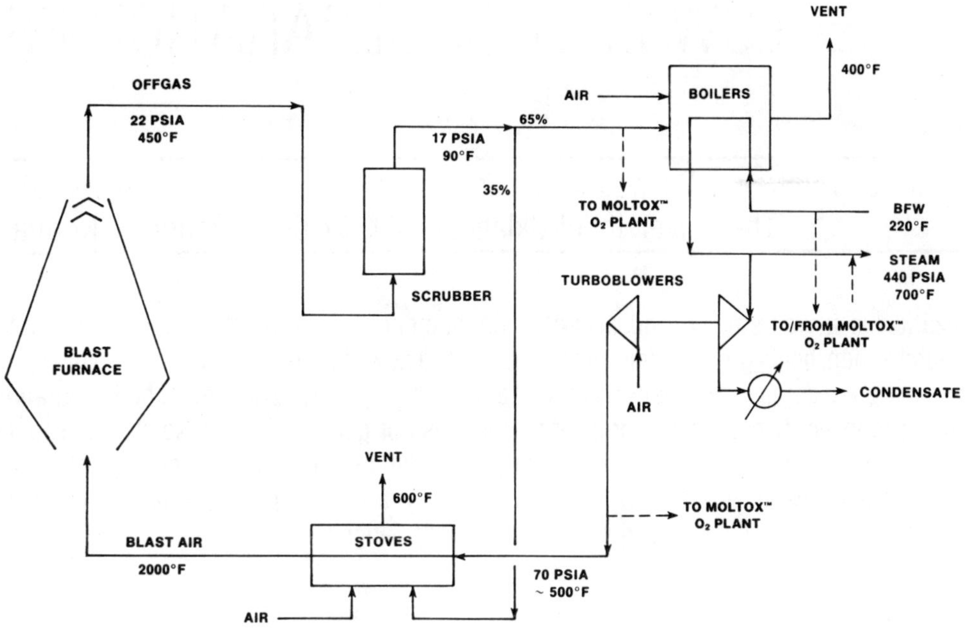

Figure 5. Blast furnace air/steam/offgas flowsheet with Moltox™ system integration.

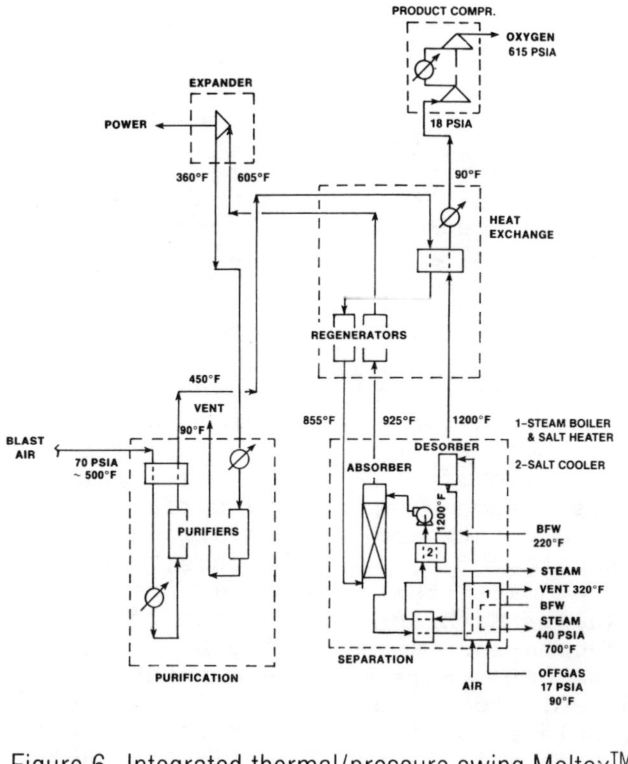

Figure 6. Integrated thermal/pressure swing Moltox™ system for BOF-steelmaking.

ADMICELLAR CHROMATOGRAPHY: A NEW LOW ENERGY SEPARATION PROCESS

Timothy P. Fitzgerald and Jeffrey H. Harwell ■ School of Chemical Engineering and Materials Science
The University of Oklahoma, 202 W. Boyd, Room 23, Norman, OK 73019

Micellar solubilization is a well-studied and accepted phenomenon of surfactant systems. The analogous surface phenomenon, adsolubilization, holds great promise as the basis for a new class of separation techniques, although it has not been investigated in any great detail. In this paper, admicellar chromatography, a new fixed bed separation process based on adsolubilization is proposed. N-heptanol, representing the class of general organic solutes, is separated from dilute aqueous solution. Over two-hundred percent increases in concentration per pass are demonstrated, and the effects of process parameters (solute concentration, flow rate, pH) on recovery are examined. Advantages over existing techniques include low energy, ease of solute recovery, and stability over wide ranges of pH.

Over the past decade, or so, new technologies have emerged which require the removal of organic compounds from dilute aqueous solutions. In the production of fuel alcohol from biomass, for example, the final step involves recovering ethanol from solutions where its concentration is typically 5-16% by volume(1,2). Other examples include coal gassification plants, which generate large amounts of wastewater with toxic phenolic compounds, and the emerging genetic engineering and recombinant DNA industry, where several desirable biological compounds are produced in dilute aqueous solution(3).

With the advent of these technologies, the need has arisen for low energy separation processes for removing dilute organics dissolved in aqueous media. The traditional process, distillation, has become increasingly expensive because of the exorbitant energy costs associated with it. Most of these costs are related to the heat capacity of the usually large volumes of water that must be treated as well. Membrane processes, which have become increasingly popular in recent years, usually do not perform well for the low molecular weight solutes of interest. Adsorption on carbon is not ideal because of the difficulty of economically recovering the solute, or the cost of regenerating the bed.

In this paper, a new low energy separation process called "Admicellar Chromatography" is introduced. The process is based on the phenomenon of adsolubilization, which is "the incorporation of compounds into surfactant aggregates adsorbed on surfaces, which compounds would not be in excess at the solid /solution interface in the absence of the surfactant." Several advantages of this process over existing techniques will be illustrated.

BACKGROUND

The adsorption of single component ionic surfactants on mineral oxides is a well established phenomena, and has been studied extensively (4-7). The surface of the adsorbent can be visualized as being composed of patches, each patch having its own energy of adsorption. At low surfactant concentrations, surfactants adsorb as individual molecules on each patch of the surface. At some specific concentration, the surfactants form an aggregate on the most energetic patch, and the adsorption on this patch increases sharply. At higher concentrations, surface aggregates form on successively less energetic patches. At a specific surfactant concentration, the adsorption becomes constant. This usually corresponds to the onset of micelle formation in the bulk, and the concentration where this occurs is known as the Critical Micelle Concentration, or CMC. Constant adsorption also occurs when the surfactant attains complete bilayer coverage of the surface, at a concentration below the CMC.

Mineral oxides such as alumina complete the bonds of the atoms near the surface by chemically adsorbing H_2O molecules, resulting in a surface of close packed oxygens, some of which carry hydrogen atoms. Because of this, the alumina surface can become charged by the protonation or deprotonation of the chemically adsorbed water (8-9). If each -OH on the surface can exist as either $-O^-$ or $-OH_2^+$ in equilibrium with the surface $[H^+]$, then the charge on the surface can be altered by manipulating the solution pH. In alumina the surface has a net positive charge at low pH's and a net negative charge at high pH's. The pH at which the surface has a net charge of zero is called the point of zero charge (PZC).

The aggregates formed by adsorption of ionic surfactants from aqueous solution at higher electrolyte concentrations on oppositely charged mineral oxide surfaces have been shown to be micelle-like structures and are called "admicelles"(4) or "hemimicelles"(6). The concentration where admicelles first form on the most energetic surface patch is called the "Critical Admicelle Concentration" (CAC), indicating that admicelles are the surface analogue of micelles. Again, the CAC is often referred to in the literature as the "Hemimicellar Concentration" (HMC).

As mentioned previously, this separation process is based on the phenomenon of adsolubilization, which can be considered a surface phenomenon analogous to micellar solubilization, a well accepted property of surfactant systems. As defined by Rosen, it is

> the spontaneous dissolving of a substance (solid, liquid, or gas) by reversible interaction with the micelles of a surfactant in a solvent to form a thermodynamically stable isotropic solution with reduced thermodynamic activity of the solubilized material(10).

The term adsolubilization is intended to convey that while the surfactant aggregate is adsorbed onto a solid substrate, so that the system cannot be referred to as isotropic, the mechanisms of solute incorporation in the aggregate are identical to those in solubilization, and not simply a coadsorption process.

OVERVIEW OF PROCESS

The proposed process has three steps (See Figure 1). In the first step, the bed preparation step, a dilute surfactant solution above the CMC (Stream A) is equilibrated with a

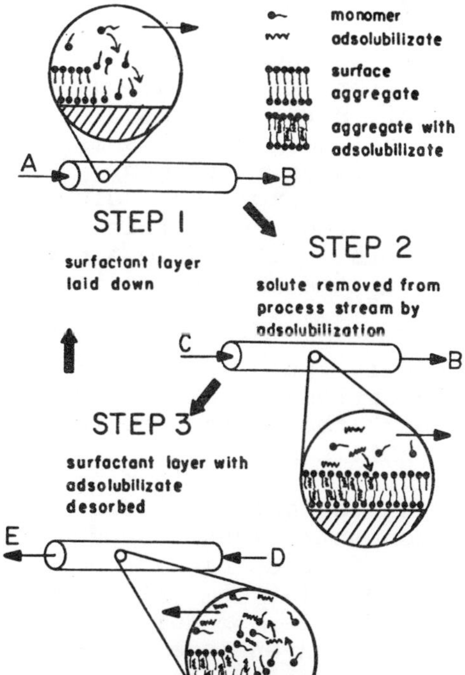

Figure 1. Schematic of Fixed Bed Admicellar Chromatography Process to Remove Dissolved Organics from Water.

fixed bed of an adsorbent with an amphoteric surface, such as α-Al_2O_3. The pH of this stream is adjusted so that the surfactant will adsorb onto the surface. The concentration of the stream is above the CMC to minimize equilibration time. After the surfactant bilayer has propagated sufficiently from the bed inlet, Stream A is switched off from the column, and Stream C is introduced. This initiates step 2, the adsolubilization step.

Initially, stream B is devoid of surfactant, but sometime during the process, either in step 1 or step 2, surfactant will begin to appear in the effluent as the wave forming the bilayer reaches the outlet. This stream can either be recycled, or because of the biodegradability of most commercial surfactants, dumped directly into the environment.

Stream C contains the solute we wish to separate and some surfactant at the system CMC. The surfactant is included to maintain the stability of the bilayer, which can be thought of as being in equilibrium with the surfactant monomer in the bulk. If surfactant were not present, the bilayer would begin to decompose to establish this equilibrium. Again, as in the first step, Stream B is initially devoid of solute, but at some point, the solute begins to show up at the outlet, indicating the column has become saturated. When this occurs, Stream C is switched to a new bed, and Step 3, the bed stripping step, begins.

At this point, a new influent stream is introduced at a pH which will cause the surfactant to desorb (Stream D). As the surfactant desorbs, the solute is flushed out with the surfactant, but now at much higher concentrations. Depending on the nature of the target molecule, the effluent (Stream E) could be disposed of as is, or further treated with a more conventional technique. An important point here is that if further treatment is necessary, the solute is now at a much higher concentration. Any downstream equipment could be scaled down potentially by a factor of 10 or even 100. This could be very significant in the overall economics of removing the solute. If the solute is not to be recovered, but disposed of--i.e. if the adsolubilization unit is serving as a substitute for a carbon adsorption bed--then step 3 can be thought of as a low energy bed regeneration step that does not require dumping the bed. In either application, when the concentration of the solute in Stream E falls below the point where it is either no longer economical or no longer necessary to continue recovering it, the influent stream is switched back to Stream A, the column begins re-equilibrating with surfactant, and a new bilayer begins forming.

The process as presented here has several advantages over conventional techniques. The reversibility of bilayer formation, the ease at which the bed can be regenerated, and the ease of recovering the target molecule are obvious advantages over carbon bed adsorption. Compared to membrane techniques, the surfactant bilayer is stable over a wide pH range; membranes are often sensitive to pH changes. This process can be applied to both high and low molecular weight compounds, a clear advantage over membranes, which typically do not perform well on low molecular weight solutes. With the wide availability of commercial surfactants, and the many substrates available, one can tailor a surfactant/substrate system to fit the range of pH's encountered in any given process; a surfactant/substrate combination could be designed for a specific solute or pollutant as well. Also, commercially available surfactants are relatively inexpensive for the low concentrations required (μ-moles/l) and as mentioned previously, biodegradable.

The process also has some potential disadvantages as well. Depending on the column porosity, the surface area of the adsorbent, and the desorption of the bilayer with the solute, phase boundaries could be crossed during the Bed Stripping Step (Step 3, Figure 1). Precipitate formation and even liquid/liquid phase separation are all potential hazards. Formation of a precipitate or an extremely viscous phase, for example, might be sufficient to plug the column, making it necessary to dump the bed.

EXPERIMENTAL

Materials Used

In both the static adsorption isotherms and in the column experiments, Sodium Dodecyl Sulfate (Fisher Scientific 95-98%) and 1-Heptanol (Aldrich Chemical 98%) were used as received. HCl and NaOH (Fisher), 0.1N - 1.0N , were used to adjust the pH, and the water was distilled and deionized.

For the isotherms, Aluminum Oxide C (Degussa Inc), a primarily gamma alumina, with a surface area of 100 m^2/g was used. For the column experiments, a mixture of 75% low surface area alumina (Alfa Products #87354) and 25% high surface area (Alfa Products #89369, 320 m^2/g) was used.

Methods

For the heptanol adsorption isotherms 10 milliliters of pH 12 feed solution was added to 0.5 grams of alumina in screw top centrifuge tubes. The tubes were centrifuged at 1000 RPM for one hour, and then were placed in a water bath at 30 °C for four days to equilibrate. The liquid on the top of the alumina was decanted. The concentration of SDS was measured using a high performance liquid chromatograph with a conductivity detector, and the heptanol concentration was measured on a gas chromatograph using a flame ionization detector. The equilibrium pH was measured using a combination pH electrode.

For the isotherms done at pH 3 feed, 15 milliliters of solution was added to 0.5 grams of alumina, and the above procedure followed.

For the column runs, a stainless steel column was dry packed with a 75% low surface area/25% high surface area alumina mixture. This mixture was chosen because it gave the best flow properties through the column.

Table 1 summarizes the experimental conditions for the runs presented. Each run consisted of four steps. First, the alumina bed was equilibrated with water at pH 8.2 to 8.4. Then, a 50,000 mole/liter SDS solution at pH 8.2 to 8.4 was injected into the column, and the SDS concentration in the effluent was monitored. When the concentration in the effluent approached that of the inlet, this stream was switched off, and a new stream at pH 8.2 to 8.4 containing SDS at 10,000 mole/liter and heptanol at 2,000 to 3,200 -mole/liter was introduced. The concentrations of both SDS and heptanol in the effluent were monitored. When the concentration of heptanol in the effluent approached that at the inlet, this stream was switched off and the final stream containing water at pH 11.5 to 12.0 was injected. When the concentrations of both SDS and heptanol at the outlet approached zero, this stream was switched off.

It is important to notice that

Table 1. Summary of Run Conditions.

	Stream A	Stream C	Stream D
Run I			
[SDS]*	52,800	9,694	-
[C_7OH]	-	2,785	-
pH	8.20	8.20	11.85
Pore Vol. Injected	9.47	17.10	7.13
Flowrate**	0.30	0.30	0.30 - 0.10
Run II			
[SDS]	49,575	9,671	-
[C_7OH]	-	3,214	-
pH	8.37	8.28	11.56
Pore Vol. Injected	4.89	10.83	7.99
Flowrate	0.10	0.10	0.10
Run III			
[SDS]	49,416	9,998	-
[C_7OH]	-	2,069	-
pH	8.27	8.37	11.96
Pore Vol. Injected	5.52	10.35	6.32
Flowrate	0.30	0.30	0.10

*Concentrations given in μmole/liter
**Flowrates in ml/min

the experimental runs were not performed according to the three step procedure described in the introduction for a possible industrial application. The experimental runs were designed to make the results more interpretable by minimizing wave/wave interactions within the column; in an actual application the experimental procedure would be unnecessarily wasteful of surfactant.

Burets were used to record the amount of each stream injected, and the pH of the effluent streams were measured using a flow cell fitted with a pH electrode. In line sampling valves allowed the SDS and heptanol concentrations at the outlet to be automatically measured using HPLC and GC, respectively, as shown in Figure 2.

RESULTS

1-Heptanol was chosen as a model solute representing the class of general organic that might need to be removed from an aqueous process stream.

Static adsorption isotherms for heptanol at various pH and SDS concentrations were prepared to simulate expected conditions in the column. From these isotherms, the conditions presented earlier were selected for the column runs.

For the column experiments, the flow rate was varied in order to determine at what conditions mass transfer effects became significant, and the heptanol feed concentration was varied to determine how efficient the process was in removing it.

Figures 3 to 7 present the results for a run at a high flow rate (0.3 ml/min), Figures 8 to 12 a low flow rate run (0.1 ml/min), and Figures 13 to 17 a run at the extreme conditions of low heptanol feed concentration (2,026 mole/l, 37.2 ppm) and less than ideal pH conditions. In all of these, effluent concentrations are plotted as a function of pore volumes of solution injected. A pore volume is the void space in the fixed bed, (i.e. the space not taken up by the packing). A useful visualization of a pore volume is the volume of liquid required to propel an imaginary slug of solution from the inlet to the outlet of the column. For example, assuming steady state, for a step change in concentration at the inlet for a nonadsorbing solute, at least one pore volume of this new solution would have to be injected before any noticeable change in outlet concentration was detected.

In all of these experiments, SDS and heptanol mass balances were calculated to see if there were any losses due to irreversible adsorption on the column, and if so how much

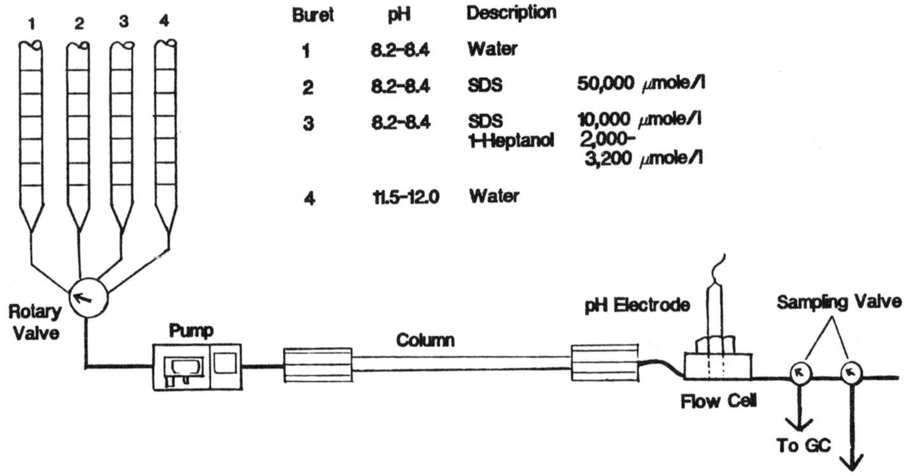

Figure 2. Schematic of Experimental Apparatus.

heptanol and SDS could be recovered. Integrating the area under each effluent curve gave the amount of that component coming out of the column during that step. Summing these gave the total amount of each component eluted during the entire run. These values were subtracted from the amounts injected, giving a total SDS and heptanol mass balance for the entire process. The SDS mass balances all closed within 10 to 12%, and the heptanol balances closed within 10 to 19%.

DISCUSSION OF RESULTS

Figure 3 shows the effluent SDS profile during Bed Preparation (Step 1, Stream B of Fig.1) for Run I. The inlet SDS concentration is indicated by the horizontal line on the figure. No SDS was detected until 2.5 pore volumes of surfactant solution were injected into the column. The SDS breakthrough was sharp, and the effluent approached but never reached the inlet concentration.

Figures 4 and 5 illustrate the

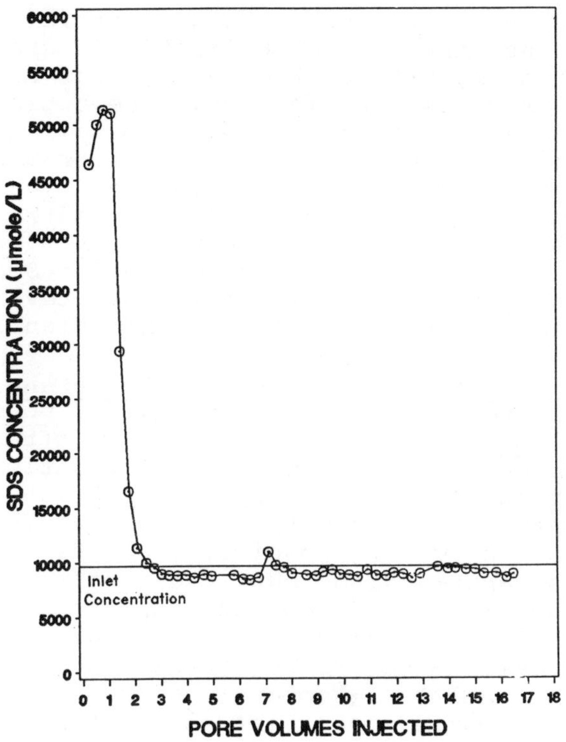

Figure 4. Run I; Effluent SDS Profile for Adsolubilization Step.

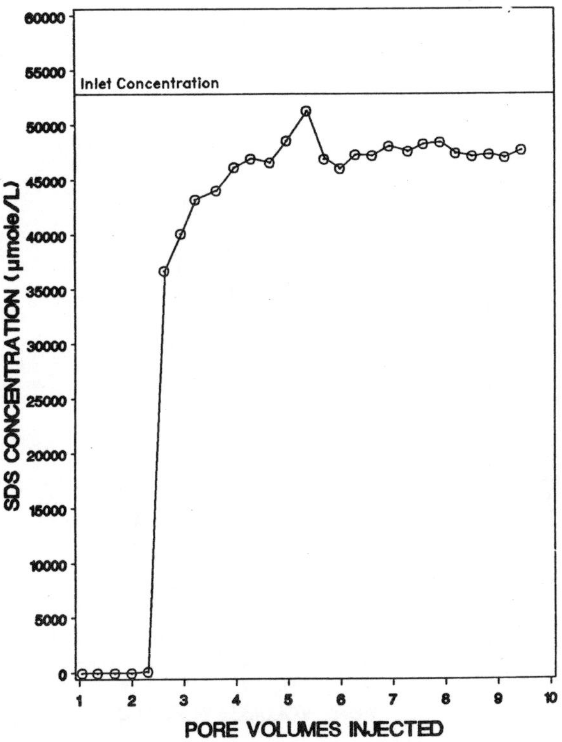

Figure 3. Run I; Effluent SDS Profile for Bed Preparation Step.

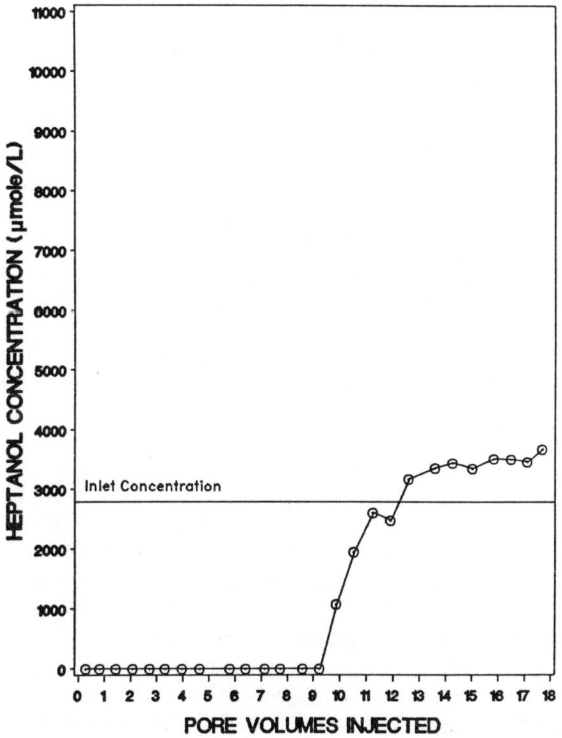

Figure 5. Run I; Effluent 1-Heptanol Profile for Adsolubilization Step.

effluent SDS and 1-Heptanol profiles during the adsolubilization step (Step 2, Stream B of Fig 1) for Run I. As shown in Figure 4, the outlet SDS concentration began to fall to the new inlet concentration after one pore volume was injected, and had almost reached the inlet concentration after two pore volumes. As in the first step, the new outlet SDS concentration approached but never reached the new inlet concentration.

Complete removal of 1-Heptanol was achieved for over 9.5 pore volumes as shown in Figure 5. After heptanol breakthrough, the outlet concentration rose to the inlet concentration, but not as sharply as the SDS, indicating possibly that the bed was approaching but had not yet reached saturation and heptanol was still being removed from the process stream. A surprising observation was the outlet 1-Heptanol concentration overshooting the inlet concentration after breakthrough occurred. This is an occurrence often observed in binary adsorption studies, where one component adsorbs more strongly than the other, but it was not expected here.

Figures 6 and 7 show the effluent profiles during Bed Stripping (Step 3, Stream E of Fig 1) for Run I. The effects of desorbing SDS appeared in the effluent after about 1.5 pore volumes of the flush solution were introduced as shown in Figure 6. The SDS desorption remained constant over several pore volumes, fell off sharply after about 4.5 pore volumes, then slowly approached zero. The SDS did not desorb in a sharp peak over a few pore volumes as was expected.

The effects of desorbing heptanol also began to appear in the effluent after 1.5 pore volumes of flush were injected as shown in Figure 7. The heptanol effluent concentration remained relatively constant over the first 1.5 pore volumes, then increased significantly over the next three pore volumes. During this time, the heptanol concentration rose steadily from 3,594 to 4,920 µmole/l (64.7 to 88.6 ppm). It decreased slightly, then rose sharply as a wave

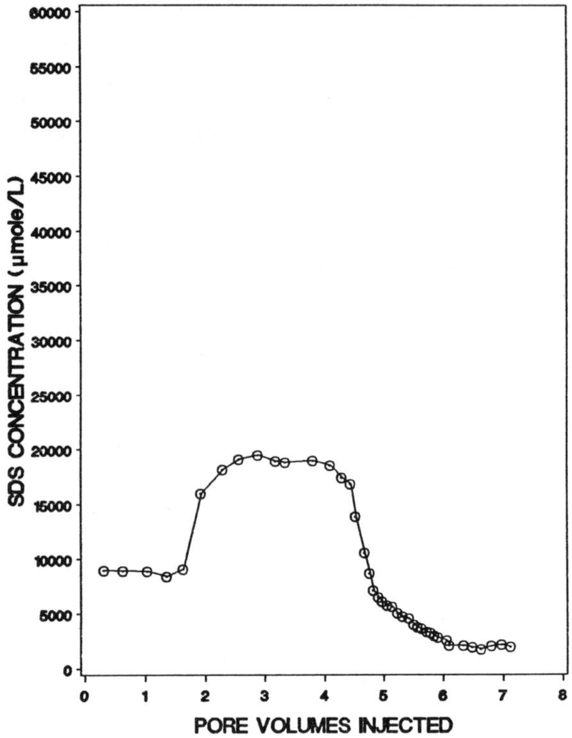

Figure 6. Run I; Effluent SDS Profile for Bed Stripping Step.

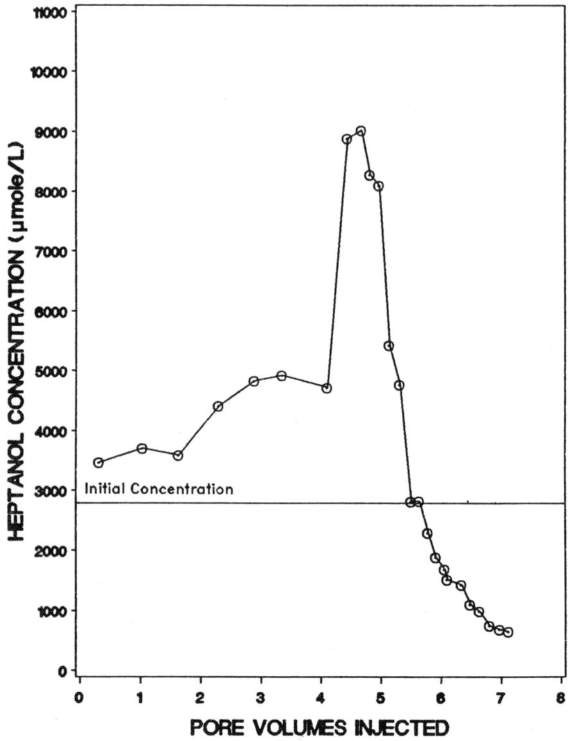

Figure 7. Run I; Effluent 1-Heptanol Profile for Bed Stripping Step.

of concentrated heptanol eluted out over the next 1.5 pore volumes. The heptanol concentration rose from 4,720 to 9,020 mole/l (85.0 to 162 ppm). This represents an over three fold increase in concentration from the initial feed of 2,785 mole/l (50.1 ppm). After this wave had eluted, the heptanol concentration at the outlet fell quickly and slowly approached zero.

It should be pointed out that it was necessary to decrease the flow rate during the Solute Recovery step from 0.3 to 0.1 ml/min at 4.5 pore volumes. During this last step, the pressure drop across the glass column increased from about 70 psig to near the column's limit of 500 psig. Subsequent runs were done in a stainless steel column, with a presssure limit of 9,000 psig.

Effluent stream profiles for the Bed Preparation and Adsolubilization steps of Run II are presented in Figures 8, 9, and 10. Comparing these to the same steps in Run I, Figures 3, 4, and 5, one finds very

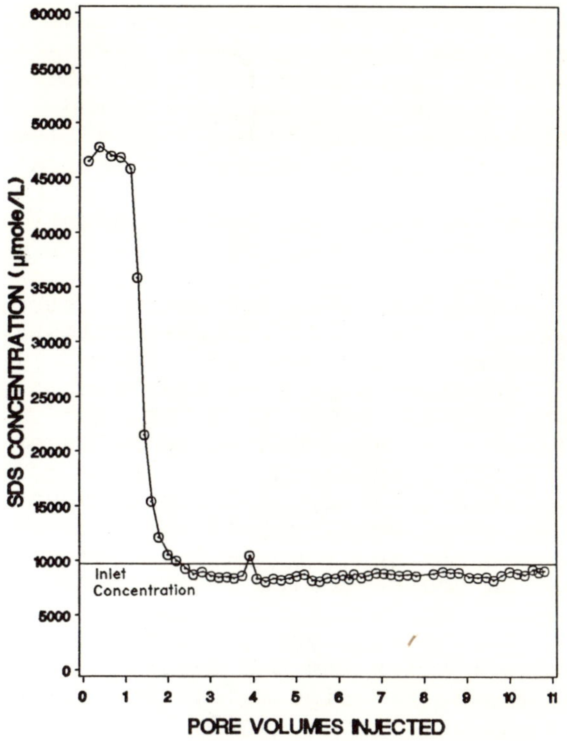

Figure 9. Run II; Effluent SDS Profile for Adsolubilization Step.

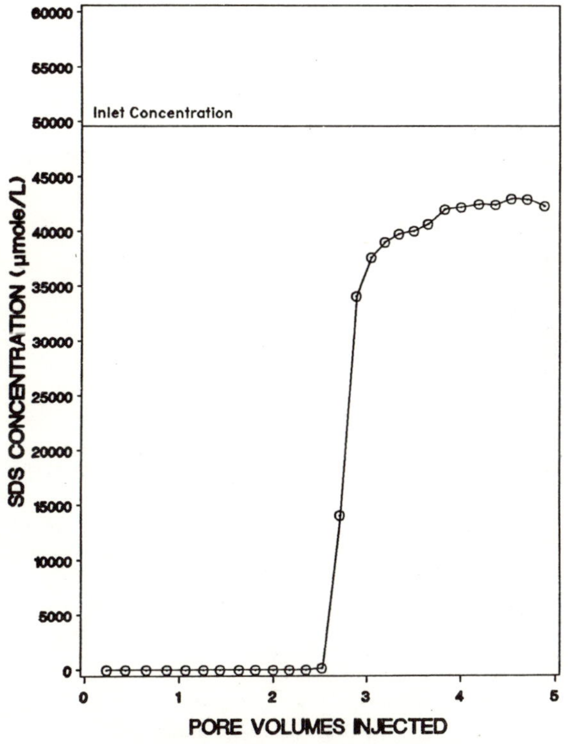

Figure 8. Run II; Effluent SDS Profile for Bed Preparation Step.

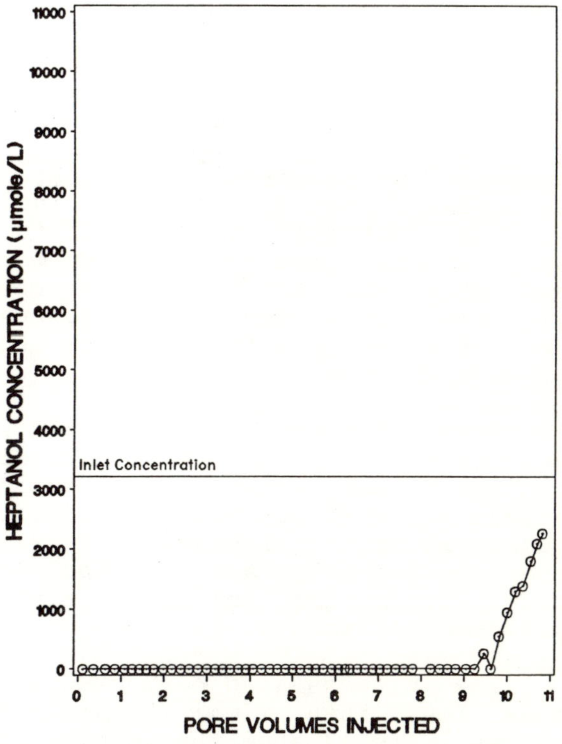

Figure 10. Run II; Effluent 1-Heptanol Profile for Adsolubilization Step.

little difference between the two; in fact, Runs I and II were virtually identical for these first two steps. Since the two runs were performed at different column flow rates (see Table 1), the results would indicate that mass transfer effects at the flow rates examined were not significant.

For the Bed Stripping step, however, the results were slightly different. As shown in Figure 11, the SDS eluted in a broader and less concentrated wave than in Run I (Figure 6). Although it took longer, the SDS appeared to be more completely eluted, a definite advantage, especially if surfactant recovery is an important consideration in the economics of the process.

For Run II, the effluent heptanol profile for the Bed Stripping step is shown in Figure 12. The outlet concentration rose slowly from 2,275 to 4,115 μmolar (41.0 to 74.1 ppm) after 2.5 pore volumes of flush solution were injected, remained relatively constant for the next three pore volumes, then rose sharply when a very concentrated wave of heptanol appeared at the outlet after 5 pore volumes. The heptanol increased from an initial 3,214 to 10,360 μmolar (57.9 to 186.5 ppm), again, an over three fold increase from its initial concentration! Compared to the effluent heptanol profile in Run I (Figure 7), the heptanol in Run II took longer to elute out, but eluted more completely and in a much sharper wave; moreover, unlike Run I, the heptanol eluted completely from the column after two more pore volumes.

From the results presented so far, it is clear that this process can produce substantial concentration increases for organics in dilute aqueous solution. Increases of 225% were consistently achieved for both runs, with minimal losses of either surfactant or solute.

The process also performed well under less than ideal conditions. Figures 13 to 17 show the results for Run III where the inlet heptanol concentration was very low (See Table 1), and the bed was not completely

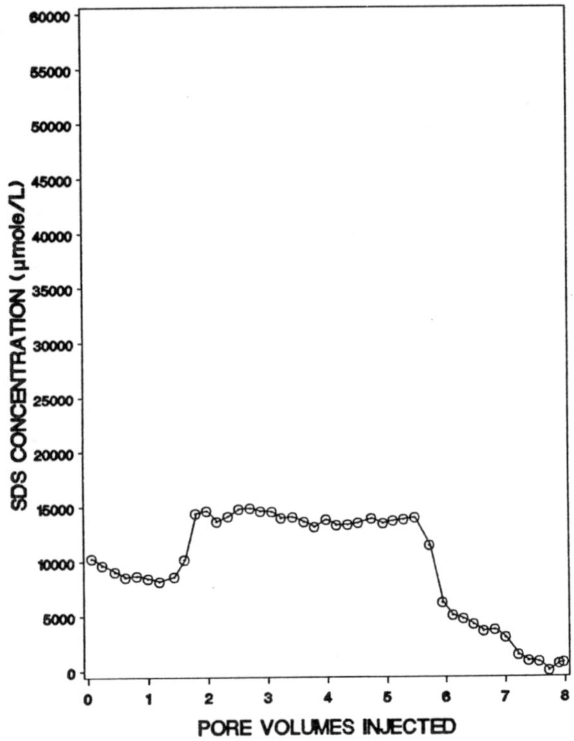

Figure 11. Run II; Effluent SDS Profile for Bed Stripping Step.

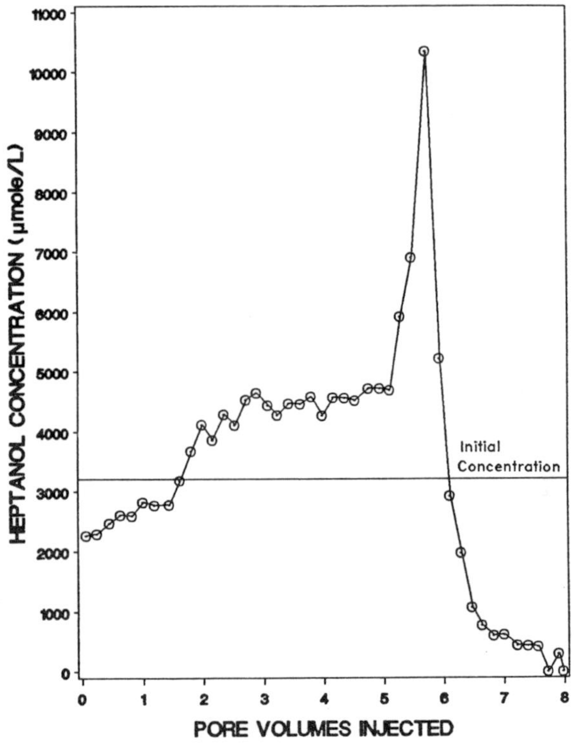

Figure 12. Run II; Effluent 1-Heptanol Profile for Bed Stripping Step.

equilibrated to the desired pH range of 8.2 to 8.4.

When the run was begun, the effluent pH of the stream used to condition the bed prior to each run was 9.78. The PZC of alumina is 9.0-9.5. Above this pH, one expects very little to no adsorption of SDS on alumina. Figure 13 shows the effluent SDS profile for this run. Substantial SDS adsorption was observed , with the surfactant appearing in the effluent after 1.5 pore volumes. The SDS breakthrough was not as sharp as in previous runs, but it was quite surprising to see this much adsorption at all.

Figures 14 and 15 show the effluent SDS and heptanol profiles during the Adsolubilization step. As in the previous runs, the effluent SDS concentration reached the new inlet concentration after about 2.5 pore volumes. Surprising was the heptanol profile shown in Figure 15. Complete heptanol removal was achieved for almost seven pore volumes. This is quite amazing when compared to Runs I and II, where hep-

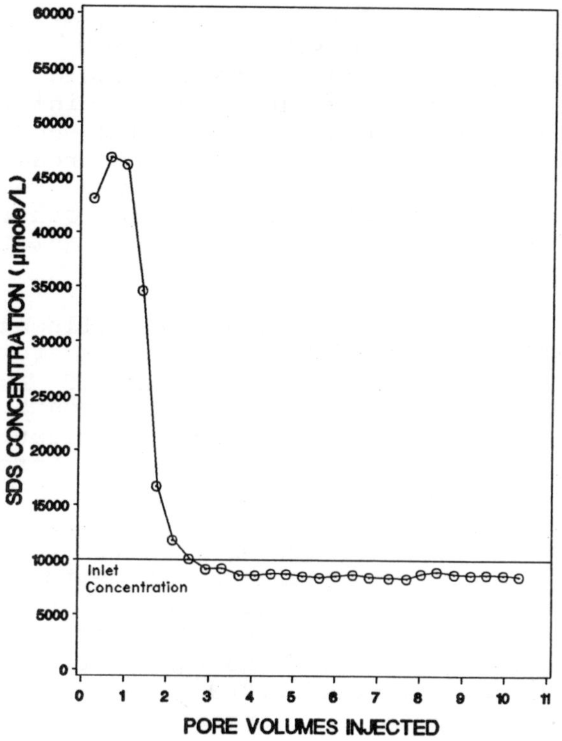

Figure 14. Run III; Effluent SDS Profile for Adsolubilization Step.

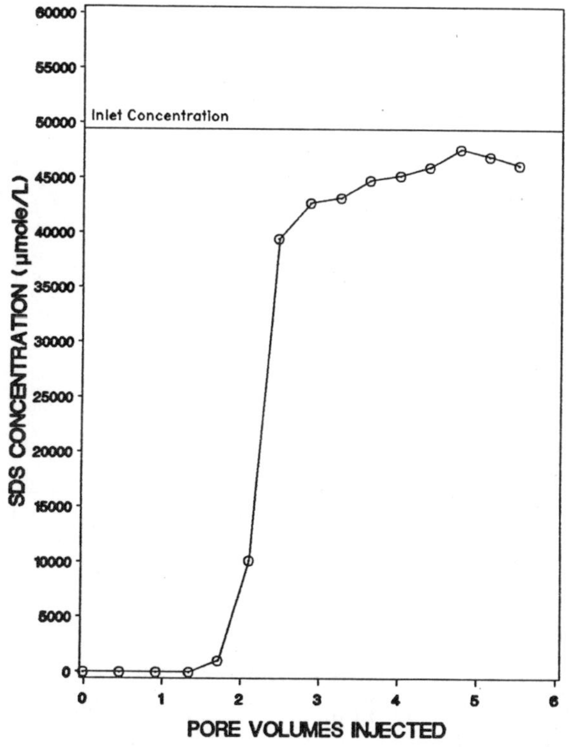

Figure 13. Run III; Effluent SDS Profile for Bed Preparation Step.

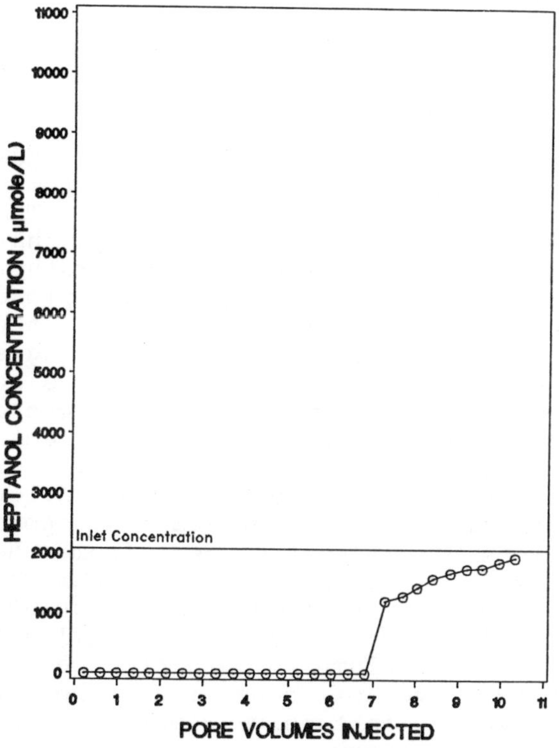

Figure 15. Run III; Effluent 1-Heptanol Profile for Adsolubilization Step.

tanol appeared in the effluent after 9.5 pore volumes. It might be suggested that this behavior could be the result of the low heptanol inlet concentration used, which might cause the bed to take longer to reach saturation; however, this type of behavior was not observed in other runs done at different heptanol inlet concentrations.

The SDS and heptanol effluent profiles for the Bed Stripping step are shown in Figures 16 and 17. As in Runs I and II, the SDS desorbed in an almost square wave. The effluent SDS concentration approached zero after 6.5 pore volumes.

Figure 17 shows once again that a significant increase in heptanol concentration was achieved. In this instance, the concentration of heptanol was more than doubled, rising from an initial 2,069 μmole/l to 4,431 μmole/l (37.2 ppm to 79.7 ppm). This last run gives strong support to the argument made earlier that the process can perform over a wide range of pH's with satisfactory results.

CONCLUSIONS

The results reported here constitute a preliminary feasibility study of a new separation process, admicellar chromatography. No attempt was made to optimize the system or to systematically vary process parameters. The runs were not performed in ways that might improve performance, but rather in ways that were designed to improve their interpretability. All of the separations reported here were single pass experiments, i.e. there was no recycle. Incorporating a recycle stream could have significant effects on the recovery of the solute. For example, during the Adsolubilization step, in all cases at least 4.0 pore volumes of effluent eluted from the column before the wave of concentrated solute reached the outlet. Recycling this part of the effluent into the column might produce a second wave of concentrated solute and would certainly improve the overall recovery.

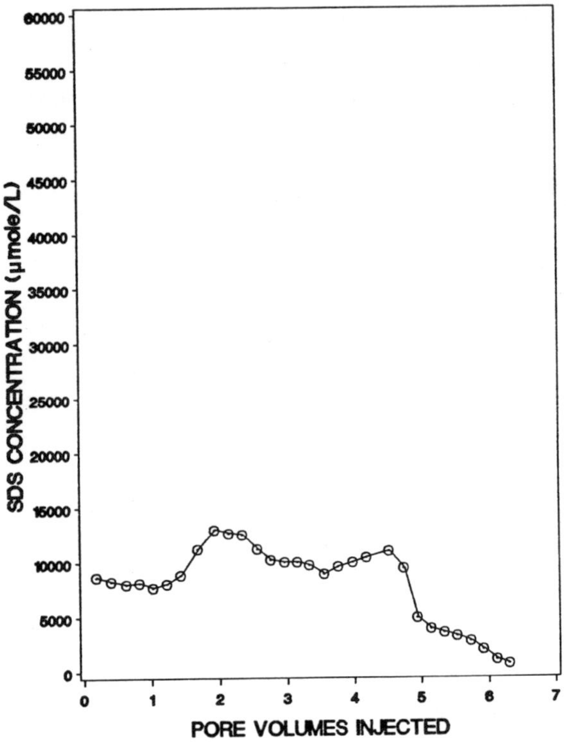

Figure 16. Run III; Effluent SDS Profile for Bed Stripping Step

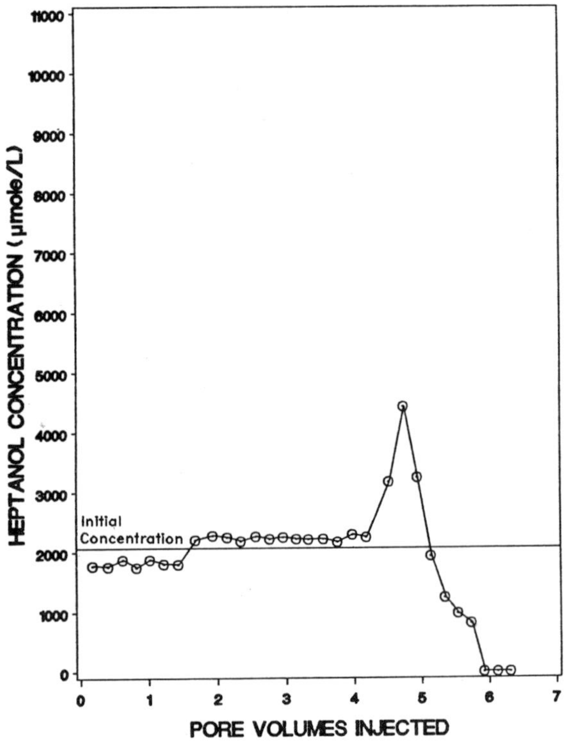

Figure 17. Run III; Effluent 1-Heptanol Profile for Bed Stripping Step

Further increases in concentration might also be achieved by making this a stagewise separation, as is done in extraction systems. Say the heptanol, as in Run II, is initially at 3,214 µmole/l (57.9 ppm). Using this process, the first run through the column would increase the heptanol concentration to 10,360 µmole/l (186.5 ppm). If the process was repeated on this effluent, the heptanol could be brought to near its solubility limit in water (at the conditions studied here) of 14,150 µmole/l (254.7 ppm). This second stream could then be cooled slightly, causing the heptanol to phase separate. The final result would be a pure heptanol phase obtained using a process that required little external energy input.

Plans for future work include the application of this process to more commercially valuable solutes such as biological compounds and pharmaceuticals. Other work hopes to examine ways to sharpen the desorption wave of the solute and surfactant. This would allow faster turn around times on the fixed beds, and might possibly chieve greater increases in conce.tration as well, an important consideration for potential industrial application.

LITERATURE CITED

1. Westby, C.A. and W.R. Gibbons, Biotechnology and Bioengineering, 24, 1681, (1982).
2. Westby, C.A. and W.R. Gibbons, Biotechnology and Bioengineering, 25, 2127, (1983).
3. Cooney, C.L., Science, 219, 728, (1983).
4. Harwell, J.H., J.C. Hoskins, R.S. Schechter, and W.H. Wade, Langmuir, 1, 251, (1985).
5. Cases, J.M. and B. Mutaftshiev, Surf. Sci., 9, 57-72, (1983).
6. Gaudin, A.M. and D.W. Fuerstenau, Min. Eng., 7, 66, (1955).
7. Scamehorn, J.F., R.S. Schechter, and W.H. Wade, J. Colloid Interface Sci., 85, 463, (1982).
8. Yopps, J.A. and D.W. Fuerstenau, J. Colloid Interface Sci., 19, 61, (1964).
9. James, R.O. and G.A. Parks, Surface Colloid Sci., 12, (1982).
10. Rosen, M.J., Surfactants and Interfacial Phenomena, Wiley, New York, Chapter 4, (1978).

A MODEL OF OPEN-GRADIENT MAGNETIC SEPARATION FOR COAL CLEANING USING A SUPERCONDUCTING QUADRUPOLE FIELD

R. D. Doctor, C. B. Panchal and C. E. Swietlik ■ Argonne National Laboratory
Energy and Environmental Systems Division
9700 South Cass Ave., Argonne, IL 60439

The application of a 3.6-T superconducting quadrupole magnet for the open-gradient magnetic separation of ash-forming minerals and pyritic sulfur from coal is presented. The results of modeling in support of the experimental program are described, together with specific parameters to be investigated (such as coal particle-size distribution and particle magnetic susceptibility).

A novel magnetic-separation technique is considered for cleaning pulverized coal to separate the ash and pyritic fractions, which in combination with strongly paramagnetic trace minerals have a net weak magnetic (paramagnetic) susceptibility, from the non-magnetic (diamagnetic) maceral fractions. This work focuses on results for paramagnetic particles only; the equations presented in this paper must be modified in order to consider diamagnetic materials.*

SCOPE

Open-gradient magnetic separation (OGMS) deflects paramagnetic ash and pyrite particles falling under the influence of gravity and draws them toward the bore wall; at the same time, the diamagnetic macerals that constitute the clean coal are repulsed from the field toward the center of the bore. By using the high magnetic field strengths available from superconducting magnets to act on pulverized coal falling through the bore, practical spatial separations of these two fractions can be achieved in a short distance. In the model developed here, the system uses a 1-m-long superconducting quadrupole magnet with a highly isotropic field having a gradient of 60 Wb/m^3. Such a system is potentially attractive for three reasons: the power cost for the separation is low, the separator has no moving parts, and the pulverized-coal product has no added surface moisture. The equations governing the movement of paramagnetic coal fractions in this system are presented.

CONCLUSIONS AND SIGNIFICANCE

The main findings of this study are:

1. An OGMS system using a superconducting quadrupole magnet could effectively separate the ash and pyritic coal fractions from the macerals.

2. The process is designed for operation in a continuous, on-line mode.

3. The system would employ a modest-size superconducting quadrupole magnet, 1 m long, with a gradient of 60 Wb/m^3. The technology for mass-producing these magnets has been

*Work performed for the Illinois Center for Research on Sulfur in Coal under subgrant no. 84-142 under an agreement with the U.S. Department of Energy.

developed by the national laboratories for their accelerator programs.

4. The proposed OGMS system is basically a static separator with no moving parts. The components required for the process (fluidized coal-handling equipment and superconducting magnets) can be expected to operate with a high degree of reliability and availability and with a minimum of maintenance.

5. Parasitic power costs for maintaining the superconducting magnetic field are low.

6. The optimal size range for the pulverized coal is 50 to 150 µm.

7. This novel separation technology could be used in conjunction with commercially available coal-cleaning systems.

BACKGROUND

Coal beneficiation is well established commercially for upgrading run-of-mine (ROM) coals for both metallurgical and steam-raising markets; at present, more than 522 coal-cleaning plants are operating in the U.S. Conventional coal-beneficiation processes encounter inherent difficulties in the cleaning of fine-size coals, because the traditional definition of clean coal as the low-ash product of a separation does not meet the optimum economic condition. This is especially true of processes that successfully clean fine coal fractions to yield a relatively low-ash product but have poor thermal recoveries because of the surface moisture that clings to fine coal (1).

The requirements for separating, handling, and dewatering fine-size coal are so expensive that a significant fraction is usually rejected to the waste pond. Processes (such as magnetic beneficiation) that operate on a dry, rather than wet, coal feed are favored candidates for study as advanced coal-cleaning systems.

It has been observed experimentally that when pulverized coal passes freely through a magnetic-field gradient, the mineral components tend to separate from the organic components due to inherently different magnetic characteristics. The pyrites and ash-producing components are paramagnetic, whereas the organic components (including the "clean" coal) are diamagnetic. A paramagnetic particle becomes slightly magnetized in the presence of a magnetic field so that, if the field is nonuniform, the particle will tend to move toward the region of higher field intensity. A diamagnetic particle in a nonuniform magnetic field tends to move in the direction of lower field intensity. Open-gradient magnetic separation takes advantage of this principle to separate the organic fraction of the coal from the ash and pyritic fractions.

Recent interest in magnetic coal-beneficiation techniques has focused on the application of high-gradient magnetic separation (HGMS), which has been used commercially for beneficiation of mineral ores (2-8). In HGMS, the high field gradients are induced locally around ferromagnetic filaments placed in a uniform magnetic field, and paramagnetic minerals are trapped on the filaments and removed from the fluid stream containing the organic coal.

Technically, HGMS is a semibatch-type process, inasmuch as the filaments are moved out of the magnetic field so that the particles will fall off. In contrast, an OGMS process can operate continuously, because the mineral particles are spatially separated as they move through the magnetic field instead of being captured inside the separator. Superconducting magnets can produce the necessary field volumes and gradients at low power costs. Moreover, the production and operation of such magnets has been well established through research projects in high-energy physics. However, very little experimental work has been done on OGMS, except for experiments that have been carried out at Oak Ridge National Laboratory (ORNL) (9). Larger-scale testing of this coal-beneficiation process using a system configured as shown in Figure 1 is a logical next step in the overall research program.

COAL PREPARATION

In the ORNL work, run-of-mine coal (Western Kentucky #9) was pulverized until it could pass through a 600-μm sieve (28 mesh) while being retained on a 45-μm sieve (325 mesh) (10,11). Sufficient sieving data are available to reconstruct the size distribution of the coal. The ORNL investigators chose this size range so that the associated costs of the coal work-up would closely parallel conventional gravity-separation coal-cleaning technologies.

This size range may not be optimal for magnetic beneficiation, and Oder (Bechtel) has suggested grinding down the coal to 10 μm (12,13). A grind to the 10-μm range would presumably reduce the various crystalline structures within the coal matrix to a point where all the pyritic and other mineral fractions within the run-of-mine coal could be considered as containing a single mineral or carbonaceous phase. If the coal were ground to the 10-μm range, there would be sufficient data to simulate the system. However, it is likely the economic, materials-handling, and safety considerations would dictate less severe grinding.

If the size remains in the 150-600-μm range, sufficient data to conduct this

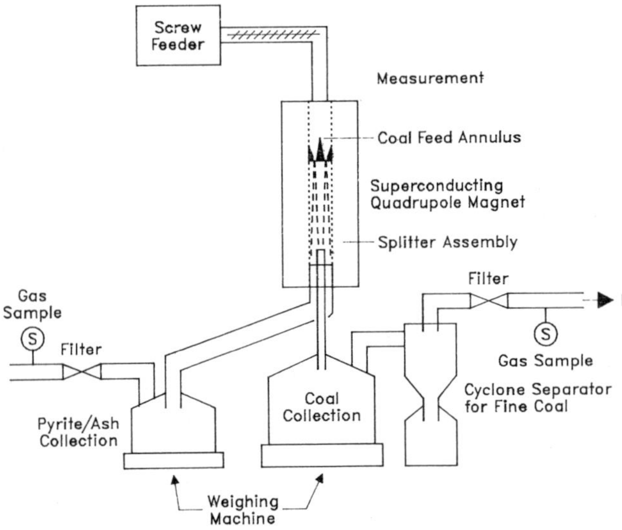

FIGURE 1 Test Configuration for Open-Gradient Superconducting Magnetic Coal Separator

assessment will be considerably more difficult to acquire, because for each size range (150-200 μm etc.) the particles may represent one to several separate crystalline phases of the following: carbonaceous matter, illite, kaolinite, quartz, calcite, dolomite, siderite, pyrite, feldspar, muscovite, apatite, bassanite, and others.

PARAMAGNETIC SUSCEPTIBILITY

In reviewing the bulk susceptibility of the ash and pyritic fractions of the coal, we find that our values are within the range of susceptibilities appropriate to Illinois No. 6 coal (14). The work of Liu and Bluhm will serve to illustrate the reasonable and conservative nature of our work (15,16). Using our model, we examined the movement of a particle that is partially pyrite and other mineral matter with moisture, and we used a conservative value of 3.1×10^{-6} emu/g·Oe. The work done on HGMS has shown that grinding to the 50-μm size range will liberate ash and pyrite with nearly three orders of magnitude of variation in magnetic susceptibility values. The reason for the large measured variation in the field samples, as explained by Wechsler, is that once liberated, natural pyrite has a small but appreciable magnetic susceptibility of approximately 3.0×10^{-6} emu/g·Oe (17). Monoclinic pyrrhotite, which is believed to occur together with pyrite in some coals, has a much higher susceptibility, 280×10^{-6} emu/g·Oe. Even in small concentrations, the presence of pyrrhotite within a pyrite inclusion can greatly increase the probability of trapping the inclusion in the matrix of a high-gradient separator. As an additional benefit, many of the ash-forming inorganic minerals in coal also have appreciable magnetic susceptibilities and should be amenable to capture by magnetic means, provided that they are first liberated from the organic matter by size reduction.

SINGLE-PARTICLE BALLISTIC MODEL IN OGMS SYSTEM

A ballistic model that can be used for single particles passing through an intense magnetic-field gradient is presented. Particle-particle interactions and complex behavior due to particle-size distribution are considered later.

Quadrupole magnets are capable of producing a very uniform magnetic field gradient throughout a relatively large working volume in a cylindrical bore (18). Furthermore, superconducting magnets, operating at temperatures on the order of 4K, are able to generate intense magnetic-field gradients with very low power consumptions. Once energized, the electricity consumption in the magnet is negligible, and only the refrigeration power is significant. One estimate for a magnetic-separation process puts the energy savings over conventional magnets at about 75% (19). The quadrupole magnet modeled is based on a proven design employed by particle accelerators; several hundred of these magnets have been deployed and reliably operated.

We make the following assumptions:

1. The magnetic-field axis (z-axis) is vertical, and the gradient is constant in the radial direction (r-axis) from the bore center to the wall and is isotropic (invariant) in the Θ-axis.

2. Particle motion along the r-axis and z-axis will use either Stokes's Law (Re <0.3) or an empirical correlation (0.3 < Re <1000), where "Re" signifies Reynolds number.

3. Macerals, pyrite, and ash can be approximated as spherical particles.

MAJOR FORCES ON COAL PARTICLES IN THE MAGNETIC FIELD

The major forces on the particles in the magnetic separator are magnetic force, fluid-drag force, gravitational force, and inertial forces.

Magnetic Force

The magnetic force acting on a weakly magnetic particle is proportional to the magnitude and gradient of the magnetic field. The induced magnetic moment of the particle (μ) is proportional to the field and the magnetic susceptibility (χ) of the particle. The direction of the force is the direction of increasing magnetic-field strength for paramagnetic particles (they are attracted to the magnet) and decreasing field strength for diamagnetic particles (they are repelled from the magnet). Since the carbonaceous particles in the pulverized coal are diamagnetic, they will be expelled from high magnetic fields to lower magnetic fields at the bore center, while the pyrites and other minerals will be accelerated to the high fields near the bore wall.

Mathematically, the magnetic force acting on a spherical particle for a multi-dimensional case will be:

$$[F]\ \text{magnetic} = \frac{\pi}{6} D_p^{\ 3} \cdot \chi \cdot \frac{[B]}{\mu_o} \cdot \nabla[B] \qquad (1)$$

For angular motion, the field gradient for the quadrupole magnet to be used is constant for all Θ values. For radial motion, this magnet possesses the special property of having a constant r-gradient (60 T/m), which leads to [B] = (dB/dr) ·r for the geometry of this quadrupole magnet. Hence, Equation (1) may be simplified to yield the following:

$$[F]\ \text{magnetic} = \frac{\pi}{6} D_p^3 \cdot \chi \cdot \frac{[B]}{\mu_o} \cdot \frac{dB}{dr} = \frac{\pi D_p^{\ 3} \chi}{6\ \mu_o} \cdot \left(\frac{dB}{dr}\right)^2 \cdot r \qquad (2)$$

Fluid-Drag Force

The fluid-drag force acting on the particle will be determined by the relative motion of the particles with respect to the gas carrier fluid. For the low Reynolds number (Re <0.3) expected for the radial motion of the particle, the Stokes's Law formulation may be used. The force on a spherical particle at low Reynolds number can be expressed as follows:

$$[F]\ \text{fluid drag} = 3\pi\eta D_p\left([v_p]-[v_f]\right) \qquad (3)$$

Alternative formulations for the drag force that employ empirically derived drag coefficients will be used for those flow regimes where Reynolds numbers greater than 0.3 are found. These drag forces will be represented as follows:

$$[F]\text{ fluid drag} = \frac{\pi}{8} C_D D_p^2 \rho_f ([v_p] - [v_f])^2 \tag{4}$$

Equation (4) is applicable until Re >1000, at which point Newton's Law for drag force would be employed. However, experimental work is not envisioned for this regime.

Gravitational Force

The coal particles will experience the effects of gravity while entrained in the carrier fluid. Mathematically, the action of this force on a spherical particle may be expressed as:

$$[F]\text{ gravity} = (\rho_p - \rho_f) \frac{\pi D_p^3}{6} \cdot [g] \tag{5}$$

Because the basis of the magnetic separation is the radial magnetic force exerted by the magnet, and because this force is so much smaller than the gravitational force, it is clear that the magnet must be so arranged that the two forces are orthogonal (i.e., the magnet bore must be parallel to the gravitational force).

Inertial Force

This is the rate of change in velocity for the particle. It may be expressed mathematically as follows:

$$[F]\text{ inertia} = \rho_p \cdot \frac{\pi D_p^3}{6} \cdot \frac{d^2 r}{dt^2} \tag{6}$$

EXPRESSIONS FOR THE RADIAL MOTION OF PARTICLES

For the radial motion of particles, the force balance in the radial direction is derived from the following equation:

$$[F]\text{inertia} = [F]\text{magnetic} + [F]\text{fluid drag} \tag{7}$$

On the assumption that the magnet is vertical, the radius is orthogonal to gravity, and no gravitational-force term appears. Additionally, the fluid will have no initial velocity component in the r-direction.

Stokes's Law Regime

For the flow regime in which Stokes's Law is applicable (Re <0.3), Equation (7) can be expressed as follows:

$$\rho_p \cdot \frac{\pi D_p^3}{6} \cdot \frac{d^2 r}{dt^2} = \frac{\pi D_p^3 \chi}{6\,\mu_o} \cdot \left(\frac{dB}{dr}\right)^2 \cdot r - 3\pi\eta\, D_p \frac{dr}{dt} \tag{8}$$

Since dB/dr is constant, Equation (8) can be rearranged to constitute a second-order linear differential equation, for which the solution is given by the following:

$$r(t) = r_o\,[\exp(\lambda_1 t) - \exp(\lambda_2 t) \cdot \lambda_1/\lambda_2]/(1 - \lambda_1/\lambda_2) \tag{9}$$

where:

$$\lambda_{1,2} = -\frac{9\eta}{\rho_p D_p^2} \pm \frac{1}{2}\left[\left(\frac{18\eta}{\rho_p D_p^2}\right)^2 + \frac{4\,\chi\left(\frac{dB}{dr}\right)^2}{\rho_p\,\mu_o}\right]^{1/2} \tag{10}$$

This equation can be conveniently solved (by computer) for the time required to reach the wall, given the particle's initial position, using the values in Table 1. (For further discussion, see App. A.)

Intermediate-Flow Regime

The drag force on the particle in Equation (7) will follow the formulation in Equation (4) for the flow regimes intermediate between Stokes's Law and Newton's Law (0.3 < Re <1000):

Table 1. Parameters for Simple Ballistic Model, Considering a 150 μm Pyrite Particle

Identifier	Value	Units
χ_{pyrite}	19×10^{-6}	Dimensionless (formulated in terms of MKS units)
B_o	3.6	Wb/m^2 (at the cold bore wall)
dB/dr	60	$(Wb/m^2)/m$
μ_o	$4\pi \times 10^{-7}$	(Wb/A·m)
η	175×10^{-6} 1.75×10^{-5}	P (poise) $N \cdot s/m^2$ (nitrogen, 60°F, 1 atm)
D_p	150 1.5×10^{-4}	μm m
r	0-0.05	m (variable)
D_{Bore}	0.064	m
ρ_p	4.9 4900	g/cm^3 kg/m^3
ρ_f	1.2	kg/m^3 (nitrogen, 60°F, 1 atm)
ρ_{coal}	1300-1600	kg/m^3

$$[F]r\text{-inertia} = \frac{\pi D_p{}^3 \chi}{6\mu_o} \cdot \left(\frac{dB}{dr}\right)^2 \cdot r - \frac{\pi}{8} C_D D_p{}^2 \rho_f v_p{}^2 \quad (11)$$

This leads to a nonlinear differential equation. Since this equation has no explicit solution, it must be solved numerically. (For further discussion, see App. B). It is assumed that the drag coefficient will depend on the velocity along the vertical axis as well as the radial axis.

EXPRESSIONS FOR THE VERTICAL MOTION OF PARTICLES

For the vertical motion of particles, Stokes's Law may or may not be applicable. Therefore, a general formulation is used for the drag force:

$$[F]z\text{-inertia} = [F]\text{gravity} + [F]\text{fluid drag} + [F]z\text{-inertia}(o) \quad (12)$$

Stokes's Law Regime

A general solution to Equation (12) in the flow regime governed by Stokes's Law is

$$z = \frac{gD_p{}^2\rho_p}{18\eta} \cdot \left(t + \left[\frac{D_p{}^2\rho_p}{18\eta} - \frac{v_o}{g}\right] \cdot \left[\exp\left(18\eta\, t/D_p^2\, \rho_p\right) - 1\right]\right) \quad (13)$$

For details of the derivation, see App. C.

Intermediate-Flow Regime

A general solution (see App. D) in the regime intermediate between Stokes's and Newton's Laws yields the following:

$$z = \frac{8D_p\rho_p}{C_D\,\rho_f} \cdot \left(\ln[1 + k_1 \cdot \exp(t/\theta)] - \ln(1 + k_1)\right) - \gamma t \quad (14)$$

where:

$$k_1 = \frac{\gamma + v_o}{\gamma - v_o} \text{ (dimensionless)} \quad (15)$$

$$\theta = \left(\frac{3C_D\rho_f}{D_p\rho_p} \cdot \left(g + \frac{v_o{}^2}{2}\right)\right)^{-1/2} \quad (16)$$

$$\gamma = \left(\frac{4D_p\,\rho_p}{3C_D\,\rho_f} \cdot \left(g + \frac{v_o{}^2}{2}\right)\right)^{1/2} \quad (17)$$

This establishes the basic residence time and required bore length for the separator in the intermediate-flow regime.

RESULTS

The equations presented in this paper were programmed in FORTRAN to provide guidance in the development of the experimental program for open-gradient magnetic separation. The examples presented here are for completely liberated pyrite, employing conservative susceptibility data. The travel time for a falling pyrite crystal to hit the bore wall is indicated in Figure 2, while the required length of the superconducting quadrupole with this gradient to draw the pyrite crystals to the wall is shown in Figure 3. The calculated trajectories of the pyrite crystals in the bore appear in Figure 4. It can be seen that even a short bore length should be adequate for the removal of small pyrite crystals.

ACKNOWLEDGMENTS

This work was supported by funding from the Illinois Center for Research on Sulfur in Coal, Champaign, Illinois, as well as discretionary funding from the Energy and Environmental Technology branch of Argonne National Laboratory. The computer code was developed by the authors with assistance from Karen Foote and John Anderson, Argonne National Laboratory (ANL). The authors wish to acknowledge Barbara Rogowski and Roberta Walker, ANL, for word-processing effort and Floyd Bennett, ANL, for editing the manuscript.

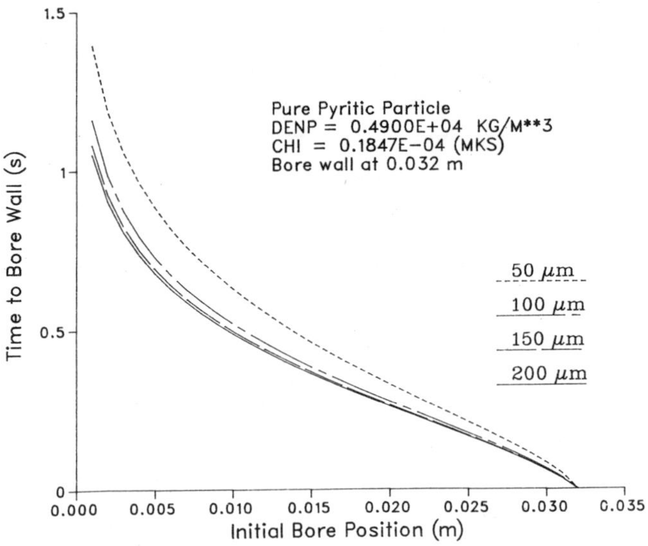

FIGURE 2 Travel Time in the Bore (The pyrite particles in an open-bore magnetic separator will encounter the wall at the times calculated from the model. The drag forces considerably reduce the velocity of 50-μm and smaller-diameter particles, which should lead to collisions and agglomeration.)

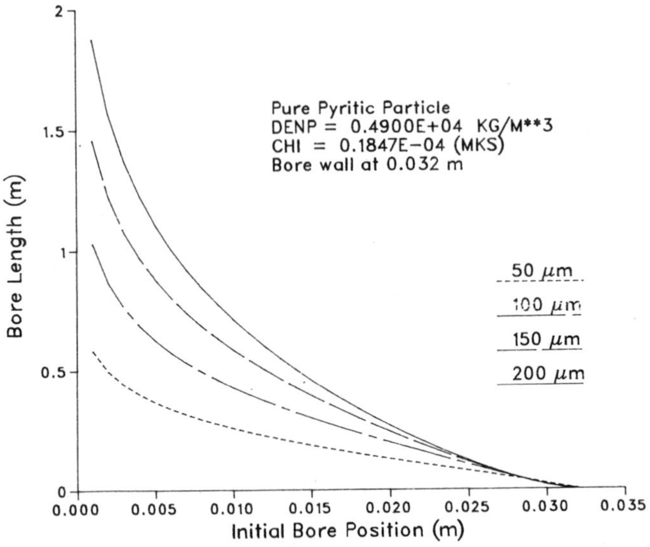

FIGURE 3 Required Bore Length for Motion of Pyrite Particles in an Open-Bore Magnetic Separator (The 1-m superconducting quadrupole magnet available should be flexible enough to permit the testing and separation of a wide range of particles.)

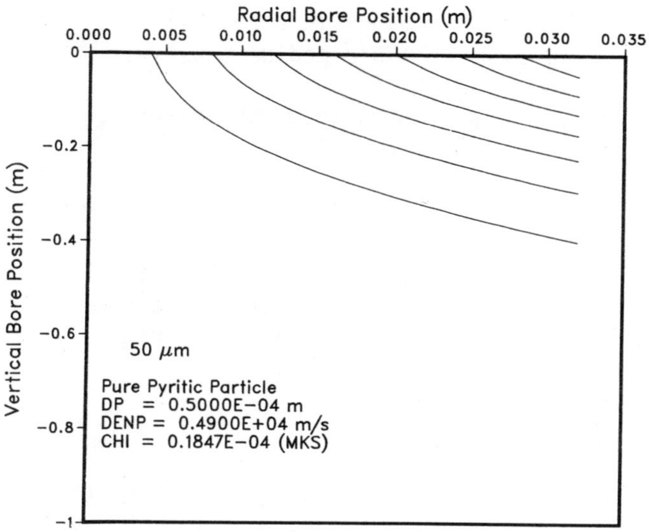

4(a) 50-μm Particles

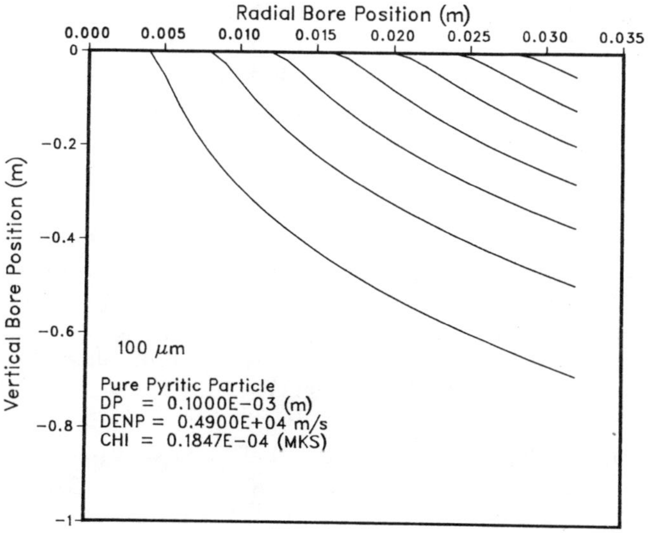

4(b) 100-μm Particles

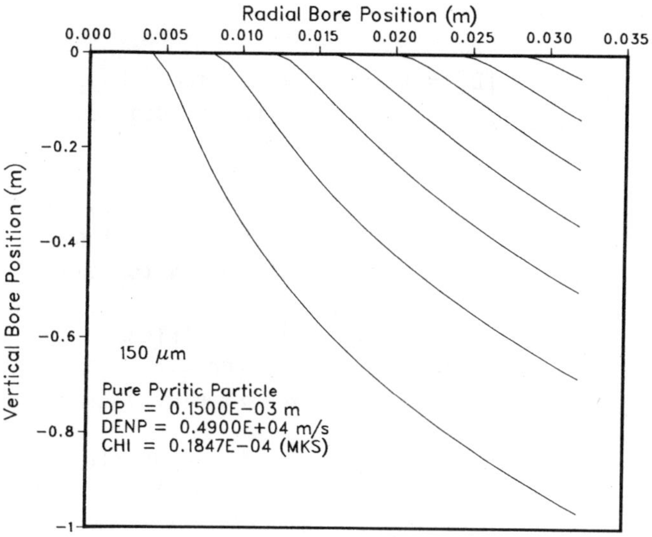

4(c) 150-μm Particles

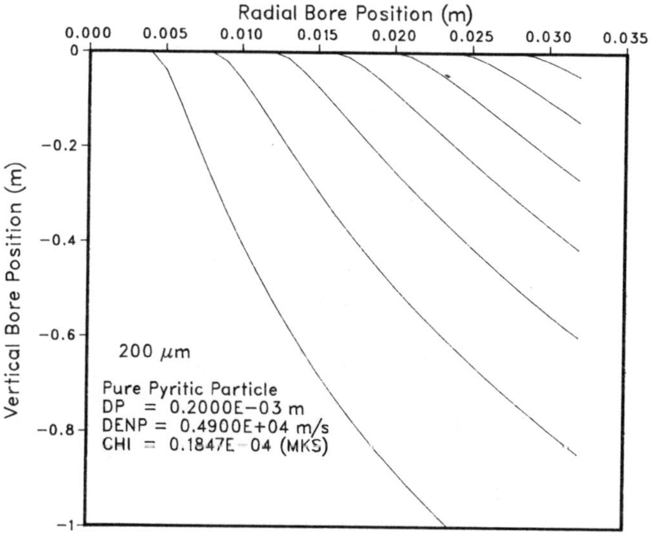

4(d) 200-μm Particles

FIGURE 4 Trajectories of Pyrite Particles (The trajectories of pyrite particles falling through the open-gradient magnetic separator show that even a relatively short field could effectively remove fine pyrite crystals.)

NOTATION

[B] = magnetic field strength in a vacuum with coordinates r, Θ, z (T = Wb/m^2)

C_D = 24/Re (1 + 0.15 $Re^{0.687}$) (an empirical drag coefficient for Re <0.3; dimensionless)

C_D = 18/$Re^{0.6}$ (an empirical drag coefficient for 0.3 < Re <1000; dimensionless)

dB/dr = gradient of the magnetic field in the radial direction (T/m = Wb/m^3)

D_p = particle diameter (m)

d[v_p]/dt = acceleration experienced by the particle (m/s^2)

[F] fluid drag = fluid-drag force (N)

[F] gravity = gravitational force (N)

[F] inertia = inertial force on a particle (N)

[F] magnetic = magnetic force (N)

[g] = gravitational acceleration (9.81 m/s^2)

r = magnet bore radius (m)

t = time (s)

[v_f] = velocity vector of the fluid (m/s)

[v_p] = velocity vector of the particle (m/s)

v_o = initial velocity of the particle in the vertical direction (z-axis) (m/s)

η = fluid viscosity ($N\cdot s/m^2$)

λ = terms for differential equations (1/s)

μ_o = magnetic permeability of the fluid medium (Wb/A·m)

ρ_p = density of the particle (kg/m^3)

ρ_f = density of the fluid (kg/m^3)

χ = volume magnetic susceptibility (dimensionless; formulated in the MKS system)

LITERATURE CITED

1. Norton, G., Coal Mining 21(9):38 (Sept. 1984).

2. Hise, E.C., I. Wechsler, and J.M. Douhlin, "Separation of Dry Crushed Coals by High-Gradient Magnetic Separation," Oak Ridge National Laboratory Report ORNL-5571 (Oct. 1979).

3. Hise, E.C., I. Wechsler, and J.M. Douhlin, "The Continuous Separation of Dry Crushed Coal at One Ton per Hour by High-Gradient Magnetic Separation," Oak Ridge National Laboratory Report ORNL-5763 (Dec. 1981).

4. Oder, R.R., IEEE Transactions on Magnetics, MAG-12 (5):428-435 (Sept. 1976).

5. Liu, Y.A., and M.J. Oak, AIChE Journal 29(5):780 (Sept. 1983).

6. Cohen, E., et al., IEEE Transactions on Magnetics, MAG-12 (5):503-506 (Sept. 1976).

7. Cohen, E., et al., IEEE Transactions on Magnetics, MAG-12 (5):552-555 (Sept. 1976).

8. Parker, M.R., Physical Technology 12:263-268 (1981).

9. "Impact of Coal Cleaning on the Cost of New Coal-Fired Power Generation," Bechtel National, Inc., Electric Power Research Institute Report EPRI-CS-1622 (March 1981).

10. Hise, E.C., et al., "Separation of Dry Crushed Coals by High-Gradient Magnetic Separation," Oak Ridge National Laboratory Report ORNL-5571 (Oct. 1979).

11. "Environmental Control Implications of Generating Electric Power from Coal," Argonne National Laboratory Report ANL/ECT-3, App. A, Part 1, pp. 293-303 (Dec. 1977).

12. Oder, R.R., "Engineering Aspects of Dry Magnetic Separation," Proc. of the Workshop on Novel Concepts, Methods and Advanced Technology in Particulate-Gas Separation, University of Notre Dame, South Bend, Ind. (Aug. 1978).

13. Bogner, E., et al., "A 12-cm-Bore Superconducting DC Quadrupole Lens," presented at the Third International Conf. on Magnet Technology, Hamburg, Federal Republic of Germany (1970).

14. Ergun, S., and E. Bean, "Magnetic Separation of Pyrite from Coals," U.S. Department of the Interior, Bureau of Mines Report 7181 (1968).

15. Liu, Y.A., and C.J. Lin, IEEE Transactions on Magnetics, MAG-12 (5):539 (1976).

16. Bluhm, D., et al., "Selective Magnetic Enhancement of Pyrite in Coal by Dielectric Heating at 27 and 2450 MHz," Ames Laboratory, Report IS-M-284, Ames, Iowa (1980).

17. Wechsler, I., J.M. Douhlin, and R. Eddy, "Coal Preparation Using Magnetic Separation," Electric Power Research Institute Report EPRI CS-1517, Vol. 3, pp. 3-10 (July 1980).

18. Holman, A.S., et al., "Initial Investigation of Open-Gradient Magnetic Separation," Oak Ridge National Laboratory Report ORNL-5764 (April 1982).

19. "Cutting the Power Needs of Magnetic Separators," Chemical Week, pp. 22-23 (March 16, 1983).

APPENDIX A:

EXPRESSIONS FOR THE RADIAL MOTION OF PARTICLES FOR LOW-REYNOLDS-NUMBER FLOW

For the radial motion of particles, the force balance in the radial direction is derived from the following equation:

$$[F]\text{inertia} = [F]\text{magnetic} + [F]\text{fluid drag} \qquad (A.1)$$

Because of the assumption that the magnet is vertical, the radius is orthogonal to gravity and no gravitational-force term appears. Additionally, the fluid will have no initial velocity component in the r-direction. Hence, for the flow regime in which Stokes's Law is applicable (Re <0.3), Equation (7) in the text can be expressed as follows:

$$\rho_p \cdot \frac{\pi D_p^3}{6} \cdot \frac{d^2 r}{dt^2} = \frac{\pi D_p^3 \chi}{6\,\mu_o} \cdot \left(\frac{dB}{dr}\right)^2 \cdot r - 3\pi\eta\, D_p \frac{dr}{dt} \qquad (A.2)$$

Rearranging terms gives:

$$\frac{d^2 r}{dt^2} + \frac{18\eta}{\rho_p D_p^2} \cdot \frac{dr}{dt} - \frac{\chi \frac{d[B]^2}{dr}}{\rho_p\,\mu_o} \cdot r = 0 \qquad (A.3)$$

Since dB/dr is constant, this is a second-order linear differential equation. The solution is given by the following:

$$r(t) = k_1 \exp(\lambda_1 t) + k_2 \exp(\lambda_2 t) \qquad (A.4)$$

where:

$$\lambda_{1,2} = -\frac{9\eta}{\rho_p D_p^2} \pm \frac{1}{2}\left[\left(\frac{18\eta}{\rho_p D_p^2}\right)^2 + \frac{4\chi\left(\frac{dB}{dr}\right)^2}{\rho_p\,\mu_o}\right]^{1/2} \qquad (A.5)$$

Constants k_1 and k_2 are evaluated by using the following boundary conditions:

I. At t = 0, radial velocity is $\frac{dr}{dt} = 0$. (A.6)

II. At $t = 0$, radial position of the particle is r_o. (A.7)

The resulting equation is given by:

$$r(t) = r_o\,[\exp(\lambda_1 t) - \exp(\lambda_2 t) \cdot \lambda_1/\lambda_2]/(1 - \lambda_1/\lambda_2) \quad (A.8)$$

This equation can be conveniently solved (by computer) for the time required to reach the wall, given the particle initial position, using the values in Table 1. The k_2 term in Equation (A.4) is significant for small-diameter particles near the center of the bore.

APPENDIX B:

EXPRESSIONS FOR THE RADIAL MOTION OF PARTICLES FOR INTERMEDIATE-REYNOLDS-NUMBER FLOW

For the intermediate-flow regime ($0.3 < Re < 1000$), the basic force balance on the particle is as follows:

$$[F]\text{inertia} = [F]\text{magnetic} + [F]\text{fluid drag} \quad (B.1)$$

Substituting Equations (2) and (4) from the text, one obtains

$$[F]r\text{-inertia} = \frac{\pi D_p^3 \chi}{6\mu_o} \cdot \left(\frac{dB}{dr}\right)^2 \cdot r - \frac{\pi}{8} C_D \cdot D_p^2 \cdot \rho_f \cdot v_p^2 \quad (B.2)$$

Substituting for the inertial force [Equation (6)], expressing velocities and accelerations in terms of r, and collecting terms yields the following:

$$0 = \rho_p \cdot \frac{\pi D_p^3}{6} \cdot \frac{d^2 r}{dt^2} - \frac{\pi D_p^3 \chi}{6\mu_o} \cdot \left(\frac{dB}{dr}\right)^2 \cdot r + \frac{\pi}{8} C_D D_p^2 \rho_f \left(\frac{dr}{dt}\right)^2 \quad (B.3)$$

where a calculated value will be used for C_D and then reexamined through subsequent iterations. Because v_p dependence is to a low exponential power, this introduces little error.

$$C_D = \frac{18.5}{Re^{0.6}} = \frac{18.5}{\left(\frac{D_p v_p \rho_f}{\eta}\right)^{0.6}} \quad (B.4)$$

Let the following terms be defined:

$$\alpha = -1 \quad (B.5)$$

$$\beta = -\frac{3}{4}\frac{C_D \rho_f}{D_p \rho_p} \quad (B.6)$$

$$\gamma = \frac{\chi}{\rho_p \mu_o} \cdot \left(\frac{dB}{dr}\right)^2 \quad (B.7)$$

Then Equation (B.3) can be rearranged as follows:

$$0 = \alpha d^2r/dt^2 + \beta(dr/dt)^2 + \gamma r \quad (B.8)$$

This nonlinear differential equation has no explicit solution and must be solved by numerical techniques. By designating $r'' = f(r, r')$, where time t is the independent variable, perform a substitution of variables. If we let $v = r'$, we obtain $v' = f(r, v)$.

$$v' = \beta v^2 + \gamma r \quad (B.9)$$

It is possible to eliminate the variable t by considering r as the independent variable. Then, by the chain rule, one obtains

$$\frac{dv}{dt} = \frac{dv}{dr} \cdot \frac{dr}{dt} = v\,\frac{dv}{dr} \quad (B.10)$$

Hence, the original differential equation may be written as follows:

$$v\,\frac{dv}{dr} = \beta v^2 + \gamma r \quad (B.11)$$

This is now a first-order differential equation, which has a solution with v as a function of r. At this point, a test for an exact differential equation can be performed. Rearranging the terms of the equation yields the following:

$$(-\beta v^2 - \gamma r) + (v)\,\frac{dv}{dr} = 0 \quad (B.12)$$

This constitutes an expression of the following form:

$$M(r,v) + N(r,v)\,\frac{dv}{dr} = 0 \quad (B.13)$$

This is an exact differential equation if

$$\frac{\partial M}{\partial v} = \frac{\partial N}{\partial r} \qquad (B.14)$$

However, the following condition applies:

$$-2\beta v \neq 0 \qquad (B.15)$$

Since the differential equation is not exact, one can attempt to compute an integrating factor μ. The integrating factor may be found, if it is a function of only one variable, from the following differential equations:

$$\frac{d\mu}{dr} = \frac{M_v - N_r}{N}\mu \qquad (B.16)$$

$$\frac{d\mu}{dr} = \frac{-2\beta v - 0}{v}\mu \qquad (B.17)$$

$$\frac{d\mu}{dr} = -2\beta\mu \qquad (B.18)$$

This differential equation for the integrating factor has an exponential solution

$$\mu(r) = e^{-2\beta r} \qquad (B.19)$$

Multiplying Equation (B.11) by this integrating factor yields

$$(\beta v^2 e^{-2\beta r} - \gamma r e^{-2\beta r}) + (v e^{-2\beta r})\frac{dv}{dr} = 0 \qquad (B.20)$$

The integrating factor μ should have transformed Equation (B.11) to an exact equation. The test for an exact equation can be applied once again. The differential equation of form

$$M(r,v) + N(r,v)\frac{dv}{dr} = 0 \qquad (B.21)$$

is exact if

$$\frac{\partial M}{\partial v} = \frac{\partial N}{\partial r} \qquad (B.22)$$

which becomes in this case:

$$-2\beta v e^{-2\beta r} = 2\beta v e^{-2\beta r} \qquad (B.23)$$

Thus, the transformed equation is exact, and a solution Ψ may be given in implicit form. If a solution Ψ can be found for the exact equation, it must satisfy both of the following conditions:

$$\frac{\partial\psi}{\partial r} = M = -\beta v^2 e^{-2\beta r} - \gamma r e^{-2\beta r} \qquad (B.24)$$

and

$$\frac{\partial\psi}{\partial v} = N = v e^{-2\beta r} \qquad (B.25)$$

Integrating the first of these equations with respect to r yields

$$\psi = \frac{-\beta v^2}{-2\beta} e^{-2\beta r} - \gamma\left[\frac{e^{-2\beta r}}{-2\beta}\left(r - \frac{1}{-2\beta}\right)\right] \qquad (B.26)$$

$$\psi = \frac{v^2}{2} e^{-2\beta r} + \frac{\gamma}{2\beta} e^{-2\beta r}\left(r + \frac{1}{2\beta}\right) \qquad (B.27)$$

An integration in the second equation with respect to v yields another expression for Ψ:

$$\psi = \frac{v^2}{2} e^{-2\beta r} \qquad (B.28)$$

Reconciling these two expressions for Ψ, we obtain the following implicit solution for the differential equation:

$$\psi = \frac{v^2}{2} e^{-2\beta r} + \frac{\gamma}{2\beta} e^{-2\beta r}\left(r + \frac{1}{2\beta}\right) = C_1 \qquad (B.29)$$

where C_1 is a constant of integration. This constant can be determined from the initial conditions.

Recalling that $v = r' = \frac{dr}{dt}$, we can substitute this into the implicit solution to obtain

$$\frac{(r')^2}{2} e^{-2\beta r} + \frac{\gamma}{2\beta} e^{-2\beta r}\left(r + \frac{1}{2\beta}\right) = C_1 \qquad (B.30)$$

Rearranging this first-order nonlinear differential equation through algebraic manipulations, we obtain the following:

$$\frac{(r')^2}{2} e^{-2\beta r} = C_1 - \frac{\gamma}{2\beta} e^{-2\beta r}\left(r + \frac{1}{2\beta}\right) \qquad (B.31)$$

$$(r')^2 = 2e^{2\beta r} C_1 - \frac{\gamma}{\beta}\left(r + \frac{1}{2\beta}\right) \qquad (B.32)$$

$$r' = \pm [2e^{2\beta r} C_1 - \frac{\gamma}{\beta}(r + \frac{1}{2\beta})]^{1/2} \quad (B.33)$$

Equation (B.33) is a first-order non-linear differential equation that may be solved numerically in time.

A host of numerical techniques exists for solving an equation of the form

$$r' = f(t,r), \; r(t_o) = r_o \quad (B.34)$$

The simplest technique is the Euler or tangent-line method. This technique is not highly accurate; however, it is sufficient for preliminary investigations if the time step $\Delta t << 1$. In this context, the method assumes the form:

$$\begin{aligned} r_{n+1} &= r_n + \Delta t \; f(t_n, r_n) \quad (B.35) \\ &= r_n + \Delta t \; r_n', \quad n = 0,1,\ldots \end{aligned}$$

Given an initial position r(t) and initial velocity (r'(t), this equation can be used to step through time by Δt, calculating an approximate position r_n at each time step.

The Euler method, however, computes each slope at the beginning of the time interval, and as a result the algorithm becomes inaccurate and unstable as time progresses. An improved method based upon a fourth-order Taylor expansion is the Runge-Kutta method. The slope in this formulation is a weighted average of derivatives throughout the time interval.

For the problem

$$r' = f(t,r), \quad r(t_o) = r_o \quad (B.36)$$

the Runge-Kutta method employs the following interaction equation:

$$r_{n+1} = r_n + \frac{\Delta t}{6}(K_{n1} + 2K_{n2} + 2K_{n3} + K_{n4}) \quad (B.37)$$

where:

$$\begin{aligned} K_{n1} &= f(t_n, r_n) \\ K_{n2} &= f(t_n + \frac{1}{2}\Delta t, \; r_n + \frac{1}{2}\Delta t \; K_{n1}) \quad (B.38) \\ K_{n3} &= f(t_n \frac{1}{2}\Delta t, \; r_n + \frac{1}{2}\Delta t \; K_{n2}) \\ K_{n4} &= f(t_n + \Delta t, \; r_n + \Delta t \; K_{n3}) \end{aligned}$$

$n = 0,1,\ldots$

These equations are then used to step through time with $\Delta t << 1$. Since the error is proportional to $(\Delta t)^5$, these equations are highly accurate for most differential equations.

APPENDIX C:

EXPRESSIONS FOR THE VERTICAL MOTION OF PARTICLES FOR LOW-REYNOLDS-NUMBER FLOW

For the flow regime governed by Stokes's Law, the basic force balance on the particle is as follows:

$$[F]\text{z-inertia} = [F]\text{gravity} + [F]\text{fluid drag} \quad (C.1)$$

Expressed in terms of the mass of the particle (m), this becomes:

$$m\frac{dv}{dt} = mg - 3\pi\eta D_p v \quad (C.2)$$

Note that the velocity (v) is the relative velocity between the fluid and particle:

$$v = v_p - v_f \quad (C.3)$$

Now, separating variables, one obtains

$$dt = \frac{mdv}{mg - 3\pi\eta D_p v} \quad (C.4)$$

Integrating yields the following:

$$t = \frac{m}{-3\pi\eta D_p} \ln(mg - 3\pi\eta D_p v) + k_o \quad (C.5)$$

The logarithmic term in this expression can only be evaluated if we integrate again:

$$mg - 3\pi\eta D_p v = k_1 \exp(-3\pi\eta D_p t/m) \quad (C.6)$$

Hence,

$$v = \frac{mg - k_1 \exp(-3\pi\eta D_p t/m)}{3\pi\eta D_p} \quad (C.7)$$

From the initial boundary condition at t = 0, $v = v_o$; then

$$k_1 = mg - 3\pi\eta D_p v_o \quad (C.8)$$

and

$$v = \frac{mg}{3\pi\eta D_p}\left(1 - \exp\left(\frac{-3\pi\eta D_p t}{m}\right)\right) + v_o \cdot \exp\left(\frac{-3\pi\eta D_p t}{m}\right) \quad (C.9)$$

Taking v to be equal to dz/dt and integrating for z yields

$$z = \frac{m}{3\pi\eta D_p}\left[gt + \left(\frac{mg}{3\pi\eta D_p} - v_o\right) \cdot \exp\left(\frac{-3\pi\eta D_p t}{m}\right)\right] + k_2 \quad (C.10)$$

To find k_2, assume the following boundary condition at t = 0, z = 0; then

$$k_2 = v_o \cdot m/3\pi\eta D_p - g \cdot (m/3\pi\eta D_p)^2 \quad (C.11)$$

This leads to the following:

$$z = \frac{mgt}{3\pi\eta D_p} + g\left(\frac{m}{3\pi\eta D_p}\right)^2\left(\exp\left(\frac{-3\pi\eta D_p t}{m}\right) - 1\right) - v_o\, m/3\pi\eta D_p \left(\exp\left(\frac{-3\pi\eta D_p t}{m}\right) - 1\right) \quad (C.12)$$

Assuming the particles are spherical, this expression can be further simplified by noting that

$$m = \frac{\pi}{6}\rho_p D_p^3 \quad (C.13)$$

Substituting this value and collecting terms gives the following:

$$z = \frac{gD_p{}^2\rho_p}{18\eta}\left(t + \left(\frac{D_p{}^2\rho_p}{18\eta} - \frac{v_o}{g}\right) \cdot \left[\exp\left(-18\eta\, t/D_p{}^2\rho_p\right) - 1\right]\right) \quad (C.14)$$

APPENDIX D:

EXPRESSIONS FOR THE VERTICAL MOTION OF PARTICLES FOR INTERMEDIATE-REYNOLDS-NUMBER FLOW

In the intermediate-flow regime (0.3 < Re < 1000), the force balance on the particle is as follows:

$$[F]\text{z-inertia} = [F]\text{gravity} + [F]\text{fluid drag} + [F]\text{z-inertia(o)} \quad (D.1)$$

Expanding this expression, using Equation (4) for the fluid-drag force, yields the following:

$$m \cdot dv/dt = mg - \frac{\pi}{8} C_D D_p^2 \rho_f v^2 + \frac{1}{2} m v_o{}^2 \quad (D.2)$$

where:

$$C_D = \frac{24}{Re}(1 + 0.15\, Re^{0.687}) \quad (D.3)$$

Only short distances (where C_D is approximately constant) will be covered.

Let

$$A = \frac{\pi}{8} C_D D_p{}^2 \rho_f \quad (D.4A)$$

$$B = m(g + \frac{1}{2} v_o{}^2) \quad (D.4B)$$

$$v = v_p - v_f \quad (D.5)$$

Then one obtains the following:

$$dt = \frac{m\,dv}{B - Av^2} \quad (D.6)$$

Now integrate:

$$t = \frac{m}{2(A \cdot B)^{1/2}} \cdot \ln\left(\frac{(B/A)^{1/2} + v}{(B/A)^{1/2} - v}\right) \quad (D.7)$$

Let

$$\theta = m/[2\,(A \cdot B)^{1/2}] \quad (D.8)$$

$$\gamma = (B/A)^{1/2} \quad (D.9)$$

Then:

$$\frac{\gamma + v}{\gamma - v} = k_1 \exp(t/\theta) \quad (D.10)$$

Assume the following initial boundary condition:

I. At t = 0, $v = v_o$.

Then:

$$k_1 = \frac{\gamma + v_o}{\gamma - v_o} \quad (D.11)$$

The following rearrangement of Equation (D.10) can now be solved for velocity:

$$v = \gamma \cdot \frac{k_1 \exp(t/\theta)-1}{k_1 \exp(t/\theta)+1} \tag{D.12}$$

This is integrated to obtain

$$z = \gamma \left(2\,\theta \cdot \ln\,[1 + k_1 \exp(t/\theta)] - t\right) + k_2 \tag{D.13}$$

where for $t = 0$, $z = 0$:

$$k_2 = -2\gamma \cdot \theta \cdot \ln(1 + k_1) \tag{D.14}$$

Hence:

$$z = \gamma\left(2\theta \cdot \ln\,[1 + k_1 \exp(t/\theta)] - t\right) - 2\gamma\theta \cdot \ln\,(1 + k_1) \tag{D.15}$$

This expression has been simplified and used as Equation (14).

REVERSING THE RELATIVE VOLATILITY BY EXTRACTIVE DISTILLATION

An-I Yeh and Lloyd Berg ■ Montana State University, Bozeman, MT 59717

Extractive distillation has shown that the components of azeotropes can be separated by rectification. By the proper selection of agents, either of the components can be recovered overhead as the more volatile component. This reversing of the volatile is reported for the separation of the ethanol - water, methyl acetate - methanol and acetone - methanol azeotropes. The selection of suitable agents can be correlated to some extent by dispersion, polar, induction, assymetry, acidity and basicity parameters and hydrogen bonding assymetry factor.

Extractive distillation is the method of carrying out a rectification in the presence of a higher boiling liquid or mixture of liquids. It is most commonly used to separate two close boiling compounds or to break up azeotropes. When a separation is accomplished, the usual expectation is that the more volatile compound, that is, the one with the lowest normal boiling point, will come out as overhead product while the less volatile compound and the extractive agent come out as bottoms products. That, however, is not always the result, as has been reported a number of times. Buell (1) used extractive distillation to reverse the volatility of cis-butene-2 and trans-butene-2 relative to 1,3-butadiene. Berg (2) reversed the volatility of ethyl acetate from ethanol and Berg & Yeh (3) brought out isopropyl ether (b.p.=68.5°C.) overhead from acetone (b.p.=56.2°C.). It is the object of this paper to report on several extractive distillation systems which by the proper selection of agents, can bring out overhead either component of an azeotropic mixture.

DEHYDRATION OF ETHANOL

The dehydration of ethanol is on of the earliest applications of extractive distillation. In 1923, Schneible (4) used glycerine to take water (100°C.) out the stillpot and purified ethanol (78.6°C.) as overhead product. Smith (7) employed ethoxyethanol and butoyx-ethanol as extractive agents. Drout (8) used glycols, glycol ethers and glycol esters. Catterall (9) reported butyl, amyl and hexyl alcohols as effective extractive agents. All of these investigators described extractive distillation systems where the ethanol is taken off overhead and the water accompanys the extractive agent to the stillpot. We discovered that by the proper selection of extractive distillation agents the less volatile water can be recovered in the pure state as column overhead by extractive distillation. A considerable number of potential agents were evaluated in an Othmer type vapor-liquid equilibrium still. The most attractive performers were then employed in a 2-inch diameter glass perforated plate rectification column which had been calibrated with a test mixture of ethylbenzene - p-xylene and found to have 4.5 theoretical plates. The results are presented as the relative volatility of the overhead product to the bottoms product on an agent-free basis.

Some effective agents as determined in the vapor-liquid

Table 1 - Relative Volatility of Water to Ethanol for Several Concentrations

	Relative Volatility			
	Concentration of Ethanol			
Agent	95%	90%	85%	85%
Trimellitic anhydride, Ethyl salicylate	2.1	1.1	0.8	
Hexahydrophthalic anhydride	1.4	1.1	0.9	0.7
" , Ethyl salicylate	2.1	1.4	1.0	
" , " Resorcinol	1.4	1.4	1.1	0.9
Methyl benzoate, Trimellitic anhydride	1.8	1.4	1.2	0.9
Dipropylene glycol dibenzoate	1.2	1.0	0.9	
" , Pentanol-1	1.7	1.2	1.1	

equilibrium still are listed in Table 1.

The data in Table 1 shows that as the concentration of water is increased, the relative volatility of every extractive agent decreases. The best performer was always with the ethanol-water azeotrope, 95% ethanol.

Several of the agents which had been evaluated in the vapor-liquid equilibrium still were then tested in the rectification column. The results are shown in Table 2.

The operating procedure for the semi-batch extractive distillation was as follows. The stillpot was charged with 400 grams of the mixture to be studied and heat applied. When the condensate reached refluxing and the system stabilized, agent was added at the desired rate and temperature. The reflux rate was controlled by adjusting the heat input to the stillpot. The temperature of the incoming agent, condenser, stillpot, vapor and liquid phases on the top plate was recorded every ten minutes. The system was operated at total reflux and samples taken. Three sets of samples were taken from the overhead and stillpot after one hour, 1-1/2 and two hours of steady operation. Analyses were by gas chromatography. During the operation the agent was pumped into the system and accumulated in the stillpot until the operation was completed. The "blank" run is the system operated without adding any agents.

Pentanol-1 caused water to be more volatile than ethanol when the initial binary mixture contained less than 40% water. The mixture of pentanol-1 and dipropylene glycol dibenzoate caused water to be more volatile than ethanol when the initial binary mixture contained less than 35% water. Dipropylene glycol dibenzoate reduced slightly the effect of pentanol-1.

The physical forces of attraction are usually higher for homologs than for non-homologs. Pentanol-1 is a homolog of ethanol. Another reason could be the molecular size difference between water and ethanol. Ethanol is about three times as large as water. The agent would collide more frequently with the molecule of larger size and thus the larger molecules could be brought down to the stillpot by the agent. However water also has very strong hydrogen bonding capability and this could offset the effects mentioned above. Therefore which key component that would be brought down to the stillpot by the pentanol-1 would depend upon the relative strength of the hydrogen bonding of water molecules. When the water content is small, less water and less hydrogen bonding capacity are present. Thus pentanol-1 brings its homologous compound, ethanol, down to the stillpot. Once water has sufficient hydrogen bonding capacity to compete with the molecular

Table 2 - Effect of Initial Ethanol Composition on Relative Volatility

Agent: 1-Pentanol

Initial composition wt.% of water	wt.% of water Overhead	Bottoms	Relative Volatility
5.0	22.1	3.8	1.55
19.25	31.9	14.3	1.26
29.94	30.3	21.8	1.10
40.63	33.5	33.5	1.00
80.05	40.7	70.3	0.75

Agent: 1-Pentanol + Dipropylene glycol dibenzoate

Initial composition wt.% of water	wt.% of water Overhead	Bottoms	Relative Volatility
5.0	18.4	3.2	1.53
19.25	20.8	14.9	1.09
29.94	28.6	21.6	1.09
40.63	25.1	32.9	0.92
80.05	31.4	83.8	0.58

interaction between the two homologous compounds, water will interact more strongly with pentanol-1 than with ethanol. Thus ethanol is obtained as overhead product up to some water content. This shows that water's different behavior in solution is due to strong hydrogen bonding.

The data in Tables 1 and 2 show that by proper selection of extractive distillation agents, the higher boiling constituent, water, can be removed as overhead product from the lower boiling ethanol. One advantage of this method appears to be that water, the compound with the highest heat of vaporization, 970 Btu/lb., needs to be vaporized only once. A disadvantage is that when the concentration of ethanol is below 85%, the relative volatility becomes close to or below one and the process ceases.

Methanol - Methyl Acetate Azeotrope

Methanol (b.p.-64.5°C.) and methyl acetate (b.p.=56.3°C.) form a binary minimum azeotrope boiling at 53°C. and containing 81.3 wt.% methyl acetate. The potential extractive distillation agents were evaluated in the vapor-liquid equilibrium still and the more attractive performers listed in Tables 3 and 4.(12)

Table 3 shows that when dimethylsulfoxide is used as the extractive distillation agent, methyl acetate is removed as overhead product with a relative volatility of methyl acetate to methanol is 7.7. Glycols are also effective extractive distillation agents in bringing methyl acetate out as overhead product. Table 4 shows that when nitroparaffins or nitrobenzene are used as the extractive distillation agents,

Table 3 - Relative Volatility of Methyl Acetate to Methanol

Agent	Relative Volatility
Dimethylsulfoxide	7.7
Ethylene glycol	7.2
Ethylene glycol + Dimethylsulfoxide	6.6
Ethylene glycol + Dimethylsulfoxide + Glycerine	6.2
Propylene glycol	5.4
Diethylene glycol	5.3
1,4-Butanediol	5.2

Table 4 - Relative Volatility of Methanol to Methyl Acetate

Agent	Relative Volatility
2-Nitropropane	2.8
1-Nitropropane	2.7
Nitroethane	2.4
Nitrobenzene	2.1
Nitromethane	2.1
Propylene carbonate + 1-Nitropropane	1.6

Table 5 - Relative Volatility of Agents in Which Acetone is More Volatile

Agents	Relative Volatility
Glycerine	3.3
Ethylene Glycol	2.8
Dimethylsulfoxide, Ethylene glycol	3.3
Ethylene glycol, 1,5-Pentanediol	2.6
Glycerine, Ethylene glycol, Diisobutyl phthalate	3.1
Glycerine, Dimethylsulfoxide, 1,4-Butanediol	3.0

methanol comes out overhead. The best agent appears to be 2-nitropropane with a relative volatility of methanol to methyl acetate of 2.8.

Acetone - Methanol Azeotrope

Acetone, b.p.=56.1°C. and methanol, b.p.=64,5°C. form a minimum azeotrope boiling at 55.7°C. containing 88 wt.% acetone. Thus any mixture of these two commonly employed solvents will always produce the azeotrope upon rectification. Extractive distillation has been applied to this separation and effective agents have been discovered which will not only break the azeotrope but will bring either the acetone or the methanol out as overhead (10,11). Agents which are effective in recovering acetone as overhead product are listed in Table 5, those which bring out methanol as overhead are in Table 6.

To confirm the validity of the data obtained in the vapor-liquid equilibrium still, several runs were made in the 4.5 theoretical plate rectification column. Table 7 presents the data for several extractive distillation agents which bring acetone out as overhead product.

Table 6 - Relative Volatility of Agents in Which Methanol is More Volatile

Agent	Relative Volatility
Acetophenone	2.3
Ethyl acetoacetate	2.2
2,4-Pentanedione	1.9
3-Heptanone	1.8
Ethyl butyl ketone	1.8
Benzophenone, Acetophenone	1.8

Table 7 - Data From Runs in Rectification Column - Acetone as Overhead

Agent	Overhead % Acetone	Bottoms % Acetone	Relative Volatility
Dimethylsulfoxide	99.7	25	4.6
1,4-Butanediol	96.7	25	2.7
Ethylene glycol, 1,5-Pentanediol	98.6	25	3.3
Dimethylsulfoxide, Ethylene glycol	97.8	25	3.0
Glycerine, Ethylene glycol, Triethylene glycol	99.3	50	3.0
Glycerine, Ethylene glycol, Diethylene glycol	98.4	50	2.5

Table 8 lists the results with agents which bring methanol out as overhead product.

The data presented in Tables 5, 6, 7 and 8 show that by the proper selection of extractive distillation agents, either acetone or methanol can be recovered in a pure state as overhead product by rectification and that extractive distillation has effectively broken the azeotrope that normally occurs when mixtures of acetone and methanol are distilled.

Table 8 - Data From Runs in Rectification Column - Methanol as Overhead

Agent	Overhead % Methanol	Bottoms % Methanol	Relative Volatility
Methyl ethyl ketone	31.1	6.2	1.53
2-Pentanone	43.6	7.3	1.66
3-Pentanone	43.9	5.8	1.76
Methyl isobutyl ketone	24.7	5.3	1.48
Methyl isoamyl ketone	30.9	4.6	1.64
2,4-Pentanedione	51.6	5.5	1.91
Diisobutyl ketone	31.8	4.8	1.64
Acetophenone	38.3	5.3	1.71
Ethyl acetoacetate	34.8	5.9	1.77
Methyl ethyl ketone + Benzil	43.2	6.8	1.68

DISCUSSION & CONCLUSIONS

In extractive distillation, if the effect of the agent is mainly dependent upon the attractive intermolecular forces occurring between the agent and key component, the molecule with more attractive intermolecular forces with the agent will be the bottoms product. Thus the potentiality of the key component to interact with the agent can be an indication of the effect of the agent. Solubility parameter which include dispersion force, polar interaction and hydrogen bonding can be used to explain the intermolecular forces occurring between molecules. However, the intermolecular forces should be attractive. This indicates that the molecules must have some similar characteristics. We attempted to categorize the molecule and find a way to select an effective agent by using solubility parameter. Table 9 (13) lists the three solubility parameters and specific volume of the key components studied.

The difference in the three solubility parameters between key components in each system is mainly polar interaction, δ_p, and hydrogen bonding, δ_h. For the acetone-methanol system, the difference in dispersion force is only 1.4 $(J/cm^3)^{1/2}$ which is a very small value compared with the differences in polar interaction, 3.2 $(J/cm^3)^{1/2}$ and the difference in hydrogen bonding, 13.0 $(J/cm^3)^{1/2}$.

Table 9 - Solubility Parameters and Specific Volume of Some Key Components

Component	v, cm^3/mol	δ_d $(J/cm^3)^{1/2}$	δ_p $(J/cm^3)^{1/2}$	δ_h $(J/cm^3)^{1/2}$
Acetone	74.0	13.0	9.8	11.0
Methanol	40.7	11.6	13.0	24.0
Ethanol	58.7	12.6	11.2	20.2
Water	18.1	12.2	22.8	40.4
Methyl acetate	79.9	13.3	9.5	10.4

comparing these three numbers shows that the difference in hydrogen bonding is the most significant. Also the difference in specific volume, 34 cm^3/mol, is appreciable.

This discussion can be applied to the ethanol-water and methanol-methyl acetate as well. This observation indicates that the dispersion force can be ignored when discussing the molecular interaction in extractive distillation. Figures 1, 2 and 3 show the plot of δ_p versus δ_h for the systems of acetone-methanol, ethanol-water, and methanol-methyl acetate respectively. In Figure 1, the location of methanol is referred to as the point A and the location of acetone is referred to as the point B. We can draw a segment AB to connect these two points. There is one point C which makes the ratio of AC to CB equal to the ratio of specific volume of acetone to that of methanol. Drawing a vertical line through the point C divides the plot into two parts. One is the methanol side and another is the acetone side. Agents such as glycerine and ethylene glycol, on the methanol side have a greater affinity for methanol than for acetone. Those agents bring methanol down to the stillpot and acetone will be the overhead product. This prediction is confirmed by the data reported in Table 9. The agents, such as methyl ethyl ketone and acetophenone, on acetone side have a greater affinity for acetone than for methanol. Therefore acetone will be the bottoms product. The results presented in Tables 5 through 8, with the exception of DFMA, are in accord with this theory. DMFA is on the acetone side but bring methanol down to the stillpot. This indicates that there is a greater molecular interaction between these agents and methanol. Ritchie (14) has shown that DMSO and DMFA can induce the acidity of alcohols. DMSO and DMFA behave as bases. Thus there exists a strong molecular interaction between these agents and methanol and thus acetone is obtained as the overhead product. The acid-base molecular interaction could cause the presence of a chemical complex.

Applying the same technique in Figures 2 and 3 shows about the same results as the previous observation. The values of solubility parameters for various compounds have been edited by Barton (13). Thus the method presented above might be considered as the simplest way to predict which key component will be the overhead product. However the reversal of the relative volatility in the ethanol-water system was limited by the concentration of ethanol. This could be due to two reasons: (i) water molecule has very strong hydrogen bonding (δ_h = 40.4 $(J/cm^3)^{1/2}$ (ii) water has small specific volume (v = 18.1 cm^3/mol.). The agent collides more frequently with the molecule with larger size. The hydrogen bonding becomes more explicit as the concentration of water increases and so would have enough intermolecular forces to attract the molecules of the extractive agent. The reversing of the relative volatility becomes more difficult when more water molecules are present. Eventually ethanol becomes more volatile when there is enough water to interact with the molecules of the agent.

The method presented above can be an initial estimation of the performance of an agent. However, a more accurate predication of the performance of an agent might be dependent upon the expression of the activity coefficient. Thomas and Eckert (15) have developed an expression, Equation (1), to calculate the infinite activity coefficients by a modified separation of cohesive energy density model and UNIFAC.

Equation (1)

$$\ln\gamma_2^{\infty} = v_2[(\lambda_1-\lambda_2)^2+q_1^2q_2^2(\tau_1-\tau_2)^2/\psi_1+(\alpha_1-\alpha_2)(\beta_1-\beta_2)/\xi]/RT+d_{12} \quad (1)$$

In equation 1, subscript 2 indicates that the solute 2 in solvent 1. The values of the liquid molar volume, v, and the dispersion parameter, λ, were assumed to be temperature independent. The polar parameter, τ, acidity parameter, α, and basicity parameter, β, were temperature dependent. Thomas and Eckert (15) presented a detailed

Table 10 - The Comparison of Relative Volatility of the Methanol-acetone System

Agent	Experimental α_{12} Still	Experimental α_{12} Column	Calculated α_{12}
MEK	1.2	1.53	1.46
2-Pentanone	1.3	1.66	1.60
3-Pentanone	1.4	1.76	1.60
Methyl isobutyl ketone	1.5	1.48	1.69
Diisobutyl ketone	1.4	1.64	1.65
Acetophnone	2.4	1.71	1.33

Table 11 - The Comparison of Relative Volatilities For The Methanol-Methyl Acetate System

Agent	Experimental α_{12}	Calculated α_{12}
Nitromethane	2.06	1.72
Nitroethane	2.40	2.41
1-Nitropropane	2.65	2.91
2-Nitropropane	2.77	2.96
Nitrobenzene	2.04	2.83

discussion of the parameter estimation. The term d_{12} accounts for the differences in molecular size and is defined as:

$$d_{12} = \ln(v_2/v_1) + 1 - (v_2/v_1) \quad (2)$$

In this study, the MOSCED model was used to calculate the relative volatilities. The values of parameters in MOSCED were reported by Thomas et al.(15). The calculated relative volatilities were computed from Equation (6) below. P_i^{sat}'s were calculated from the Antoine equations reported by Reid et al. (16). Tables (10) and (11) show the calculated results compared with the experimental data.

$$y_1P = \gamma_1x_1P_1^{sat} \quad (3)$$

$$y_2P = \gamma_2x_2P_2^{sat} \quad (4)$$

$$\gamma_1/\gamma_2 = y_1x_2P_2^{sat}/x_1y_2P_2^{sat} = \alpha 12 \; P_2^{sat}/P_1^{sat} \quad (5)$$

$$\alpha_{12} = \gamma_1P_1^{sat}/\gamma_2P_2^{sat} \quad (6)$$

From the results shown in Tables (10) and (11), the calculated α's fit the experimental α's fairly well except for the aromatic compounds, nitrobenzene and acetophenone. This could be due to the electron withdrawal of the functional group on the benzene ring in addition to other characteristics of the molecule. the advantage of the method presented is that only the properties of pure components need be known to calculate the relative volatility of the key components. Unfortunately the model is inapplicable to aqueous systems.

NOTATION

Nomenclature

α_{12}: Relative volatility of component 1 to component 2.

x, y: Composition in liquid and vapor phase, respectively.

P: Total pressure of the system in mmHg.

P^{sat}: Saturated vapor pressure of component i in mmHg.

γ: Activity coefficient of component i.

v: Molar volume in cm^3/mol.

λ: Dispersion parameter in MOSCED $(cal/cm^3)^{1/2}$.

τ: Polar parameter in MOSCED $(cal/cm^3)^{1/2}$.

q: Induction parameter in MOSCED model.

ψ: Polar asymmetry parameter in MOSCED model.

α: Acidity parameter in MOSCED model.

β: Basicity parameter in MOSCED model.

ξ: Hydrogen bonding asymmetry factor for MOSCED.

R: Gas constant.

T: Temperature in Kelvin.

LITERATURE CITED

1. Buell, C.K. & R.G. Boatright, IND.ENG.CHEM., 39, 695 (1947).

2. Berg, L. & P. Ratanapupech, U.S. Pat. 4,379,028, Apr. 5, 1983.

3. Berg, L. & A. Yeh, AIChE Journal, 31, 504 (1985).

4. Schneible, U.S. Pat. 1,469,447, Oct. 2, 1923.

5. Berg, L. & A. Yeh, U.S. Pat. 4,543,164, Sept. 24, 1985.

6. Berg, L. & A. Yeh, U.S. Pat. 4,549,938, Oct. 29, 1985.

7. Smith, P.V., Jr. & C.S. Carlson, U.S. Pat. 2,559,519, July 2, 1951.

8. Drout, W.J., Jr. & J.W. Dowling, Fr. Pat. 1,020,351, Feb. 5, 1953.

9. Catterall, W.E., U.S. Pat. 2,591,671, Apr. 8, 1952.

10. Berg, L. & A. Yeh, U.S. Pat. 4,501,645, Feb. 26, 1985.

11. Berg, L. & K.J. Warren, U.S. Pat.

12. Berg, L. & A. Yeh, CHEM.ENGR.COMM., 30, 113 (1984).

13. Barton, A.E.M., Editor, Handbook of Solubility Parameter, CRC Press (1983).

14. Ritchie, C.D. & J.E. Coetzee, Solute-Solvent Interactions, Marcel Dekker, Inc. (1969).

15. Thomas, E.R. & C.A. Eckert, IND.ENG.CHEM., 23, 194 (1984).

16. Reid, R.C., J.M. Prausnitz & T.K. Sherwood, "Properties of Gases & Liquids", 3rd. Ed., McGraw-Hill Book Co., New York (1977).

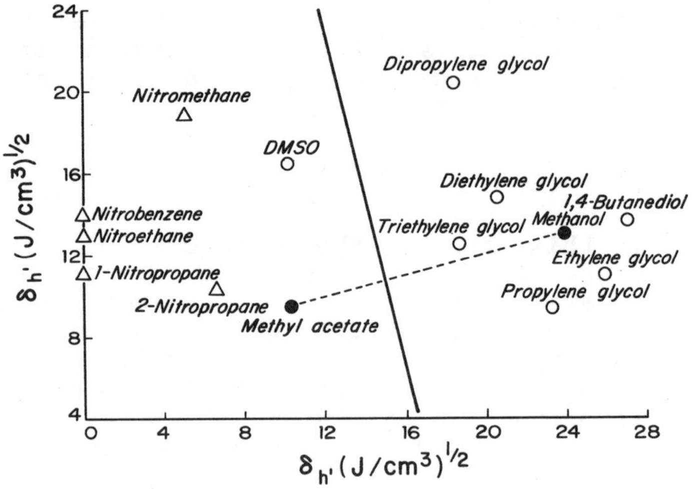

Figure 2. The polarity diagram for the methyl acetate-methanol system.

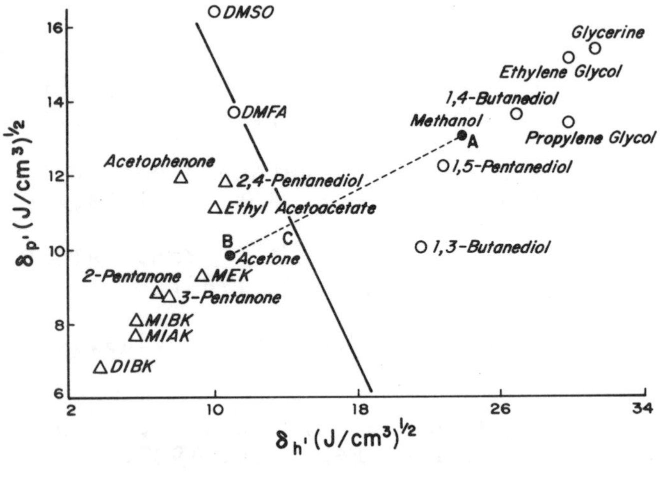

Figure 1. The polarity diagram for the acetone-methanol system.

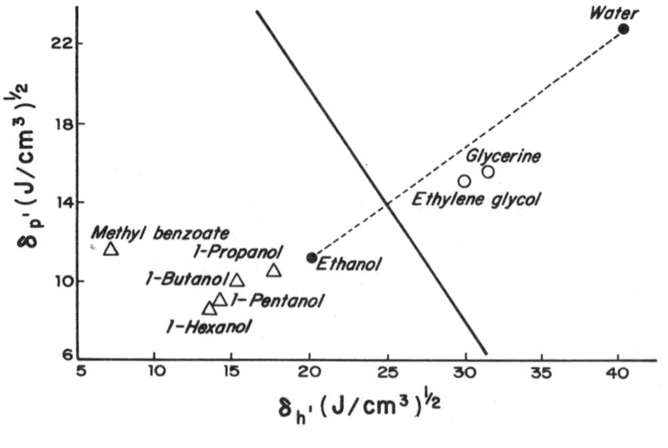

Figure 3. The polarity diagram for the uthanol-water system.

DEVELOPMENT OF A THEORETICAL MODEL FOR CONTINUOUS ROTATING ANNULAR-BED ELECTROPHORESIS COLUMN FOR BIOCHEMICAL SEPARATIONS

Ravindra Datta, Randall A. Yoshisato
and Gregory R. Carmichael ■ Department of Chemical and Materials Engineering
University of Iowa, Iowa City, IA 52242

A comprehensive theoretical framework is developed for the Continuous Rotating Annular-Bed Electrophoresis (CRAE) column recently proposed by us (*1*). This new device appears to have considerable potential for industrial application for the separation of biochemicals. Here, simplified analytical results are obtained for the purpose of preliminary analysis and design of the device. A more rigorous numerical study is being undertaken to also study the effects of hydrodynamic interactions such as thermal convection, non-uniform velocity profile in bed, and electroosmosis, on zone spreading and resolution. In addition, the effects of finite rates of adsorption on the packing and axial conduction of heat on the temperature profile in the column will be investigated.

INTRODUCTION

Electrophoresis is an extremely versatile method for the separation of biochemicals. It exploits the different electrophoretic migration velocities of charged species under the influence of an electric field. Since practically all substances acquire a positive or negative charge as a result of either adsorption onto colloidal particles or by the dissociation of soluble species in an aqueous solvent, there is hardly a class of compounds which has not been separated by electrophoresis. For analytical applications, electrophoresis and its variants are extremely popular. The various techniques developed include moving-boundary electrophoresis, zone electrophoresis, isotachophoresis and isoelectric focusing (2,3). On the preparative scale, however, progress has been much slower because of some engineering problems faced in scaling up. The major problem in the design of large scale continuous electrophoresis devices is the effective removal of the heat generated by the passage of electric current through the bed. This is necessary in order to avoid excessive temperatures and consequent thermal degradation of many biochemicals. Increased temperatures also result in thermal convection that may destroy product resolution. If anti-convectant packing is used to reduce thermal convection, increased dispersion and electroosmosis result. Of course, packed columns are also not entirely suitable for separating particulate materials.

In order to achieve increased throughput, two different continuous free-flow zone electrophoresis systems were developed by Philpot (4) and by Hannig (5) and Strickler (6). Both methods have eluent flowing laminarly in a 0.5 to 1.0 mm gap between flat parallel plates. In the Philpot method, the voltage gradient is applied across the narrow gap and hence species with different electrophoretic mobilities are collected across the narrow width of the gap. The Hannig method uses a voltage gradient applied, instead, along the length of the gap and hence requires fraction collection along the length of the gap. The number of species that can be separated in the Philpot device is limited by the narrowness of the gap.

The throughput of these instruments is rather modest, typically less than 10 ml/h for protein separations. Larger throughputs, of course, necessitate the use of larger gap free-flow devices. However, larger gaps mean increased production of heat, higher temperatures, increased thermal convection and consequent loss of resolution. Several innovative techniques have, hence, been devised to stabilize laminar flow in larger gap free-flow devices such as increased viscosity of eluent, application of a density gradient, and a velocity gradient (7). A large scale free-flow device (10 ml/min) based on the Philpot concept has been developed by Harwell in the UK (7,8) in which the flow is stabilized by a velocity gradient. It consists of an annular chamber of approximately 5 mm width between two concentric cylinders. The outer cylinder rotates at approximately 150 RPM while the inner cylinder is stationary, resulting in a radial velocity gradient. Heat generation causes a 20 to 25°C temperature rise. However, since the residence time is only 30 to 60 seconds, thermal degradation is minimized. The electric field, and hence, the separation is in the radial direction and separated components are collected as a series of concentric rings. The unit is commercially available.

Vermeulen and coworkers (9,10,11) developed an annular packed-bed column with an inner radius of 1.25 cm, an outer radius of 10 cm and a height of 122 cm. The column was capable of processing up to 1000 g/h of a binary feed stream. Due to the position of the electrodes at the centerline and along the circumference, the separation occured in the radial direction, as in the Harwell unit, with fractions being collected as concentric rings thus limiting the number of components separable. Larger number of components, of course, would necessitate a thicker bed. However, larger bed thicknesses increase the amount of Joule heating and the bed temperature. Thus, the bed thickness must be limited to 5 to 10 cm in order to prevent excessive heating and thermal degradation of the product. This is the major stumbling block in further scale-up of this design.

We have recently proposed (1) a new high throughput continuous rotating annular packed-bed device that is designed to overcome the major disadvantages of the thick-bed columns such as the Vermeulen column. The new design minimizes temperature rise in the bed, is capable of being scaled-up and can handle multicomponent mixtures. A prototype is currently being constructed for testing. Our objective here is to develop a comprehensive theoretical framework for modeling the performance of this device. This would form the basis of further numerical simulation work to study rigorously the effect of various parameters on the performance of the column. Here, a first-generation analytical solution is provided for the purpose of preliminary analysis.

CONTINUOUS ROTATING ANNULAR-BED ELECTROPHORESIS COLUMN

The continuous rotating annular electrophoresis (CRAE) column is a new approach to large scale electrophoresis, different from the Hannig-type devices as well as the Philpot/Harwell unit and Vermeulen's column. In the CRAE device, an electric field is imposed axially in a packed bed annular column, along the same direction as eluent flow, while the annular bed is rotated very slowly about its axis. Sample and eluent introduction ports at the top of the column and multiple product collection ports which ring the circumference at the bottom of the column are fixed in space. As a result of this configuration, separated fractions appear as helical bands as shown in Figure 1 with each component leaving the column at a characteristic angular position. Since separation takes place in the angular direction rather than in the radial direction as in Vermeulen's column, the bed may be kept relatively thin to improve the heat transfer and reduce the peak bed temperature rise. This should also reduce the temperature gradients and, thereby, reduce the flow instabilities due to buoyancy effects. The anticonvectant packing used may be either inert or adsorptive because of opposite charge or surface activity. A rotating device of similar geometry has already been successfully used in the chromatographic separation of multicomponent mixtures (12 to 16). It should also be evident that the proposed technique may be further refined by utilizing concepts from gradient elution, isoelectric focusing, and isotachophoresis and by varying the porosity and packing size in the column.

FORMULATION OF A THEORETICAL MODEL

Various aspects of the thin slit continuous free-flow Hannig type devices have been the subject of considerable theoretical work, although a unified and comprehensive study of their overall effects on the performance does not appear to have been done. Strickler and Sacks (17) studied the effect of horizontal electroosmotic flow adjacent to the walls and the vertical parabolic velocity profile on the separated bands and termed the resulting deformation "crescent phenomena". Reis et al. (18)

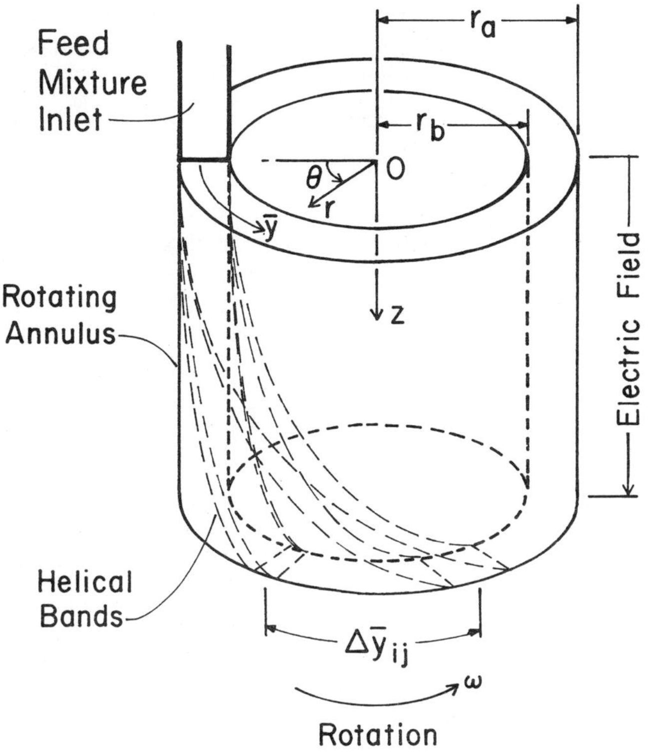

Figure 1. A schematic of the Continuous Rotating Annular-Bed Electrophoresis Column.

studied the effect of Taylor dispersion resulting from the parabolic velocity profile on band spreading and found it to be significant. The effect of heat generation and buoyancy on the flow field in free-flow devices has been studied in detail by Saville (19), Lynch and Saville (20), Turk and Ivory (21), and Naumann and Rhodes (22). Vermeulen and coworkers (9 to 11) have developed a theoretical model for their packed annulus electrophoresis column.

A Structural Model for the Packed Annulus Bed

In order to develop a hydrodynamic model that accounts for the thermal convection, wall effects and electroosmotic effects, a structural model must first be assumed for the packed bed. For instance, a parallel pore model may be assumed (23). Using such a pore model, Cohen and Metzner (24) accounted for the wall effects on the velocity profile by devising a tri-regional model with radially varying porosity. For reasons given below we follow an alternate approach in which the annular packed bed is thought to be made up of several thin concentric annular channels separated by walls of finite thickness (Figure 2). The dimensions of individual annular elements and the wall thickness would depend upon the requirements of the equality of the hydraulic radius, specific surface, and bed porosity of the structural model and those of the actual packed bed. The proposed model is conceptually simple and is capable of incorporating the wall effect by the artifice of varying the annular width of channels radially or, equivalently, by varying the bed porosity. Further, thermal convection owing to the variation of temperature in the bed can easily be incorporated into the model. It is also capable of incorporating electroosmotic effect as well as longitudinal and angular dispersion. Clearly, the pore model is not as versatile for our purposes.

If, for now, all the concentric channels are assumed to be of equal width, Δr, and the intervening walls of thickness, Δr_w, then the hydraulic radius, r_H, defined as the ratio of void volume in the bed available for flow to the wetted surface area of packing, becomes

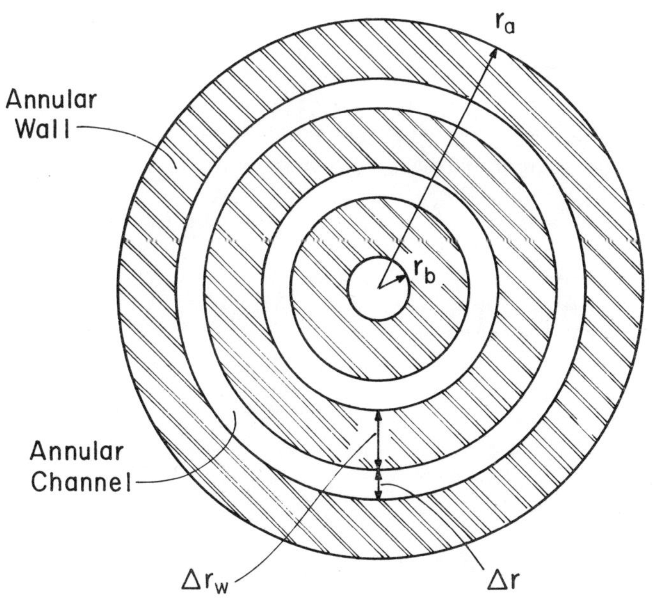

Figure 2. A structural model for the paced annulus bed consisting of N concentric annular channels of width Δr separated by walls of thickness Δr_w. Not to scale.

$$r_H = \frac{\Delta r}{2} \quad (1)$$

The hydraulic radius for a bed composed of spherical particles of diameter D_p (23) is

$$r_H = \frac{\varepsilon_B D_p}{6(1-\varepsilon_B)} \quad (2)$$

Equating these two hydraulic radii gives Δr in terms of particle size and bed porosity as

$$\Delta r = \frac{\varepsilon_B D_p}{3(1-\varepsilon_B)} \quad (3)$$

The specific surface, S_v, defined as the ratio of surface area to its volume, for spherical particles is

$$S_v = \frac{6}{D_p} \quad (4)$$

That for annular channels is

$$S_v = \frac{2}{\Delta r_w} \quad (5)$$

Equating the two specific surfaces gives Δr_w in terms of particle size

$$\Delta r_w = \frac{D_p}{3} \quad (6)$$

The number of channels, N, is calculated from

$$N = \frac{3(1-\varepsilon_B)(r_a - r_b)}{D_p} \quad (7)$$

Conservation of Momentum

The velocity profile in an annular channel with thermal convection for fully developed flow is described by (23)

$$\frac{1}{r}\frac{\partial}{\partial r}\left(r\mu \frac{\partial v_z}{\partial r}\right) = \frac{dP}{dz} - \bar{\rho} g + \bar{\rho}\beta g\,(T-\bar{T}) \quad (8)$$

where both μ and ρ are functions of temperature. The boundary conditions for the jth annular element are

at	$z = 0$	$P = P_o$	(9)
at	$z = L$	$P = P_L$	(10)
at	$r = r_{j1}$	$v_z = v_{eo}$	(11)
at	$r = r_{j2}$	$v_z = v_{eo}$	(12)

where slip is assumed at the walls due to electroosmotic flow given by the Helmholtz-Smoluchowski equation

$$v_{eo} = -\frac{\varepsilon\zeta}{4\pi\mu}\frac{dE}{dz} \quad (13)$$

The slip boundary condition is based on the assumption that the thickness of the diffuse double layer of charge at the wall is small compared with Δr (17).

Conservation of Energy

The temperature profile in the bed is obtained by the heat balance equation

$$\varepsilon_B v_z \frac{\partial(\rho C_p T)}{\partial z} = k_h\left[\frac{1}{r}\frac{\partial}{\partial r}\left(r\frac{\partial T}{\partial r}\right) + \frac{\partial^2 T}{\partial z^2}\right] + \underset{\sim}{i}\cdot\underset{\sim}{\nabla} E \quad (14)$$

where the effective thermal conductivity of the bed k_h is assumed to be isotropic and constant. The boundary conditions are

at	$z = 0$	$T = T_o$	(15)
at	$z = L$	$\frac{\partial T}{\partial z} = 0$	(16)
at	$r = r_a$	$T = T_o$	(17)
at	$r = r_b$	$T = T_o$	(18)

Conservation of Mass of species i

The flux of species i in the bed

$$\underset{\sim}{N}_i = \varepsilon_B C_i(\underset{\sim}{v} + \underset{\sim}{v}_{mig,i}) - D_i \underset{\sim}{\nabla} C_i \quad (19)$$

where the electrophoretic migration velocity of species i

$$\underset{\sim}{v}_{mig,i} = -u_i \underset{\sim}{\nabla} E \quad (20)$$

Using the flux equation in the equation of continuity, the species mass balance under steady-state becomes

$$\varepsilon_B \omega \frac{\partial C_i}{\partial \theta} + (1-\varepsilon_B)\omega\frac{\partial n_i}{\partial \theta} + \varepsilon_B(v_z + v_{mig,i})\frac{\partial C_i}{\partial z} = D_{i\theta}\frac{1}{r^2}\frac{\partial^2 C_i}{\partial \theta^2} + D_{iz}\frac{\partial^2 C_i}{\partial z^2} \quad (21)$$

This equation also includes a term for any possible adsorption on the packing as a result of either opposite charge or surface activity.

The boundary conditions are

at $\theta = 0$ for all z $\quad C_i = n_i = 0$ (22)

at $z = 0$ $0 < \theta < \theta_f$ $C_i = C_{if}$

$\theta_f < \theta \leq 2\pi$ rad $C_i = 0$ (23)

at $z = -\infty$ $0 < \theta \leq \theta_f$ $C_i = C_{if}$

$\theta_f < \theta \leq 2\pi$ rad $C_i = 0$ (24)

at $z = \infty$ all θ $C_i = 0$ (25)

This set of equations is similar to those for the annular rotating chromatographic column (16) except for the inclusion of the electrophoretic migration velocity in Equation (21).

Conservation of Charge

The current density in the bed when the packing material has zero electrical conductivity is (25)

$$\underset{\sim}{j} = \sum_{i=1}^{n} q_i \underset{\sim}{N}_i \tag{26}$$

where $\underset{\sim}{N}_i$ is given by Equation (19). It is convenient to write it in the form of Ohm's law

$$\underset{\sim}{j} = - k_m \underset{\sim}{\nabla} E \tag{27}$$

Conservation of charge yields

$$\underset{\sim}{\nabla} \cdot \underset{\sim}{j} = 0 = - \underset{\sim}{\nabla} \cdot (k_m \underset{\sim}{\nabla} E) \tag{28}$$

If the effective electrical conductivity k_m is assumed to be constant this results in the Laplace's equation

$$\nabla^2 E = 0 \tag{29}$$

The boundary conditions are

$$\text{at } z = 0 \qquad E = E_o \tag{30}$$

$$\text{at } z = L \qquad E = 0 \tag{31}$$

ANALYTICAL SOLUTION

The theoretical model formulated above is quite complex and work is underway to obtain detailed numerical results. The temperature rise within the bed due to the electric dissipation causes several complicating effects. It can significantly distort the velocity profile within the bed owing to its effect on μ and ρ. Additionally, the electrophoretic mobilities as well as the dispersion coefficients will be higher in the center region, leading to further zone distortion. Rigorously, of course, the other physical properties such as electrical and thermal conductivities and dielectric constant, etc., are also functions of temperature. Here, however, in order to obtain analytical solutions for preliminary analysis of the proposed device, several simplifying assumptions are made.

Pressure Drop and Velocity

It is assumed that the temperature variation within a channel is small enough so that constant ρ and μ may be assumed within the channel. However, these properties are allowed to vary from channel to channel as a result of the radial variation of temperature in the bed. Then Equation (8) reduces to

$$\mu \frac{1}{r} \frac{d}{dr} \left(r \frac{dv_z}{dr}\right) = \frac{d\Pi}{dz} \tag{32}$$

$$\text{where} \qquad \Pi \equiv P - \rho g z \tag{33}$$

Then for the jth annular channel the velocity profile is

$$v_z = v_{eo} + \frac{(\Pi_o - \Pi_L) r_{j2}^2}{4\mu L \tau_f} \left[1 - \frac{r^2}{r_{j2}^2} + \frac{1-\kappa^2}{\ln(1/\kappa)} \ln \frac{r}{r_{j2}}\right] \tag{34}$$

where

$$\Pi_o - \Pi_L = P_o - P_L + \rho g L \tag{35}$$

and the tortuosity factor, τ_f, accounts for the tortuous path followed by liquid compared with the column height. The bulk velocity in the jth channel is

$$\langle v \rangle_j = v_{eo} + \frac{(\Pi_o - \Pi_L) r_{j2}^2}{8\mu L \tau_f} \left[\frac{1-\kappa^4}{1-\kappa^2} - \frac{1-\kappa^2}{\ln(1/\kappa)}\right] \tag{36}$$

This equation can be used to obtain the velocity profile in the bed. Clearly, the channel width can be varied radially if necessary to simulate realistic velocity profiles in the packed bed including zero velocity at the chamber walls. If equal width, Δr, is assumed then the velocity profile is rather flat, particularly for a relatively thin annular packed bed. If each annular element is thin, the effect of curvature within a channel can be neglected and

$$\langle v\rangle_j \simeq v_{eo} + \frac{(\Pi_o - \Pi_L)\,\Delta r^2}{12\mu L\tau_f} \qquad (37)$$

which is the result for flow between parallel flat plates.

Substituting Equation (3) for Δr into Equation (37) and comparing to the Kozeny-Carman equation based on experimental values of pressure drop for laminar flow in packed beds in the absence of electroosmotic slip flow, the tortuosity factor becomes

$$\tau_f = \frac{75}{54} \qquad (38)$$

Heat Generation and Temperature Profile

Since the applied voltage gradient is only in the axial direction, the heat generation term becomes

$$\underset{\sim}{i}\cdot\underset{\sim}{\nabla}E = -k_m\left(\frac{\partial E}{\partial z}\right)^2 \qquad (39)$$

Using Equations (29) to (31)

$$\underset{\sim}{i}\cdot\underset{\sim}{\nabla}E = k_m\left(\frac{E_o}{L}\right)^2 \qquad (40)$$

Assuming a flat velocity profile in the bed, $\langle v\rangle = \langle v\rangle_j$, constant physical properties, and negligible axial thermal conduction relative to axial convection, Equation (14) becomes

$$\varepsilon_B\langle v\rangle\rho C_p\frac{\partial T}{\partial z} = k_h\frac{1}{r}\frac{\partial}{\partial r}\left(r\frac{\partial T}{\partial r}\right) + k_m\left(\frac{E_o}{L}\right)^2 \qquad (41)$$

The solution (1) in dimensionless form is

$$T^* = \left[(1 - r^{*2}) + (r_o^{*2} - 1)\frac{\ln r^*}{\ln r_o^*}\right] - \sum_{n=1}^{\infty} A_n\Phi_n(r^*)\exp(-Fo\lambda_n^2) \qquad (42)$$

where the coefficients, A_n, are given by

$$A_n = \frac{\displaystyle\int_1^{r_o^*}\left[(1 - r^{*2}) - (r_o^{*2} - 1)\frac{\ln r^*}{\ln r_o^*}\right]\Phi_n(r^*)r^*dr^*}{\dfrac{2[J_o^2(\lambda_n) - J_o^2(\lambda_n r_o^*)]}{\pi^2\lambda_n^2 J_o^2(\lambda_n r_o^*)}} \qquad (43)$$

and the eigenfunction

$$\Phi_n(r^*) \equiv Y_o(\lambda_n)J_o(\lambda_n r^*) - Y_o(\lambda_n r^*)J_o(\lambda_n) \qquad (44)$$

The eigenvalues λ_n are evaluated from the characteristic equation

$$Y_o(\lambda_n)J_o(\lambda_n r_o^*) - Y_o(\lambda_n r_o^*)J_o(\lambda_n) = 0 \qquad (45)$$

The coefficients A_n are evaluated from Equation (43) by numerical integration. Simpson's rule was used. Generally only a few terms in the summation in Equation (42) are required since the exponential term diminishes rapidly. The dimensionless temperature, T*, is thus a function of dimensionless radial position, r*, the ratio of outer to inner radius, r_o^*, and the Fourier number, $Fo = \alpha\tau/r_b^2$.

Power Consumption

The electric power dissipation in the bed is given by

$$P_e = 2\pi L\int_{r_b}^{r_a}(\underset{\sim}{i}\cdot\underset{\sim}{\nabla}E)\,r\,dr \qquad (46)$$

By use of Equation (40) and integrating, the power consumption is

$$P_e = \pi(r_a^2 - r_b^2)k_m\frac{E_o^2}{L} \qquad (47)$$

Zone Spreading and Separation

Solution of the complete mass balance

equation for species i, Equation (21) is rather difficult. See Begovich and Sisson (16) for a discussion. Here we will make several simplifying assumptions. Assuming instantaneous linear adsorption equilibrium given by

$$n_i = K_i C_i \tag{48}$$

and plug flow in the bed, Equation (21) takes the form

$$\omega\{\varepsilon_B + (1-\varepsilon_B)K_i\}\frac{\partial C_i}{\partial\theta} + \varepsilon_B(\langle v\rangle + v_{mig,i})\frac{\partial C_i}{\partial z} = D_{i\theta}\frac{1}{r^2}\frac{\partial^2 C_i}{\partial\theta^2} + D_{iz}\frac{\partial^2 C_i}{\partial z^2} \tag{49}$$

where

$$v_{mig,i} = -u_i\left(\frac{\partial E}{\partial z}\right) = u_i\left(\frac{E_o}{L}\right) \tag{50}$$

Note that the case of no adsorption can be recovered by simply substituting $K_i = 0$.

Defining (26) an effective axial velocity for i through the column by

$$v_{iz,eff} \equiv \frac{\varepsilon_B(\langle v\rangle + v_{mig,i})}{\varepsilon_B + (1-\varepsilon_B)K_i} \tag{51}$$

and effective dispersion coefficients by

$$D_{i\theta,eff} \equiv \frac{D_{i\theta}}{\varepsilon_B + (1-\varepsilon_B)K_i} \tag{52}$$

and

$$D_{iz,eff} \equiv \frac{D_{iz}}{\varepsilon_B + (1-\varepsilon_B)K_i} \tag{53}$$

Equation (49) reduces to

$$\omega\frac{\partial C_i}{\partial\theta} + v_{iz,eff}\frac{\partial C_i}{\partial z} = D_{i\theta,eff}\frac{1}{r^2}\frac{\partial^2 C_i}{\partial\theta^2} + D_{iz,eff}\frac{\partial^2 C_i}{\partial z^2} \tag{54}$$

The exit position of the peak of species i, $\bar{\theta}_i$, is obtained by writing Equation (54) without dispersion

$$\omega\frac{\partial C_i}{\partial\theta} + v_{iz,eff}\frac{\partial C_i}{\partial z} = 0 \tag{55}$$

The solution to Equation (55) yields

$$\bar{\theta}_i = \frac{\omega L}{v_{iz,eff}} = \omega t_i \tag{56}$$

and the corresponding distance travelled by the peak in the lateral direction by species i

$$\bar{y}_i = r_m\bar{\theta}_i \tag{57}$$

The circumferential separation distance between two exiting peaks is

$$\Delta\bar{y}_{ij} = r_m(\bar{\theta}_i - \bar{\theta}_j) \tag{58}$$

The resolution between two exiting peaks is, as usual, defined as the ratio of the distance between two solute peaks to their average band width

$$R_{ij} = \frac{\Delta\bar{y}_{ij}}{\frac{1}{2}(W_i + W_j)} \tag{59}$$

For a Gaussian concentration profile at the exit, 96% of a given species elutes within $\pm$ 2 σ of the peak. Thus,

$$W_i \simeq 4\sigma_i \tag{60}$$

and

$$R_{ij} = \frac{\Delta\bar{y}_{ij}}{2(\sigma_i + \sigma_j)} \tag{61}$$

A resolution of unity or greater indicates essentially complete separation.

The standard deviation at the exit, σ_i, can be estimated after obtaining the concentration profile for i at the exit by solving Equation (54). However, we follow the alternate procedure suggested by Giddings (12) and Van Deemter et al. (27). For a line source and assuming that the dispersion in θ and z directions are independent and Gaussian in nature, their variances are additive and the net standard deviation may simply be obtained from

$$\sigma_i = (\sigma_{iz}^2 \cot^2\phi_i + \sigma_{i\theta}^2)^{1/2} \qquad (62)$$

where the first term on the right hand side of Equation (62) results from the projection of the standard deviation in the z direction on the lateral direction. Further

$$\sigma_{iz}^2 = 2 D_{iz,eff} t_i \qquad (63)$$

and

$$\sigma_{i\theta}^2 = 2 D_{i\theta,eff} t_i \qquad (64)$$

The mean residence time of species i, t_i, is given by Equation (56), and

$$\cot \phi_i = \frac{r_m \omega}{v_{iz,eff}} \qquad (65)$$

These expressions should provide reasonable estimates of zone spreading and resolution.

DISCUSSION

It is evident that the design and performance of the CRAE column is affected by several factors. These include the electrophoretic mobilities of the components to be separated, the voltage gradient imposed, packing size and shape, bed porosity, any adsorption on the packing, eluent flow rate, rotation rate, heat production rate, bed thickness, and band spreading phenomena such as diffusion and hydrodynamic dispersion, electroosmosis, temperature variations in the bed and its effect on flow field, diffusion coefficients and electrophoretic mobilities. A detailed theoretical analysis of these effects and their interactions is clearly complex and must await further study. The analytical solutions obtained here, however, do provide first estimates of the effects of various parameters on the performance and design of CRAE.

Dispersion in Bed

Before embarking on a discussion of these results, some comments are in order with respect to the dispersion in the CRAE column which is the major factor limiting the degree of resolution between separated components. The hydrodynamic, or Taylor, dispersion in packed bed devices is considerably more than that in free laminar flow devices owing to the microscopic flow disturbances created in flowing around the packing. Clearly, *a priori* prediction of the dispersion coefficients in packed electrophoresis columns is difficult at this stage. For the annular channel structural model of packed beds assumed here, however, the axial dispersion coefficient can easily be calculated from the velocity profile, Equation (34). While it is not clear how this can be related to the actual dispersion coefficients in packed electrophoresis beds, it is noteworthy that at least qualitative predictions can be made. For instance, as evident from Equation (34), electroosmosis could significantly distort the velocity profile and consequently we may expect the dispersion coefficient to depend upon the strength as well as the direction of the electric field, and the nature of the packing. Thus, when electroosmotic flow is in the same direction as the bulk eluent flow, the velocity profile in channels for a given bulk flow becomes flattened and, consequently, the dispersion coefficient should be less than that for no electric field. Conversely, when electroosmotic flow is in the opposite direction, for the same bulk flow rate, the velocity variation around is the mean is now much greater and, thus, the dispersion coefficient should be higher than that for no electric field. Similarly, the effect of the temperature rise within the bed on dispersion can be qualitatively predicted when using Equation (34) for annular channels at different radial locations by incorporating the variation in local fluid properties as determined by the local temperature.

Temperature Profile in Bed

An important design constraint in the electrophoresis column is the peak temperature rise within the bed, as many proteins will denature at 50 to 60°C. The developing dimensionless temperature profiles within a CRAE column are plotted in Figure 3 using Equation (42) for different Fourier numbers. For a long column (large Fo) the developing temperature profile term in Equation (42) is small and the temperature profile can be approximated by

$$T_\infty^* = [(1 - r^{*2}) + (r_o^{*2} - 1) \frac{\ln r^*}{\ln r_o^*}] \qquad (66)$$

The location of the peak temperature in the bed, r_p^*, for this asymptotic case can be obtained by setting $dT_\infty^*/dr^* = 0$. Then

$$r_p^* = \sqrt{\frac{r_o^{*2} - 1}{2 \ln r_o^*}} \quad (67)$$

This may be substituted back into Equation (66) to determine the peak temperature. This value would also provide a conservative estimate of the peak temperature in bed for design purposes.

Illustrative Example

We will illustrate the use of the other results obtained above by using as an example the case of a binary separation in a CRAE column with dimensions similar to those of the prototype being constructed in our laboratory for testing. The values of the various parameters employed in the calculations are given in Table 1.

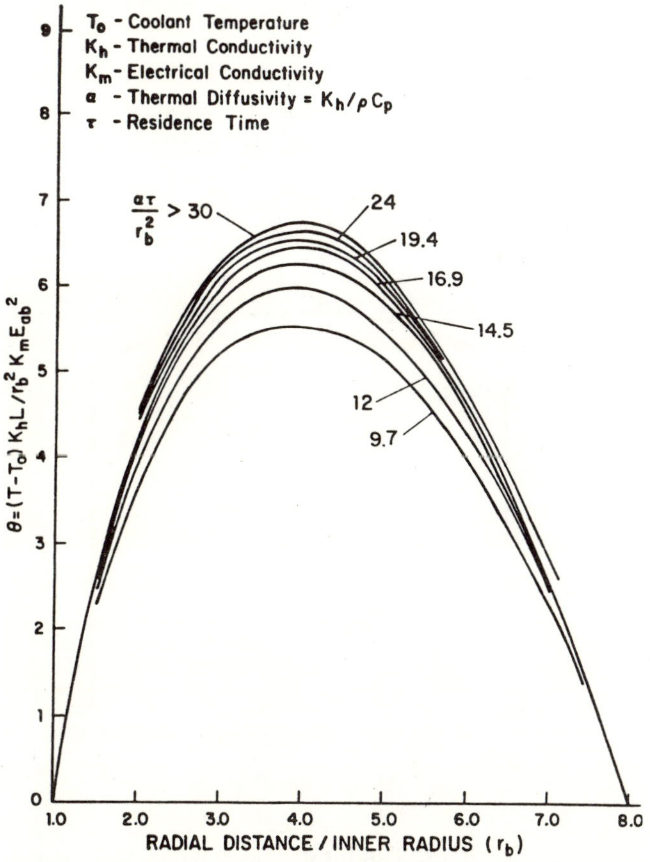

Figure 3. Calculated dimensionless temperature profiles in the CRAE column with an outer to inner radius ratio of 8.

TABLE 1 Values of Parameters Employed in the Illustrative Example

Parameter	Value	Reference
r_a	0.35 m	
r_b	0.30 m	
L	0.49 m	
ε_B	0.33	
D_p	60μm	
E_o	30 V	
k_h	0.8786 W/m•K	9
k_m	0.8 $(\Omega \cdot m)^{-1}$	9
C_p	4.1868×10^3 J/kg·K	
ρ	1.0×10^3 kg/m^3	
u_A	2×10^{-8} m^2/V.s	10
u_B	-1×10^{-8} m^2/V.s	10
T_o	278 K	
K_i	0.0	
v_{eo}	0.0 m/s	
$D_{i\theta}$	1×10^{-9} m^2/s	
D_{iz}	1×10^{-9} m^2/s	
μ	Equation (68)	28
P_o-P_L	0.0 Pa	
ω	π rad/h	

From Equation (47), the power consumption

P_e = 150 W

From Equation (3), channel width

Δr = 10.0 μm

From Equation (6), channel wall thickness

Δr_w = 20.0 μm

The number of channels, using Equation (7)

N = 1666

Assuming negligible electroosmotic velocity, from Equation (37) for thin channels

$\langle v \rangle \simeq 6.8 \times 10^{-5}$ m/s

Residence time of eluent, therefore

$\tau \simeq 7200$ s

Using these values in Equation (42), the temperature profile in the column is calculated and is plotted in Figure 4.

Assuming the viscosity of eluent (water) to be a function of temperature as given by (28)

$$\log_{10} \frac{\mu}{\mu_{20}} = \frac{1.3272\ (20-T) - 0.001053\ (T-20)^2}{T + 105} \quad (68)$$

where $\mu_{20} = 1.002 \times 10^{-3}$ Pa.s, the viscosity of water at 20°C, T is in °C, and using the temperature profile calculated above to calculate the viscosity in different channels located at different radial positions, the velocity profile in the entire bed is calculated using Equation (37) and is plotted in Figure 5. It is noteworthy that the velocity in the middle of the annulus is somewhat higher due to temperature increase within the bed. In addition, by positioning annular channel walls at r_a and r_b the boundary condition of zero velocity at the chamber walls can be satisfied. This illustrates the ease with which the effect of temperature on velocity profile in bed can be incorporated with the structural model proposed. Since the velocity variation is small, an average velocity of eluent of 6.8×10^{-5} m/s may be assumed.

From Equations (50) and (51)

$v_{Az,eff} = 6.923 \times 10^{-5}$ m/s

$v_{Bz,eff} = 6.739 \times 10^{-5}$ m/s

Therefore, from Equation (56)

$\bar{\theta}_A = 354°$

$\bar{\theta}_B = 364°$

From Equation (58)

$\Delta\bar{y}_{AB} = 55$ mm

From Equation (65)

$\phi_A = 13.72°$

$\phi_B = 13.37°$

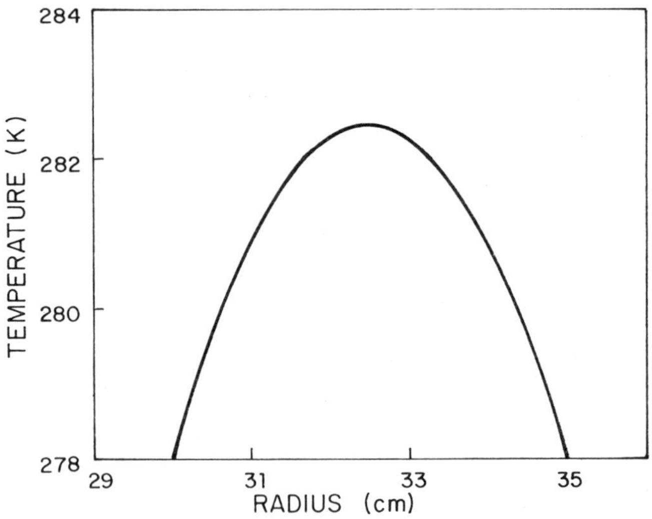

Figure 4. Temperature profile for the column described in the Illustrative Example.

Using Equation (62)

$\sigma_A = 15.9$ mm

$\sigma_B = 16.5$ mm

Therefore, resolution, using Equation (61)

$R_{AB} = 0.85$

Since this resolution is marginal, either increasing the applied voltage across the bed or increasing the rotation rate to improve separation is indicated.

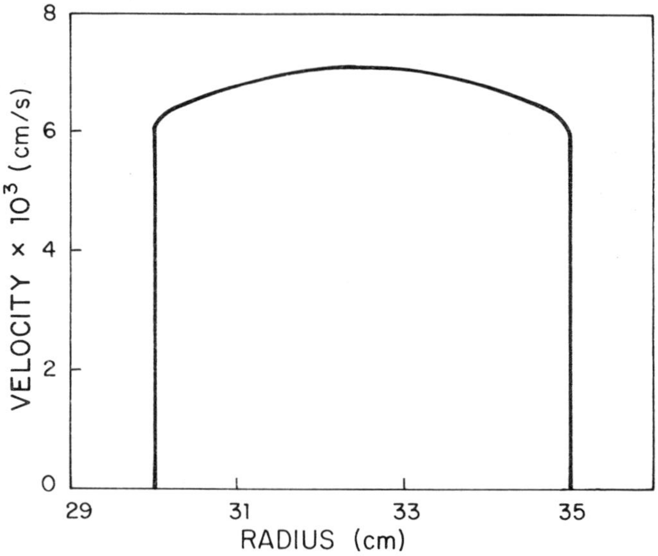

Figure 5. Velocity profile for the column described in the Illustrative Example.

CONCLUSION

A general theoretical model is formulated for the recently developed continuous rotating annular-bed electrophoresis column (1). The proposed design appears to be well suited for relatively large scale separation of biochemicals since the scaling-up in this design is not limited by bed thickness and, therefore, the temperature rise within the bed can be maintained within allowable limits. Further, since the separation takes place along the circumference, multicomponent separation is possible. In this paper we have provided simplified analytical solutions for the velocity profile and pressure drop in the column, temperature rise within the bed, power consumption, location of eluting peaks, band spreading and resolution. These analytical results represent the first step towards a more detailed numerical analysis that would incorporate thermal convection and electroosmosis and their effects on zone spreading. Other topics to be studied include consideration of axial thermal conduction in addition to convection and effects on separation of possible adsorption on the packing material. A prototype is being constructed to test the theoretical model and to study the suitability of this device for preparative scale separation of a variety of biochemicals.

ACKNOWLEDGEMENT

This work was supported in part by funds provided by the Iowa High Technology Council. Lisa Korndorf and Harry Zingher were involved in the early phase of model development. We would especially like to thank Bev Palmer for her deft typing of the manuscript.

NOTATION

A_n	eigen coefficient
C_i	concentration of i
C_{if}	feed concentration of i
C_p	heat capacity of eluent
D_{iz}	axial dispersion coefficient of species i
$D_{iz,eff}$	effective axial dispersion coefficient of species i, Eqn. (53)
$D_{i\theta}$	angular dispersion coefficient of species i
$D_{i\theta,eff}$	effective angular dispersion coefficient of species i, Eqn. (52)
D_p	diameter of particles in bed
E	electric potential
E_o	electric potential across bed
Fo	Fourier number, $\alpha\tau/r_b^2$
g	acceleration due to gravity
i	current density, species i
J_o	Bessel function of zero order, first kind
K_i	linear adsorption equilibrium constant for species i
k_h	effective thermal conductivity of bed
k_m	effective electrical conductivity of bed
L	length of column
N	number of annular channel elements in bed
$\underset{\sim}{N}_i$	flux of species i
n_i	concentration of adsorbed species i in the solid phase
P	pressure
P_e	electric power consumption
P_L	pressure at z = L in the column
P_o	pressure at z = 0 in the column
q_i	charge per mole of species i
R_{ij}	resolution between exiting peaks of species i and j
r	radial coordinate
r_a	radial position of outer cylinder
r_b	radial position of inner cylinder

r_H	hydraulic radius
r_{j1}	inner radius of jth annular channel
r_{j2}	outer radius of jth annular channel
r_m	mean radius of the annular bed
r^*	dimensionaless radial coordinate $(=r/r_b)$
r_o^*	dimensionless radial coordinate $(=r_a/r_b)$
r_p^*	dimensionless radial position of the asymptotic peak bed temperature
Δr	width of annular channel elements
Δr_w	thickness of walls of annular channels
S_v	specific surface of packing
T	temperature
T_c	characteristic temperature = $k_m(r_b E_o/L)^2/4k_h$
T_o	temperature of walls, entering eluent
T^*	dimensionless temperature = $(T-T_o)/T_c$
T_∞^*	dimensionless fully developed temperature profile
$\bar{T}$	reference temperature
t_i	retention time of species i, $L/v_{iz,eff}$
u_i	electrophoretic mobility of species i
v_{eo}	electroosmotic flow velocity
$v_{iz,eff}$	effective axial velocity of component i in the bed, Equation (51)
$v_{mig,i}$	electrophoretic migration velocity of i
v_z	axial velocity
$\langle v \rangle$	average velocity of eluent in bed
$\langle v \rangle_j$	average velocity of eluent in channel j
Y_o	Bessel function of zero order, second kind
y	lateral coordinate $\simeq r_m\theta$
$\bar{y}_i$	circumferential exit location of the peak of species i
$\Delta\bar{y}_{ij}$	separation distance between the exiting peaks of species i and j
W_i	band width of species i at the exit
z	axial coordinate

GREEK LETTERS

α	thermal diffusivity $(=k_h/\rho C_p)$
β	coefficient of thermal expansion of eluent
ε	dielectric constant of eluent
ε_B	bed porosity
θ	angular coordinate
θ_f	angle up to which feed is applied
$\bar{\theta}_i$	peak exit position for species i
ζ	zeta potential on the packing surface
κ	$\equiv r_{j1}/r_{j2}$
λ_n	eigenvalue
μ	viscosity of eluent
Π	$\equiv P - \rho gz$
ρ	density of eluent
$\bar{\rho}$	density of eluent at reference temperature

σ_i	standard deviation of Gaussian concentration profile of species i at the exit
σ_{iz}	standard deviation of species i for dispersion in the axial direction
σ_{iz}^2	variance of species i for dispersion in the axial direction
$\sigma_{i\theta}$	standard deviation of species i for dispersion in the angular direction
$\sigma_{i\theta}^2$	variance of speices i for dispersion in the angular direction
τ	residence time of eluent = $L/\langle v \rangle$
τ_f	tortuosity factor for flow in packed bed
ϕ_i	angle between the mean trajectory of species i and the horizontal plane
Φ_n	eigenvalue function, Equation (44)
ω	angular velocity of column

LITERATURE CITED

1. Yoshisato, R.A., L. M. Korndorf, G.R. Carmichael, and R. Datta, Sep. Sci. Tech., to be published, 1986.

2. Deyl, Z., F. M. Everaerts, Z. Prusik, and P.J. Svendsen, P.J., eds., Electrophoresis: A Survey of Techniques and Applications, Pt. A: Techniques, Pt. B: Applications, Elsevier Scientific, Amsterdam (1979).

3. Simpson, C.F. and M. Whittaker, eds., Electrophoretic Techniques, Academic Press, New York (1983).

4. Philpot, J. St. L., Trans. Faraday Soc., 36, 38 (1940).

5. Hannig, K., Z. Anal. Chem., 181, 244 (1961).

6. Strickler, A., Sep. Sci. Tech., 2, 335 (1967).

7. Thompson, A.R., "Free-Flow Electrophoresis", in Electrophoretic Techniques, C.F. Simpson and M. Whittaker, eds., p. 253, Academic Press, New York (1983).

8. Mattock, P., G.F. Aitchison, and A.R. Thomson, Sep. Pur. Methods, 9, 1 (1980).

9. Hybarger, R.M., C. W. Tobias, and T. Vermeulen, Ind. Eng. Chem. Process Des. Develop., 2, 65 (1963).

10. Ravoo, E., P.J. Gellings, and T. Vermeulen, Anal. Chim. Acta, 38, 219 (1967).

11. Vermeulen, T., L. Nady, J. M. Krochta, E. Ravoo, and D. Howery, Ind. Eng. Process. Des. Develop., 10, 91 (1971).

12. Giddings, J. C., Anal. Chem., 34, 37 (1962).

13. Fox, J.B., R. C. Calhoun, and W. J. Eglinton, J. Chomatog., 43, 48 (1969).

14. Fox, J.B., J. Chromatog., 43, 55 (1969).

15. Nicholas, R.A. and J. B. Fox, J. Chromatog., 43, 61 (1969).

16. Begovich, J.M. and W. G. Sisson, AIChE J., 30, 705 (1984).

17. Strickler, A. and T. Sacks, N. Y. Acad. Sci., 209, 497 (1973).

18. Reis, J.F.G., E. N. Lightfoot, and H. L. Lee, AIChE J., 20, 362 (1974).

19. Saville, D.A., Phys. Chem. Hydrodynam., 1, 297 (1980).

20. Lynch, E.D., and D. A. Saville, Chem. Eng. Commun., 9, 201 (1981).

21. Turk, R.S. and C. F.Ivory, Chem. Eng. Sci., 39, 851 (1984).

22. Naumann, R.J. and P. H. Rhodes, Sep. Sci. Tech., 19, 51 (1984).

23. Bird, R.B., W. E. Stewart, and E. N. Lightfoot, Transport Phenomena, John Wiley, New York (1960).

24. Cohen, Y. and A. B. Metzner, AIChE J., 27, 705 (1981).

25. Newman, J., Adv. Electrochem. and Electrochem Eng., 5, 87 (1967).

26. Rony, P.R., Sep. Sci., 3, 425 (1968).

27. Van Deemter, J.J., F. J. Zurderweg, and A. Klinkenberg, Chem. Eng. Sci., 5, 271 (1956).

28. Weast, R.C., and M. J. Astle, eds., CRC Handbook of Chemistry and Physics, 61st ed., CRC Press, Boca Raton (1980).

PILOT PLANT FOR CONTINUOUS DISTILLATION OF ESSENTIAL OIL-BEARING MATERIALS

G. R. Boucard and V. C Boucard ■ Texarome, Inc., Leakey, TX 78873
R. W. Serth and B. S. Ainsworth ■ Department of Chemical and Natural Gas Engineering
Texas A & I University, Kingsville, TX 78363

A pilot plant embodying a new process for the manufacture of essential oils by continuous steam distillation of natural materials is described, and the results of pilot tests for the production of Texas cedarwood oil are presented. An economic analysis of the new process is presented, including detailed cost estimates for a commercial-size continuous cedar oil plant. The environmental impact of such a plant is also considered. It is shown that the new process represents a major improvement in the production technology of large-volume essential oils.

Essential oils are aromatic liquids produced and stored in the flowers, leaves, bark, wood or roots of certain plants. They are mixtures, usually complex, of organic chemicals, the major components of which are generally terpenes, diterpenes, sesquiterpenes, and oxygenated derivatives of these compounds. Approximately 150 different oils are produced on a commercial basis. Although the total quantity of these materials produced is small by chemical-industry standards, they constitute important raw materials for the flavor and fragrance industry. The largest uses of essential oils are for flavor and fragrance formulations used in foods, beverages, pharmaceuticals, and personal care products.

A number of physico-chemical methods are employed to separate essential oils from the plant tissues in which they occur, but by far the most widely used method is that of steam distillation (1). The plant material is usually comminuted, loaded into a batch still, and contacted with steam for a period of from one hour to as long as 76 hours, depending on the material. The steam is condensed and the oil is separated from the water in a decanter. Since many essential oils are produced by small businessmen, often in under-developed parts of the world, the distillation equipment used may be extremely crude (1, 2, 3). On the other hand, the large essential oil companies operating in industrialized countries generally employ modern stainless-steel stills equipped with instrumentation to adequately control operating conditions (3). These stills may be quite large, on the order of 7 m^3 to 15 m^3. Large field distillation operations in the more industrialized countries may also use modern equipment (4). Nevertheless, the basic technology remains that of batch steam distillation as described more than thirty-five years ago by Guenther (1).

Although batch processing may be the most appropriate operation for many small-volume oils, a number of essential oils are currently produced in quantities large enough to make continuous processing extremely attractive from both an economic and a materials handling standpoint. Examples include the oils of lemongrass, the mints, cedarwood, pine and vetiver. In order to investigate the feasibility of continuously distilling such materials, bench-scale experiments were performed on a variety of essential oil-bearing materials (5). It was found that high oil yields could be obtained with short (20 to 40 seconds) residence-time distillations. Temperatures of 120°C to 150°C and particle sizes of 1.5 mm or less were required to produce high yields at essentially atmospheric pressure.

Based on the results of the bench-scale tests, a pilot plant was designed and constructed to continuously distill the oil from Texas cedarwood. The plant was designed to be flexible so that with minor modifications it is capable of processing a wide variety of essential oil-bearing materials. The purpose of this paper is to describe the operation of the pilot plant and to present the results of tests conducted on the distillation of Texas cedarwood.

PROCESS DESCRIPTION

A flow diagram for the pilot plant, which has a nominal capacity to process six metric tons per day (t/d), is shown in Figure 1. Cedar logs are manually fed to a chipper which has a nominal capacity of five t/h and is driven by a 75 kW engine that runs on propane gas. Since the chipper has a much greater capacity than is required, it operates only one to two hours per day. The wood chips are transported pneumatically to a storage bin, which consists of an agricultural forage wagon. The forage wagon has a storage volume of approximately 19 m^3 and is equipped with a live-bottom chain and a side-delivery chain for unloading the contents.

From the storage bin, the wood chips are delivered at a controlled rate to the first of two 7.5 kW hammermills, where the chips undergo further size reduction. The wood is pneumatically transported to the second hammermill (not shown) and then to the inlet of the distillation unit. The finely ground material enters the distillation unit through a plug feeder and a rotary valve. In the distillation unit, the wood is contacted with superheated steam from the boiler. The steam and wood flow through the distillation unit in a co-current flow pattern. However, other flow patterns, e.g., counter-current or cross-flow, are also feasible. The contacting occurs in discrete stages with interstage heat addition. In each stage, the wood is suspended in the steam so that intimate contact between the gas and solids is attained. The entire distillation unit is well insulated to minimize heat loss.

The distillate, which consists of steam and vaporized oil, is condensed and sent to a decanter where the oil separates from the water by virtue of its lower specific gravity. Because the oil flow rate is quite small, the oil is drawn from the decanter periodically (rather than continuously) and transferred to 0.2 m^3 drums for shipment. The condensed steam, which contains a residue of dissolved oil, is drawn continuously from the bottom of the decanter and sent to an oil separator (not shown) and then to a cooling tower (not shown) for use as condenser cooling water make-up.

The spent cedar exits the distillation unit through a rotary valve and plug discharge device, and is divided into a waste stream and a fuel stream. The latter is pneumatically transported to a specially designed wood-burning furnace which supplies all the energy for steam generation. The spent cedar comprises an excellent fuel due to its small size and low (approximately 4%) moisture content. These characteristics permit the use of an inexpensive suspension-type furnace as opposed to a dutch oven or fluidized bed furnace required to burn larger wood chips with higher moisture contents. The flow rate of wood to the furnace is controlled by a baffle whose position is controlled by the boiler pressure. The boiler is a 400 kW scotch-marine type unit. The waste wood is pneumatically carried to a waste truck for transportation to a disposal site.

Additional equipment not shown in Figure 1 includes the aforementioned cooling tower which handles the condenser cooling water, and equipment for treating the boiler-feed water.

The operation of the plant is completely automated except for the chipper feeding operation and the removal of the product oil from the decanter. In a full-scale plant, the decanter would operate in a continuous and automatic mode. Also, a hydraulic lift ("cherry picker") would be used to feed the chipper instead of manual labor. The process is entirely controlled by simple on-off control devices, with appropriate time delays to ensure proper start-up and shut-down sequences.

OPERATING EXPERIENCE

A 500 hectare tract in the hill country of south-central Texas with ample supplies of both standing and cut cedar was selected as the initial site for the pilot plant. The plant was constructed in modular fashion with the major pieces of equipment trailer mounted to make the entire operation portable. Major equipment items were constructed off-site and towed to the plant location, at which well water and electrical service were the only existing improvements.

Steady-state operation of the pilot plant was achieved after a lengthy shakedown period during which numerous modifications to the original design were made. The plant was then operated continuously for two consecutive quarters. During this time, shutdowns were made only for routine maintenance and for minor repairs and modifications. The operation of the process proved to be quite reliable, and during normal operation it ran completely unattended. However, an operator was required at the plant site at all times to take care of upset conditions, which were signalled by an alarm system. Most of the cedarwood used in the plant was harvested on site, but some wood was also purchased from local sellers. During the period of operation, a total of 6,000 kg of oil was produced and sold to a local cedar oil distillary at a competitive market price.

During the third quarter of operation the plant was shut down for major modifications, which were intended to improve the yield and increase the production rate. A number of changes were also made to eliminate trouble spots in the operation, thereby reducing the amount of maintenance and further improving the reliability of the system. The plant was restarted in the first quarter of 1984, and continues to operate as an experimental unit at the present time. The oil continues to be sold at market price, thereby permitting a financially self-sustaining operation.

Values of current operating parameters for the distillation unit are presented in Table 1. A sieve analysis of the feed to the unit is given in Table 2. The average size listed in Table 1 is a mass-average value calculated from the sieve analysis.

PRODUCT QUALITY AND YIELD

The oil produced in the pilot plant is a perfectly clear, viscous liquid with a pale yellow hue. A gas chromatographic analysis of the oil is presented in Table 3. For comparison, analyses are also given for an oil produced in a commercial (batch) plant and for the Fragrance Materials Association (FMA) Standard Texas cedarwood oil. It is obvious that the pilot-plant product is low in cedrenes and high in cedrol compared with the other two oils. This anomaly is caused by the hydration of cedrene to cedrol, which takes place under the conditions present in the distillation unit according to the reaction:

$$\underset{\text{(cedrene)}}{C_{15}H_{24}} + H_2O \xrightarrow{H^+} \underset{\text{(cedrol)}}{C_{15}H_{26}O}$$

The hydration reaction is catalyzed by the acids which occur naturally in the wood. It can be controlled by adding ammonia to the steam to adjust the pH in the system. However, it should be noted that the optimal composition of the oil depends on the end use for which the oil is intended. Since cedrol has a significantly higher market value than cedrene, a high cedrol content is desireable unless the oil is to be used specifically for making cedrene derivatives. The high thujopsene-to-cedrene ratio of the pilot plant product is also advantageous if the light fraction is to be used to manufacture thujopsene derivatives.

The average yield of oil from the pilot plant is 2.3% by weight. This value is significantly higher than the average yield of 2.0% obtained by most of the commercial plants processing Texas cedarwood. Analysis of the spent wood from the process indicates that the yield can be increased to 2.5%-3.0% by increasing the residence time in the distillation unit. Tests at the pilot plant have demonstrated that a similar improvement in yield can be obtained by increasing the operating temperature of the distillation unit. Based on the results of laboratory tests (5), a yield of 2.5% - 2.7% can also be obtained by reducing the average particle size of the feed from 0.4 mm to 0.2 mm. Yields in range of 3.0% - 3.5% can be achieved if the particle size is further reduced by use of a flour mill.

ECONOMIC ANALYSIS

Based on the pilot-plant operation, a preliminary design and cost estimate were made for a commercial-size plant having a capacity for processing 50 t/d of cedarwood. The design differs in several minor respects from that of the pilot plant. For example, steam turbines are used to drive the chipper and hammermills. Since the grinding operations account for the majority of the process power requirements, this configuration significantly reduces the operating cost. The chipper and first-stage hammermill, which is located between the chipper and storage bin, operate on one shift per day. The second-stage hammermill, which is located between the storage bin and distillation unit, operates continuously. Nearly all of the spent wood from the distillation unit is required to generate steam when all three grinding units are in operation. Hence, the entire spent wood stream is burned in the furnace and excess steam is dumped when necessary. Although this mode of operation requires additional boiler-feed water, it eliminates the need for a separate unit to dispose of the waste wood.

In addition to the above modifications, the design includes a boiler fuel bin, which serves as a surge vessel between the distillation unit and furnace. The wood stored in the bin is used to maintain boiler pressure during temporary shut-downs and for process start-ups. Screw conveyers are used to transport the wood from the distillation unit to the fuel bin and from the chipper to the first-stage hammermill. Pneumatic transport is used from the first-stage hammermill to the storage bin and from the second-stage hammermill to the distillation unit. Logs are fed to the chipper using a "cherry picker". Finally a truck scale is provided to weigh truckloads of purchased cedarwood.

A number of options are possible in the grinding train. For example, a hog could be used instead of a chipper. For fine grinding, a flour mill could be substituted for the second-stage hammermill. Since the power for grinding will be generated from the process waste, the optimum particle size will probably be the smallest size practically achievable, in order to maximize the yield. However, the cost estimates are based on essentially the same operating parameters as those of the pilot plant.

Capital and operating cost estimates for a 50 t/d plant are presented in Tables 4 and 5, respectively. The site preparation listed in Table 4 includes a water well since it is assumed that the plant will be located in an area that is not served by a municipal water system. It also includes a small wastewater pond, which is discussed in the following section.

As shown in Table 4, the required capital investment for a 50 t/d plant is approximately $860,000. From Table 5, the annual operating cost is approximately $1.1 million. This figure is based on 330 operating days per year and an annual interest rate of 15%, which is applied to the total capital investment. It is also assumed that the oil will be sold through a broker, which entails a commission charge equal to five percent of sales. These costs are to be compared with gross sales of $2.475 million, based on a yield of 2.5% and an oil price of $3.00 per pound. This gives a before-tax profit of approximately $1.36 million per year. Assuming an overall tax rate of 46 percent, the after-tax profit is $870,000 per year, and the simple payback time is one year.

The capital investment for a batch plant of the same capacity and quality construction as the continuous plant is estimated to be between $750,000 and $1.0 million. Most of the capital cost items listed in Table 4 for the continuous plant will also be incurred by the batch plant. In addition, the smaller size of the continuous distillation unit compared with batch stills tends to be offset by the added sophistication and control required in the continuous operation. Thus, there does not appear to be a significant advantage in capital cost associated with the continuous process.

On the other hand, the continuous process has a substantial advantage over the batch process with respect to operating costs. It is estimated that the continuous process will save approximately $400,000 annually in labor and overhead costs compared with the batch process. In addition, it is not clear that the batch process will permit the use of steam-driven grinding equipment, due to the larger particle size and higher moisture content of the spent wood from the batch process. Cedar oil plants do not currently use steam-driven equipment. Hence, the continuous process may also enjoy a significant advantage in power costs over the batch process. Also, the higher yield of oil obtained in the continuous process results in a correspondingly lower raw material cost per unit of product. The net result of these lower operating costs is a greatly enhanced profit margin for the continuous process.

In addition to significantly lower operating costs, the prinicipal advantages of the continuous process are reliability, ease of operation, and uniformity of product quality that result from a tightly-controlled, steady-state operation.

ENVIRONMENTAL ANALYSIS

The environmental impact from a properly controlled continuous cedar oil plant is quite modest. Estimated annual air emissions from a 50 t/d plant are given in Table 6. The emissions of cedar dust arise primarily from the pneumatic transport operations. There are two such operations in the plant: from the first-stage hammermill to the storage bin, and from the second-stage hammermill to the distillation unit. Each of the two blower exhausts is controlled by a cyclone separator followed by a fabric filter. Based on measured particle size distributions, cyclone efficiencies of 99 percent should be readily attainable. The filter efficiencies are estimated to be 99.9% based on the data given by Forrester (6). The estimated annual emissions from these sources are 0.33t, based on the above efficiencies and an operating period of 330 days per year. There will also be a certain amount of fugitive emissions of cedar dust, even with the best of housekeeping procedures. However, the fugitive dusts will occur mainly within the confines of the plant building. The exhaust from this building will be controlled with a fabric filter. Assuming a control efficiency of 99.9% for the filter and conservatively assuming that fugitive emissions amount to one percent of plant throughput, the annual emission to the environment from this source is 0.17 t. Hence, the total annual emission of cedar dust is estimated to be 0.4 t.

Based on the pilot-plant operation, fabric filters may not be required on the cyclone exhausts as assumed in the above calculations. In the pilot plant, there are no visible emissions from the cyclones on the hammermill blower exhausts. Nevertheless, the filters have been included in the design since they may be required if the commercial plant utilizes a smaller particle size than the pilot plant.

Ash, sulfur dioxide and nitrogen oxides emissions result from the combustion of the spent cedar in the boiler furnace. This unit is equipped with a cyclone separator with an estimated control efficiency of 90%. This is the average value reported for cyclones used to control fly ash from pulverized coal-fired boilers (7). The annual emission of 29.5 t of ash is based on this control efficiency, the assumption of complete combustion, and the measured ash content of Texas cedarwood, which is 1.79% by weight. In the pilot plant operation, there are no visible emissions from the furnace cyclone.

The sulfur dioxide and nitrogen oxides emissions are not controlled. The annual SO_2 emission of 3.3 t is based on an emission factor of 0.2 g SO_2/kg wood burned reported by Hall and DeAngelis (8). The NO_x emission of 90.8 t is based on an average emission factor of 5.5 g NO_x/kg wood burned (9). This emission factor is for NO_x as NO_2. Air emissions in addition to those listed in Table 6 include road dust due to truck traffic, and cooling tower drift. The amounts of these emissions are expected to be quite small and confined within the plant boundaries.

In a commercial-size plant, the condensate stream from the decanters would most likely be treated and recycled as boiler feed water. In this case, the only wastewater stream from the process would be the boiler blowdown. This stream is estimated to amount to approximately 2.0 m^3/d. A small wastewater pond is provided to handle this stream.

If the waste wood from the process is burned in the boiler furnace, then the only significant solid waste from the process is the ash collected by the furnace cyclone. The amount of ash collected is approximately 730 kg/d. This relatively small amount of material can be routed to the wastewater pond and subsequently landfilled, probably on the plant site.

Two additional environmental considerations are noise and odor generation. Noise generation is primarily associated with the grinding operations and, to a lesser extent, with truck traffic. Although these sources will be confined mainly to daytime hours, they could constitute a significant nuisance to nearby residents. The combustion operation results in a noticeable odor of burning cedar, which might be considered objectionable by some people, especially on a continuous basis. For these reasons, the plant should not be located adjacent to residential areas.

CONCLUDING REMARKS

A new process for continuously distilling essential oil-bearing materials has been successfully demonstrated in a pilot plant for the production of Texas cedarwood oil. In particular, it has been shown that a high yield of a commercial-quality oil can be obtained under field conditions. Other features of the process which have been demonstrated in the pilot plant include a high degree of automation, with a correspondingly low manpower requirement; a high degree of reliability; minimal environmental impact; and uniformity of product quality resulting from steady-state operation at tightly controlled processing conditions.

Economically, the continuous process has significantly lower operating costs than the conventional batch process for the production of cedarwood oil. At current market conditions, the simple payback time for a commercial-size cedar-oil plant utilizing the new process is less than one year.

In summary, the pilot operation described in this paper represents the first successful application of continuous processing to the production of essential oils by steam distillation. As such, it constitutes a major improvement in the technology of this field.

LITERATURE CITED

1. Guenther, E., The Essential Oils, vols. 1 to 6, Krieger Publishing Co., New York (1948 - 52).

2. Shreve, R.N. and J.S. Brink, Jr., Chemical Process Industries, 4th ed., McGraw-Hill, New York (1977).

3. White, L.J. and H.J. Eiserle, Ind. Eng. Chem., 53 (6), 421 (1961).

4. McKean, W.T., Food Eng., 43 (2), 101 (1971).

5. Serth, R.W., G.R. Boucard and B.S. Ainsworth, "Continuous partial pressure distillation of essential oil-bearing materials: part I. bench-scale tests," College of Engineering Technical Report, Texas A & I University, Kingsville (1984).

6. Forrester, W.S., Env. Sci. Tech., 8 (6), 508 (1974).

7. Jones, A.H., "Air pollution control for industrial coal-fired boilers," in Power Generation: Air Pollution Monitoring and Control, Noll, K.E. and W.T. Davis (Eds.), Ann Arbor Science, Ann Arbor (1976).

8. Hall, R.E. and D.G. DeAnglis, J. Air Poll. Cont. Assoc., 30, 862 (1980).

9. Anon., "Control techniques for nitrogen oxide emissions from stationary sources," U.S. Dept. Health, Education and Welfare, Pub. AP-67 (1970).

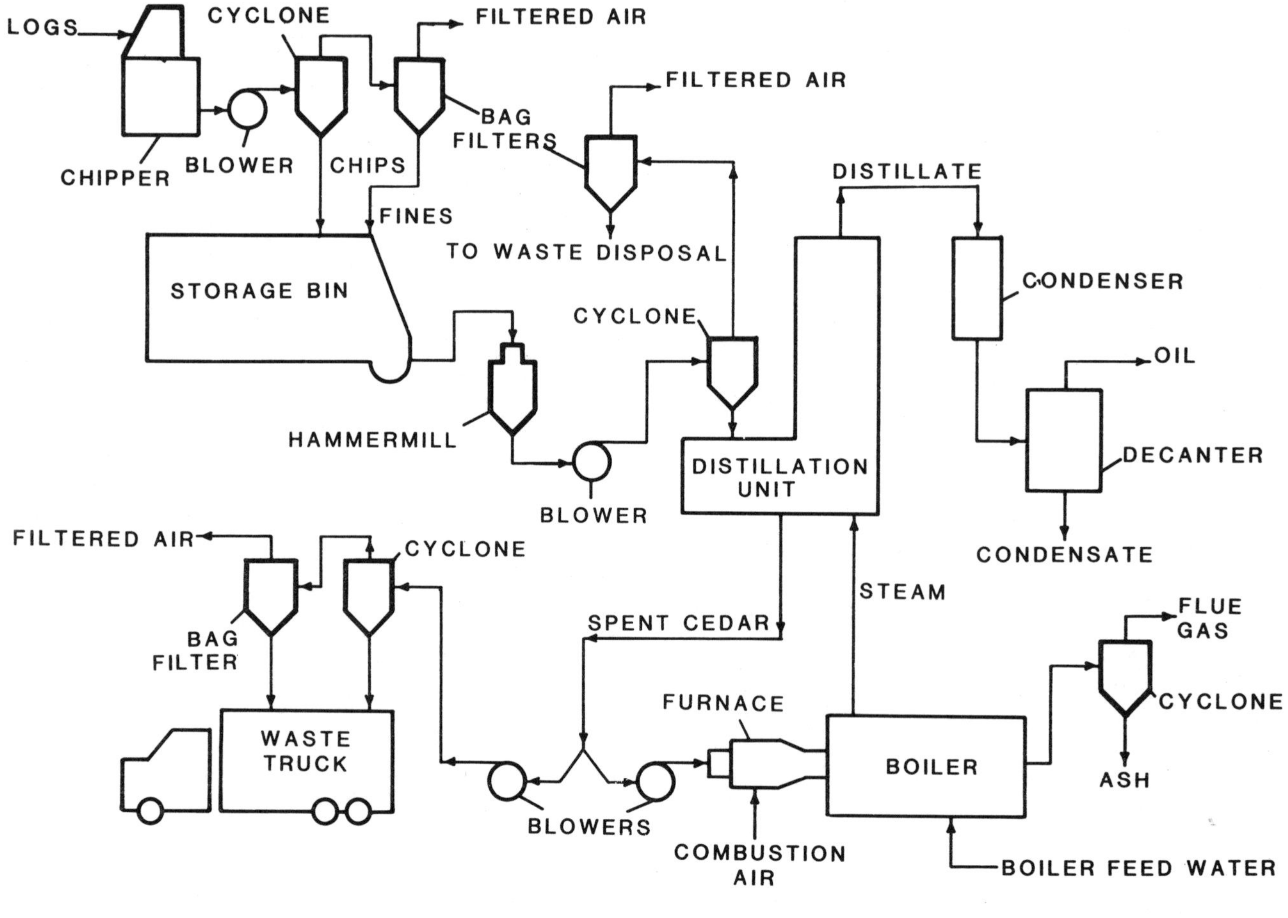

Figure 1. Schematic Flow Diagram for Pilot Plant.

Table 1. Values of Operating Parameters.

Parameter	Value
Maximum throughput (t/d)	4.0
Steam to wood ratio (kg/kg)	2.0
Temperature range (°C)	104-177
Pressure range (kPa gage)	0-25
Average Particle size (mm)	0.4
Mean residence time (s)	25

Table 2. Sieve Analysis for Feed to Pilot Plant.

Through Mesh No.	Retained On Mesh No.	Weight Percent	Average Particle Size (mm)
10	20	0.92	1.2420
20	35	40.83	0.6250
35	60	29.36	0.3315
60	Pan	28.90	0.1230

Table 3. Texas Cedarwood Oil Analyses.

Component	Pilot Plant Product (wt%)	Product from a commercial batch process (wt%)	FMA Standard (wt%)
α-cedrene	3.7	17.5	15.5
β-cedrene	-	3.7	4.1
thujopsene	39.1	33.5	33.6
cedrol	43.0	29.2	22.8
widdrol	3.0	4.6	2.6
	88.8	88.5	78.6

Table 4. Capital Costs for 50 Metric Ton Per Day Continuous Oil Plant.

Item	Cost ($)	
Installed Equipment		
Distillation unit	90,000	
Condenser	10,000	
Decanters (2)	3,000	
Boiler (8000 kW used), wood-fired furnace & accessories	70,000	
Fuel bin and accessories	10,000	
Water treatment equipment	10,000	
Storage bin (30 t, live bottom) & accessories	20,000	
Chipper	30,000	
Hammermills (2)	35,000	
Steam turbines for grinding equipment (3)	25,000	
Cooling tower	20,000	
Truck scale	20,000	
"Cherry picker"	20,000	
Screw convéyors (2)	10,000	
Pollution control equipment	30,000	
Total Installed Equipment	403,000	
Buildings		
Plant building	40,000	
Shop facility	25,000	
Office facility (portable)	10,000	
Total Buildings	75,000	
Site Preparation (including water well & evaporation pond)	40,000	
Land (6 hectare)	45,000	
Total Direct Costs		563,000
Engineering	25,000	
Construction		
Labor	115,000	
Contract work	10,000	
Crane rental	5,000	
Total Construction	130,000	
Contingency	40,000	
Total Indirect Cost		195,000
Total Fixed Capital Investment		758,000
Working Capital		100,000
Total Capital Investment		858,000

Table 5. Annual Operating Costs for 50 Metric Ton Per Day Continuous Cedar Oil Plant.

Item	Cost ($)	
Raw Material (Cedar $25/t delivered)	412,250	
Labor		
Plant Manager	40,000	
Night watchmen (2)	42,250	
Maintenance crew (2 men)	50,000	
Laborers (3)	39,600	
Clerical workers (1)	12,000	
Total Labor	183,850	
Electricity	24,000	
Maintenance & repairs	15,000	
Supplies		
Oil drums ($16/drum)	32,000	
Water treatment chemicals	10,000	
Office supplies	1,000	
Miscellaneous (shop, wearing parts, etc)	5,000	
Total Supplies	48,000	
Total Direct Production Cost		683,100
Depreciation	55,000	
Local Taxes	8,000	
Insurance	12,000	
Total Fixed Charges		75,000
Plant Overhead Costs (50% of Labor & Maint)		99,425
Administrative Costs (Legal, Accounting, etc)		3,000
Selling Costs (5% commission)		123,750
Interest on Loan (15% of Total Capital Investment)		128,700
Total Operating Cost		1,112,975

Table 6. Estimated Annual Air Emissions from 50 Metric Ton Per Day Continuous Cedar Oil Plant.

Substance Emitted	Annual Emission (t)
Cedar dust	0.4
Ash	29.5
SO_2	3.3
NO_x (as NO_2)	90.8

AN INDUSTRIAL PERSPECTIVE ON TEACHING SEPARATION SCIENCE

Norman N. Li, Edward W. Funk and Sudhir S. Kulkarni ■ Allied-Signal Research Center Inc., Des Plaines, IL

The success of an industrial researcher working on the research and development of new separation processes depends on several major factors, such as: good background in applied chemistry, ability to use mathematical and statistical techniques to guide and interpret experimental studies, and inclination and imagination to investigate new approaches. Examples were used to illustrate these factors.

INTRODUCTION

Our perspective and thus our comments are mainly directed at Ph.D. level chemical engineers who are entering industry to work in basic and exploratory research. Thus, we primarily address new areas of separation rather than areas such as optimization of separation processes, process simulation and process control.

For a Ph.D. chemical engineer to do separations research in an industrial laboratory, we believe, in general, a good knowledge of chemistry and statistics will help greatly. This is discussed below.

CHEMISTRY

Before we discuss further on the importance of chemistry knowledge for chemical engineers, it is important to point out that we don't mean to minimize the equal importance of separation or "unit operation" principles, which a Ph.D. chemical engineer can bring to the research laboratory. For example, classical separation principles teach the advantages of countercurrent flow, the disadvantages of axial diffusion and mixing, and the use of staging concepts to enhance a separation. While we recognize the importance of these "unit operations" principles, we believe strongly that a chemical engineer researcher should also know the underlying chemistry associated with the separation being investigated.

We have interviewed many chemical engineering graduate students for employment. We notice there is, in general, a weakness in their chemistry background. Very few Ph.D. chemical engineers have chemistry courses taken in their graduate study. Indeed, very few schools require examinations of chemistry as part of the chemical engineering Ph.D. qualifying examination. In one case, an employment candidate gave an eloquent seminar on his Ph.D. thesis research which involved sophisticated stage-to-stage calculations and equilibrium theories for a distillation process. However, when questioned about the molecular interactions among the compounds to be fractionated, we found that he did not really have much understanding of key factors such as hydrogen bonding.

Let's look at an example of membrane separations. Figure 1 presents the structure of a typical TFC (thin-film composite) membrane which is used for desalination. It consists of a fabric support which is coated by a porous polysulfone and then a polyamide layer is formed on the surface of the polysulfone.(1-3) These complex membranes give 99+% rejection of salt at 10 gal/ft^2-D--they are widely used throughout the world. For this type of composite membrane, there are abundant opportunities for chemical engineers to develop new membranes. The membrane casting procedure, as illustrated for polysulfone in Figure 2, requires background in traditional areas of chemical engineering (thermodynamics and mass transfer). Our point is that chemical engineers are well suited to make significant inventions in membrane materials and are not limited to only model the permeation data and design of membrane process systems. We have seen quite a few recent chemical engineering Ph.D. dissertations on separations by membranes. Most of them deal with mathematical models and/or process schemes, and very few are involved with the chemistry work in making the membrane. The reason probably is that the graduate students simply worked on the courtesy samples of films from manufacturers and therefore were not involved in the membrane chemistry. This approach leaves a student oblivious to important factors such as membrane degradation and pore compaction. Interpretation of a membrane process without consideration of the membrane chemistry can lead to erroneous conclusions.

A second example concerns adsorption. It is not uncommon that in adsorption studies, there may be significant chemical effects such as complexation, dissociation or even polymerization. In such cases, a researcher cannot just blindly measure adsorption isotherms in evaluating potential adsorbents; attention must be given to the chemistry involved in the system. Figure 3 shows the structure of the new ZSM series of molecular sieves.(4) These have potential use as adsorbents for processes such as the UOP's Sorbex process.(5-8) Again, chemical engineers can contribute in the synthesis of these types of new materials.

No doubt, mathematical contributions are important to chemical engineering and have been significant in advancing areas such as distillation, gas absorption, and extraction. But there are other areas where we believe the mathematical analysis has drowned the need for critical study of the basic chemistry of the separation. Few chemical engineers can expect to spend their careers working on chemically simple petrochemical systems. Therefore, chemical engineers must be prepared to handle the chemistry associated with separation processes for complex systems.

STATISTICS

While every Ph.D. chemical engineer had math courses in undergraduate and graduate education, it is not very common to have a background in statistics. yet, with a strong mathematical background, it would be easy for a chemical engineer to master the necessary statistics and apply it in his research. Statistics can be used to design experiments and analyze data and to apply in quality control. We can significantly reduce the number of experiments if statistics are used to carefully formulate the experimental program and strategy. The information obtained using a statistically designed experimental program allows a systematic and rigorous data analysis. In the usual way of data analysis, one tends to study X as a function of Y, or at most, as another function of Z. The results are plotted as X vs. Y or as a series of X vs. Y curves with Z as the variable among these curves. This type of analysis makes, of course, the assumption that all other variables are held constant while X, Y and Z are varying. In a real system, there are usually many variables involved. To ensure that laboratory results simulate the real system, one needs to be able to analyze all these variables when they are all varying realistically at the same time. This can only be done effectively by using statistics, which allows isolation of the key variables in a multidimensional data space.

In performing data analysis, if researchers are not knowledgeable about statistics, they often mistake "background noise" as real trends in their results. One important contribution of statistics is to estimate experimental error. The more complex the experiment, the greater the need of statistical treatment for rigorous data interpretation.

A good textbook on statistics for researchers is given by Box, Hunter and Hunter. (9) The importance of statistics in chemical engineering is illustrated by the Chemical Engineering tutorial review on statistics which began in the July 1984 edition. As an example, we find statistical techniques valuable in development of new membrane materials for gas and liquid separations. Without a set of statistically-designed experiments, the researcher can spend large amounts of time (and money) seeking to find by intuition the key variables in membrane casting and the significant interactions between the variables. Figure 4 shows an illustration of a study from our laboratory involving a membrane formulation for CO_2/CH_4 separations with seven variables in the casting procedure. From a set of 16 experiments, the magnitude of the effects of the seven variables and their second-order interactions could be estimated (see Figure 5). As shown in Figures 4 and 5, only three variables were statistically significant and that there was only one significant interaction between the variables. These statistical techniques are very valuable for complex systems which cannot reasonably be described in detailed mathematical models. In the membrane case, they help to achieve consistent and high quality of the membranes produced in our laboratory.

MATHEMATICAL MODELING

We all recognize the importance of mathematical modeling, which helps provide a better understanding of a given separation process and enables prediction of the process performance. Ph.D. chemical engineers in general have good mathematical skills. Our point here is that we should not overdo it. We have seen very complex mathematical models which contain parameters difficult or impossible to measure experimentally. For example, an elegant membrane permeation model published contains a key parameter that describes the pore tortuosity in a membrane. We doubt anyone knows how to measure experimentally this factor.

There are complex separation phenomena which can be explained by relatively simple mathematical models. This is illustrated in Figure 6 for the case of modeling the separation by liquid surfactant membranes (Li, References 10-16). As shown in Figure 6, the liquid surfactant membrane (or emulsion liquid membrane) is a complicated structure on a microscopic level. It would be very difficult if not impossible, to measure the internal circulation of the millions of the micron-size droplets within a globule while the globule is being stirred in an extractor. An "advancing boundary" model is capable of fitting the experimental results. Since the model is based on the assumption that there is no internal circulation, it is now generally accepted that the assumption is valid. This conclusion has practical importance--any extractor that can be designed to initiate and promote the internal circulation of the micron-size droplets would yield an improved separation. It is interesting to note that in terms of predicting the separation, this model only improves the accuracy of an earlier model by a few percent.(13) However, the added complexity of the later model helps to elucidate the separation mechanism whereas the earlier model, (15,16) which is a modified version of Fick's First Law on Diffusion, cannot do so. The points about this example are two: the earlier model is probably adequate for process design while the later model explains the separation mechanism.

In summary of our discussion on mathematical modeling, it is important for the researchers to nave a good sense of the physical magnitude of various effects so that a model can be only as complex as necessary. It is important for students to develop a sense of proportion in the modeling of laboratory data.

INNOVATION AND CREATIVITY

Creative ideas can lead to new separations and new separations will be needed for emerging areas such as biotechnology. Few chemical engineering programs attempt to develop creativity in their students yet research labs require new and creative ideas. In interviewing students, every industrial laboratory seeks those students who will become inventors and source of new ideas. Granted this is hard to teach, but it should be taken as a serious challenge for educators in Separation Science. There are certainly factors which can at least help a researcher become more creative in his work. For example, to possess knowledge of chemistry would broaden a chemical engineer's understanding about the separation challenge he is facing in the laboratory and would make him more creative in finding the solution.

CONCLUSIONS

With industries' emphasis on developing energy-efficient separation processes and with emerging new technology areas that require sophisticated separation techniques, researchers in the field of separation science and technology are facing strong challenges. Ph.D. chemical engineers entering industry to work on separations research are generally well prepared in mathematical modeling skill but could have a much stronger background in chemistry and applied statistics. We also see a need for them to learn in their education how to isolate key problems in a complex separation system under study and to be creative in their approach in solving these problems.

LITERATURE CITED

1. Li, N.N., "Membrane Processes," p. 17-24, Perry's Chemical Engineers' Handbook, McGraw Hill,(1984)

2. Edwards, V.H., "Origins of Separation Systems," Chapter 11, p. 269 in Recent Developments in Separation Science," Ed. by N.N. Li, Vol. II, CRC Press (1972).

3. Podall, H.E., "Reverse Osmosis," Chapter 8 p. 171 in Recent Developments in Separation Science, Ed. by N.N. Li, Vol. II, CRC Press (1972).

4. Flanigan, E.M., Bennett, J.M., Grose, R.W. Cohen, J.P., Patton, R.L., Kinchner, R.M., and Smith, J.V., "Silicalite, a New Hydrophobic Crystalline Silicon Molecular Sieve," Nature, 271, 512 (1978).

5. Broughton, D.B., Neuzil, R.W., Pharis, J.M and Brearley, C.S., "The Parex Process for Recovering Paraxylene," Chem. Eng. Process 66, 70 (1970).

6. deRosset, A.J., Neuzil, R.W. and Korous, D.J., "Liquid Column Chromatography as a Predictive Tool for Continuous Countercurrent Adsorptive Separations," I&EC, Proc. Des. Dev., 15, 261 (1976).

7. deRosset, A.J., Neuzil, R.W., Tajbl, D.G., Braband, J.M., "Separation of Ethylbenzene from Mixed Xylenes by Continuous Adsorptive Processing," Separation Science and Technology, 15, 657 (1980).

8. Neuzil, R.W., Rosback, D.H., Jensen, R.H., Teague, J.R., and deRosset, A.J., "An Energy-Saving Separation Scheme," Chemtech 498, (1980).

9. Box, G.E.P., Hunter, W.G., Hunter, J.S., Statistics for Experimenters, John Wiley and Sons, New Yori, New York, (1978).

10. Li, N.N., "Separating Hydrocarbons with Liquid Membranes," U.S. Patent 3,410,794, (1968).

11. Li, N.N., "Permeation through Liquid Membrane Permeation," AIChE Journal, Vo.1.17, No, 2, p. 459 (1971).

12. Ho, W.S., Lightfoot, E.N., Li, N.N., "A Diffusion-Controlled Model with Advancing Reaction Front," AIChE Journal, Vol. 28, No. 4, p. 662, (1982).

13. Hatton, T.A., Lightfoot, E.N., Li, N.N., "The Mathematical Modeling of Liquid Surfactant Membranes," Proceedings of Joint Meeting of AiChE and Chemical Industry and Engineering Society of China, Volume 2, 558 (1982).

14. Li, N.N., Cahn, R.P., Naden, D., and Lai, R.W., "Liquid Membrane Processes for Copper Extraction," Hydrometallurgy, 9, p.277 (1983).

15. Li, N.N., "Encapsulation and Separation by Liquid Surfactant Membranes," The Chemical Engineers (U.K.), p. 325, July (1981).

16. Li, N.N., and Frankenfeld, J.W., "Liquid Membrane Separations," a chapter in Encyclopedia of Chemical Processing and Design, Ed. J.J. McKetta, to be published by Marcel Dekker.

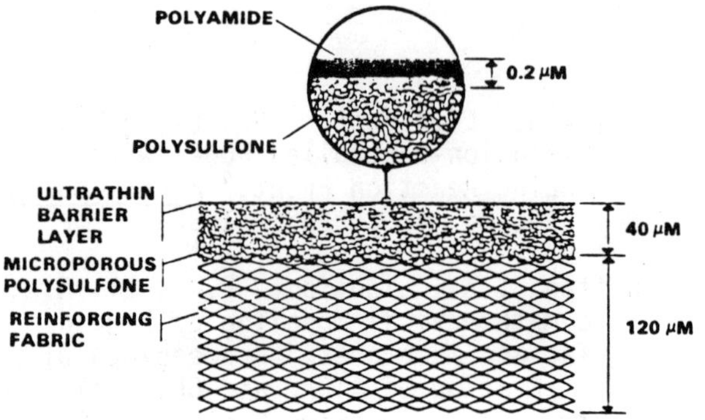

Figure 1. Thin-film composite membrane.

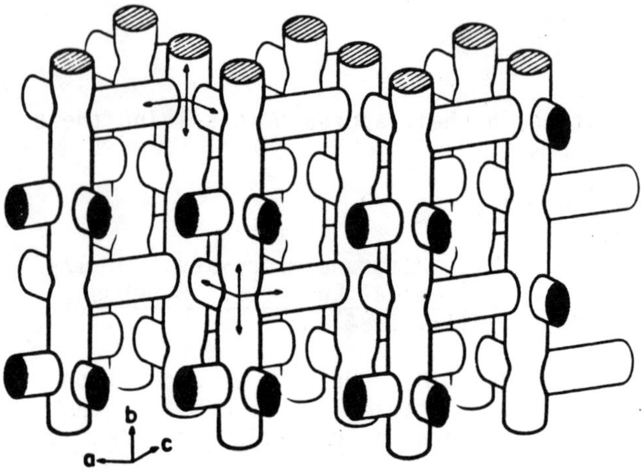

Figure 3. Structure of silicalite-1.

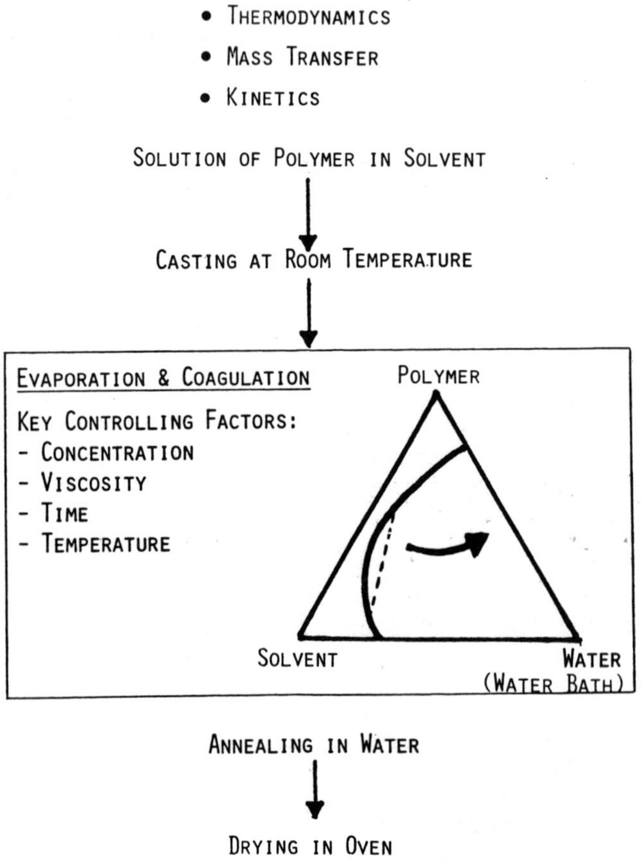

Figure 2. Typical membrane casting sequence.

	Additive 1 Conc. % X1	Additive 2 % X2	Additive 3 % X3	Time Before Drain X4	Wait Before Oven X5	Oven Time X6	Polymer Purity X7	Flux mL/min	$\bar{P}_{CO_2}/\bar{P}_{CH_4}$
1	-	-	-	+	+	+	-	20.0	5.11
2	+	-	-	-	-	+	+	12.0	5.85
3	-	+	-	-	+	-	+	4.4	7.23
4	+	+	-	+	-	-	-	1.0	11.53
5	-	-	+	+	-	-	+	~100.	1.0
6	+	-	+	-	+	-	-	18.0	5.58
7	-	+	+	-	-	+	-	1.7	9.31
8	+	+	+	+	+	+	+	5.0	8.83
9	+	+	+	-	-	-	+	6.0	9.13
10	-	+	+	+	+	-	-	2.5	11.45
11	+	-	+	+	-	+	-	3.5	4.56
12	-	-	+	-	+	+	+	61.3	1.38
13	+	+	-	-	+	+	-	1.17	15.94
14	-	+	-	+	-	+	+	1.9	17.0
15	+	-	-	+	+	-	+	3.7	7.9
16	-	-	-	-	-	-	-	33.6	2.02

Figure 4. Factorial design to guide membrane casting.

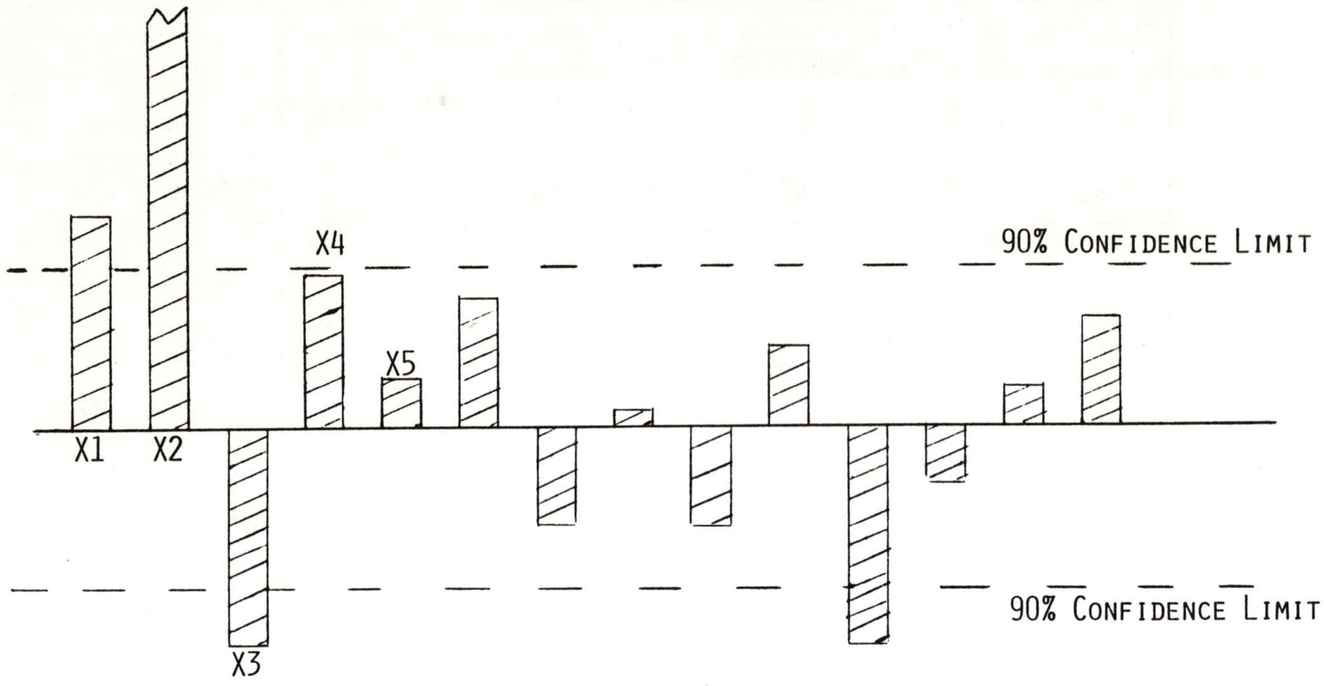

Figure 5. Results of statistical design on membrane casting.

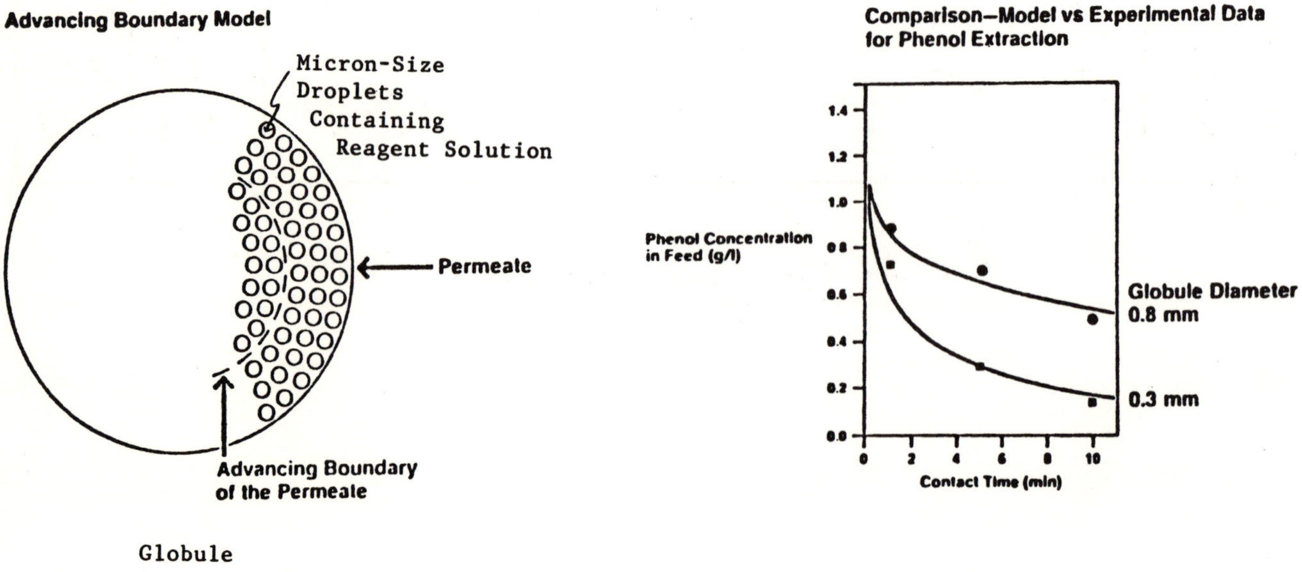

Figure 6. Liquid membrane modeling.

INDEX

SYMPOSIUM SERIES

ADSORPTION

96 **Developments in Physical Adsorption**
117 **Adsorption Technology**
219 **Recent Advances in Adsorption and Ion Exchange**
230 Adsorption and Ion Exchange—'83
233 Adsorption and Ion Exchange—Progress and Future Prospects
242 Adsorption and Ion Exchange: Recent Developments

AEROSPACE

33 **Rocket and Missile Technology**
52 Chemical Engineering Techniques in Aerospace

BIOENGINEERING

69 Bioengineering and Food Processing
84 The Artificial Kidney
86 Bioengineering ... Food
93 Engineering of Unconventional Protein Production
99 Mass Transfer in Biological Systems
108 Food and Bioengineering—Fundamental and Industrial Aspects
114 Advances in Bioengineering
163 Water Removal Processes: Drying and Concentration of Foods and Other Materials
172 Food, pharmaceutical and bioengineering—1976/77
181 Biochemical Engineering Renewable Sources of Energy and Chemical Feedstocks

CRYSTALLIZATION

110 Factors Influencing Size Distribution
193 Design Control and Analysis of Crystallization Processes
215 Nucleation, Growth and Impurity Effects in Crystallization Process Engineering
240 Advances in Crystallization From Solutions

DRAG REDUCTION

111 Drag Reduction
130 Drag Reduction in Polymer Solutions

ENERGY

Conversion and Transfer

5 Heat Transfer, Atlantic City
57 Heat Transfer, Boston
59 Heat Transfer, Cleveland
75 Energy Conversion Systems
79 Heat Transfer with Phase Change
87 Advances in Cryogenic Heat Transfer
113 Convective and Interfacial Heat Transfer
118 Heat Transfer—Tulsa
119 Commercial Power Generation
138 Heat Transfer—Research and Design
162 Energy and Resource Recovery from Industrial and Municipal Solid Wastes
174 Heat Transfer: Research and Application
189 Heat Transfer—San Diego 1979
202 Transport with Chemical Reactions
208 Heat Transfer—Milwaukee 1981
216 Processing of Energy and Metallic Minerals
225 Heat Transfer—Seattle 1983
236 Heat Transfer—Niagara Falls 1984

Nuclear Engineering

53 Part XIII
56 Part XIV
94 Part XX
104 Part XXI
106 Part XXII
119 Commercial Power Generation
168 Heat Transfer in Thermonuclear Power Systems
169 Developments in Uranium Enrichment
191 Nuclear Engineering Questions Power Reprocessing, Waste, Decontamination Fusion
221 Recent Developments in Uranium Enrichment

ENVIRONMENT

78 Water Reuse
97 Water—1969
115 Important Chemical Reactions in Air Pollution Control
122 Chemical Engineering Applications of Solid Waste Treatment
124 Water—1971
126 Air Pollution and its Control
133 Forest Products and the Environment
137 Recent Advances in Air Pollution Control
139 Advances In Processing and Utilization of Forest Products
144 Water—1974: I. Industrial Wastewater Treatment
145 Water—1974: II. Municipal Wastewater Treatment
146 Forest Product Residuals
147 Air: I. Pollution Control and Clean Energy
148 Air: II. Control of NO_{xx} and SO_x Emissions
149 Trace Contaminants in the Environment
151 Water—1975
156 Air Pollution Control and Clean Energy
157 New Horizons for the Chemical Engineer in Pulp and Paper Technology
165 Dispersion and Control of Atmospheric Emissions, New-Energy-Source Pollution Potential
170 Intermaterials Competition in the Management of Shrinking Resources
171 What the Filterman Needs to Know About Filtration
175 Control and Dispersion of Air Pollutants: Emphasis on NO_x and Particulate Emissions
177 Energy and Environmental Concerns in the Forest Products Industry
184 Advances in the Utilization and Processing of Forest Products
188 Control of Emissions from Stationary Combustion Sources Pollutant Detection and Behavior in the Atmosphere
195 The Role of Chemical Engineering in Utilizing the Nation's Forest Resources
196 Implications of the Clean Air Amendments of 1977 and of Energy Considerations for Air Pollution Control
198 Fundamentals and Applications of Solar Energy
200 New Process Alternatives in the Forest Products Industries
201 Emission Control from Stationary Power Sources: Technical, Economic and Environmental Assessments
207 The Use and Processing of Renewable Resources—Chemical Engineering Challenge of the Future
209 Water—1980
210 Fundamentals and Applications of Solar Energy II
211 Research Trends in Air Pollution Control: Scrubbing, Hot Gas Clean-up, Sampling and Analysis
213 Three Mile Island Cleanup
223 Advances in Production of Forest Products
232 Applications of Chemical Engineering in the Forest Products Industry
239 The Impact of Energy and Environmental Concerns on Chemical Engineering in the Forest Products Industry
243 Separation of Heavy Metals and Other Trace Contaminants
246 Advances in Process Analysis and Development in the Forest Products Industries.

FLUIDIZATION

101 Fundamental Processes in Fluidized Beds
105 Fluidization Fundamentals and Application
116 Fluidization: Fundamental Studies Solid-Fluid Reactions, and Applications
176 Fluidization Application to Coal Conversion Processes
205 Recent Advances in Fluidization and Fluid-Particle Systems
234 Fluidization and Fluid Particle Systems: Theories and Applications
241 Fluidization and Fluid Particle Systems: Recent Advances